The Culture of Science

Routledge Studies in Science, Technology and Society

1 **Science and the Media**
Alternative Routes in Scientific Communication
Massimiano Bucchi

2 **Animals, Disease and Human Society**
Human-Animal Relations and the Rise of Veterinary Medicine
Joanna Swabe

3 **Transnational Environmental Policy**
The Ozone Layer
Reiner Grundmann

4 **Biology and Political Science**
Robert H Blank and Samuel M. Hines, Jr.

5 **Technoculture and Critical Theory**
In the Service of the Machine?
Simon Cooper

6 **Biomedicine as Culture**
Instrumental Practices, Technoscientific Knowledge, and New Modes of Life
Edited by Regula Valérie Burri and Joseph Dumit

7 **Journalism, Science and Society**
Science Communication between News and Public Relations
Edited by Martin W. Bauer and Massimiano Bucchi

8 **Science Images and Popular Images of Science**
Edited by Bernd Hüppauf and Peter Weingart

9 **Wind Power and Power Politics**
International Perspectives
Edited by Peter A. Strachan, David Lal and David Toke

10 **Global Public Health Vigilance**
Creating a World on Alert
Lorna Weir and Eric Mykhalovskiy

11 **Rethinking Disability**
Bodies, Senses, and Things
Michael Schillmeier

12 **Biometrics**
Bodies, Technologies, Biopolitics
Joseph Pugliese

13 **Wired and Mobilizing**
Social Movements, New Technology, and Electoral Politics
Victoria Carty

14 **The Politics of Bioethics**
Alan Petersen

15 **The Culture of Science**
How the Public Relates to Science Across the Globe
Edited by Martin W. Bauer, Rajesh Shukla and Nick Allum

The Culture of Science
How the Public Relates to Science
Across the Globe

**Edited by Martin W. Bauer,
Rajesh Shukla and Nick Allum**

NEW YORK   LONDON

First published 2012
by Routledge
711 Third Avenue, New York, NY 10017

Simultaneously published in the UK
by Routledge
2 Park Square, Milton Park, Abingdon, Oxon OX14 4RN

Routledge is an imprint of the Taylor & Francis Group, an informa business

© 2012 Taylor & Francis

First issued in paperback 2013
The right of Martin W. Bauer, Rajesh Shukla and Nick Allum to be identified as the authors of the editorial material, and of the authors for their individual chapters, has been asserted in accordance with sections 77 and 78 of the Copyright, Designs and Patents Act 198

Typeset in Sabon by IBT Global.

All rights reserved. No part of this book may be reprinted or reproduced or utilised in any form or by any electronic, mechanical, or other means, now known or hereafter invented, including photocopying and recording, or in any information storage or retrieval system, without permission in writing from the publishers.

Trademark Notice: Product or corporate names may be trademarks or registered trademarks, and are used only for identification and explanation without intent to infringe.

Library of Congress Cataloging-in-Publication Data
The culture of science : how the public relates to science across the globe / edited by Martin W. Bauer, Rajesh Shukla, and Nick Allum.
  p. cm. — (Routledge studies in science, technology and society ; 15)
Includes bibliographical references and index.
1. Science—Social aspects. I. Bauer, Martin W. II. Shukla, Rajesh, 1970– III. Allum, Nick.
Q175.5.C85 2011
306.4'5—dc22
2010052026

ISBN13: 978-0-415-87369-7 (hbk)
ISBN13: 978-0-415-85102-2 (pbk)

Contents

List of Figures		ix
List of Tables		xi
Acknowledgments		xv

1 Towards Cultural Indicators of Science with Global Validity   1
   MARTIN W. BAUER, RAJESH SHUKLA AND NICK ALLUM

2 The Culture of Science and the Politics of Numbers   18
   BENOÎT GODIN

PART I
Longitudinal Analysis

3 Attitudes toward Science in France: 1972–2005   39
   DANIEL BOY

4 American Public Understanding of Science: 1979–2006   55
   SUSAN CAROL LOSH

5 The Image of Science in Bulgaria and UK, 1992–2005: Does Generation Matter?   76
   KRISTINA PETKOVA AND VALERY TODOROV

6 The Changing Culture of Science across Old Europe: 1989 to 2005   92
   MARTIN W. BAUER

7 Knowledge of Science and Technology in Japan: IRT Scores for 1991 and 2001   110
   KINYA SHIMIZU AND TAKUYA MATSUURA

8	Adult Scientific Literacy and Its Surveys in China since 1992 KE WANG, FUJUN REN, WEI HE AND CHAO ZHANG	126

PART II
Cross-National Comparisons

9	Comparing the Public Understanding of Science across China and Europe XUAN LIU, SHUKUN TANG AND MARTIN W. BAUER	139
10	Information and Attitudes towards Science and Technology in Iberoamerica CARMELO POLINO AND YURIJ CASTELFRANCHI	158
11	The Science Culture Index (SCI): Construction and Validation RAJESH SHUKLA AND MARTIN W. BAUER	179
12	A Comparative View on Adolescents' Attitudes towards Science SVEIN SJØBERG AND CAMILLA SCHREINER	200

PART III
Measurement Issues

13	The Sources and Impact of Civic Scientific Literacy JON D. MILLER	217
14	Using Latent Trait Models to Assess Cross-National Scales of the Public's Knowledge about Science and Technology SALLY STARES	241
15	Statistical Modelling of Public Understanding of Science ANIL RAI AND RAJESH SHUKLA	262
16	The Cultures of Public Understanding of Science—Defining Cultural Distance GAUHAR RAZA AND SURJIT SINGH	282

PART IV
Cultural Aspects of Sensitive Topics

17 Beliefs about Astrology across Europe   301
    NICK ALLUM AND PAUL STONEMAN

18 The Human-Animal Boundary in Switzerland:
    A Cross-Cultural Perspective   323
    FABIENNE CRETTAZ VON ROTEN

19 Religious Belief and Attitudes about Science in the United States   336
    SCOTT KEETER, GREGORY SMITH AND DAVID MASCI

20 Worldviews and Perceptions of Stem Cell Research
    across Europe   353
    RAFAEL PARDO

PART V
Complementary Data Streams

21 Measuring PEP/IS, a New Model for Communicative
    Effectiveness of Science   375
    HAK-SOO KIM

22 Suggesting Cultural Indicators Derived from Exhibitory Science   385
    BERNARD SCHIELE

23 Building a Science News Media Barometer—SAPO   400
    CARLOS VOGT, YURIJ CASTELFRANCHI, SABINE RIGHETTI,
    RAFAEL EVANGELISTA, ANA PAULA MORALES AND FLAVIA GOUVEIA

24 Validating Survey Measures of Scientific Citizenship   418
    NIELS MEJLGAARD AND SALLY STARES

25 Benchmarking Climate Indicators for Science Communication
    and Public Engagement across Europe   436
    STEVE MILLER

26 **Monitoring Science in the Public Sphere: The Case of Italy** 449
MASSIMIANO BUCCHI AND FEDERICO NERESINI

Contributors 463
Index 465

Figures

2.1	The traditional model for measuring science.	22
2.2	S&T activities (UNESCO).	32
3.1a	Opinion toward scientists by generation.	42
3.1b	Effect of science on human relations: positive.	43
3.2a	Attitudes toward chemical fertilizers.	45
3.2b	Attitudes toward ethical principles.	45
3.3a	Generally speaking, do you feel that science does humanity more good than harm, more harm than good, or about as much good as harm?	46
3.3b	Overall judgment of science.	47
3.4a	Beliefs in parasciences by generations.	49
3.4b	Index of interest for scientific culture.	49
4.1	Adjusted cohort effects on science and pseudoscience measures.	64
5.1	The estimated marginal means of attitude variables.	
	a	82
	b	82
	c	83
	d	83
	e	84
	f	84
	g	85
6.1	Overall trends in knowledge, attitudes and interest presented as standard z-values.	93
6.2	Knowledge of and attitudes to science by generation cohort: knowledge, quality of life, interesting work and pace of life shown as z-values.	97
6.3	Knowledge, attitude, interest by education and age cohort (controlled for other variables).	
	a	100
	b	100
	c	101
	d	101
7.1	The ICC for all items.	119
7.2	Mean IRT scores of six age groups in Japan.	120

x *Figures*

9.1	Marginal means for rationalism (rejecting lucky numbers).	150
11.1	Conceptual framework of SCI.	183
11.2	Estimates of PUS index, STS index, and SC index.	
	a	186
	b	186
	c	186
	d	186
	e	187
	f	187
11.3	Estimates of variability—PUS and STS indicators.	
	a	190
	b	190
12.1	Hierarchical cluster analysis of residual ACE mean scores for all countries.	203
12.2	Science and technology are important for society.	206
12.3	The benefits of science are greater than the harmful effects it could have.	207
12.4	I would like to become a scientist.	208
12.5	I would like to get a job in technology.	209
12.6	Country means of interest in learning science vs. level of development measured by Human Development Index (HDI).	210
13.1	Civic scientific literacy in the United States, 1988–2005.	221
13.2	Civic scientific literacy in 34 countries, 2005.	222
13.3	A path model to predict civic scientific literacy in adults, 2005.	225
13.4	A model to predict CSL in Sweden, 2005.	226
14.1	Item characteristic curves for the 'knowledge' trait from a 2-trait discrete trait model for 3-category nominal items, British sample.	249
14.2	Item characteristic curves from 3-trait discrete trait models for 3-category nominal items, with measurement models equal for all traits, for 15 countries, biotechnology 'knowledge' trait.	253
14.3	Item characteristic curves from a 1-trait discrete trait model for binary items, with measurement models equal for 15 countries.	255
14.4	Item characteristic curves from final science literacy scale model, curves for UK.	257
16.1	Measuring cultural distance.	287
17.1	How scientific are astrology, horoscopes, and other subjects?	308
17.2	Factor loading plots for horoscopes and astrology conditions.	310
17.3	Standardised random effect estimates for 25 European countries.	315
17.4	Levels of cognitive polyphasia in Europe.	317
17.5	Engagement with science.	318

23.1	Quantitative indicators of SAPO.	
	A	407
	B	407
	C	407
	D	408
	E	408
24.1	Bi-plot from correspondence analysis of measures of participation and competence.	425
26.1	Attitudes towards investments in nuclear energy—Science in Society Monitor.	452
26.2	Patterns of coverage of the issue of nuclear power in Il Corriere della Sera and La Repubblica (no. of articles per year, 1992–2008).	457
26.3	Italian citizens in favour, against, or uncertain about investments in nuclear power and patterns of media coverage.	458

Tables

1.1	Nationally Representative Survey of PUS, Science Literacy and Attitudes from across the Globe	13
2.1	Historical Development of Statistics on Science	20
2.2	OECD Methodological Documents	22
3.1	Logistic Regression of the Overall Judgment of Science	52
4.1	Questions Used to Measure Adult Public Understanding of Science	60
4.2	Civic Science Literacy Variables by Time, Age Group and Cohort	63
4.3	Support for Pseudoscience by Time, Age Group and Cohort	65
4.4	Standardized Multiple Regression Effects on Basic Science Knowledge	67
5.1	ANOVA. Effects of Country, Cohort, and Interaction between Country and Cohort on Science Literacy	89
5.2	ANOVA. Effects of Cohort for Bulgaria and UK Separately	90
6.1	Shows the Ranking of EU12 Countries on Knowledge (know13) by Years and by Generational Cohort	95
6.2	The Results of a Binary Logistic Regression of Knowledge, Attitude and Interest	104
7.1	Classical Item Statistics and Item Parameters of the 2PL Model	116
7.2	Mean IRT Score by Year and by Selected Demographic Factors in Japan	122
8.1	Technical Parameters of Each Survey	127
8.2	Percentage of Scientifically Literate Public and Scientific Literacy Dimensions	129
8.3	Correlation of the Scientific Literacy Index and Independent Variables	131
8.4	Level of Scientific Literacy of Those Reporting the Use of Different Sources of S&T Information	132
8.5	Level of Scientific Literacy of Those Who Hold Particular Professions in High Esteem	133
9.1	Distribution of Responses on Each Knowledge Item	141

9.2	General Description of Indices (mean, median, SD; ANOVA, correlations)	144
9.3	Cluster Analysis of the Resulting Indicators	146
9.4	The Distributions of PUS Types across EU27 and China (Anhui province)	148
9.5	Equivalence Recoding for Attitude (Att2) and Interest(I) Items	155
10.1	Comparative Indicators by Country	160
10.2	S&T Information Indicators for Different Media	165
10.3	ICIC (S&T Information Index) by City	166
10.4	Balance between Future Risks and Benefits from S&T	167
10.5	"Attitudinal Typology" Considering Risks and Benefits from S&T by City	169
10.6	Model: ICIC Determinants' Coefficients a	172
11.1	Percentage Contribution of Indicators to Subindices	188
11.2	The Index Scores for STS, PUS and SC to the Basis of EU=100	196
12.1	Pearson Product Moment Correlation Coefficient of Mean Agreement with Statements with HDI	210
13.1	Comparison of Total Effect of Selected Variables on Civic Scientific Literacy.	229
13.2	Confirmatory Factor Analysis of Science Knowledge Items, 2005	236
14.1	Distribution of Responses to Knowledge Questions across Fifteen EU Countries	243
14.2	Qualitative Summaries of Unusual Item Characteristic Curves (ICCs) on Biotechnology 'Knowledge' Traits, from 2-trait Models, 15 Countries	251
14.3	Percentages of Respondents in Each Level of the Final Joint Model of Biotechnology Knowledge Items	254
15.1	Definition of Variables Used in the Analysis	270
15.2	Results of Stepwise Logistic Regression: All India	271
15.3	Logistic Regression Estimation Results: All India	272
15.4	Measure of Association for Different Statistics	273
16.1	Questions Posed to the Respondents	289
16.2	Statistical Properties and Cultural Distance	290
16.3	Relative Cultural Distance and Ranking of Selected Indian States	292
16.4	Magnitude and Polarity of Shift in Cultural Distance	294
17.1	Random Effects Regression Estimates (individual characteristics)	312
17.2	Variance Components for Intercept-only and Full Models	314
17.3	Formal and Informal Engagement in Science (unstandardised coefficients)	319
18.1	Linguistic Regions' Difference in Agreement with Science-related Items (mean, standard deviation in parenthesis, Anova F statistic, and significance)	327
18.2	Linguistic Regions' Difference in Type of Governance (% by criteria, square-test statistic, and p-value)	328

18.3	Linguistic Regions' Difference in Agreement with Attitudes toward Nature and Animals (mean, standard deviation in parenthesis, Anova F statistic, and p-value)	329
18.4	Modelization of Approval of Animal Cloning for Medical Research	330
19.1	Opinions on Evolution and Climate Change, by Religious Affiliation	339
19.2	Predicted Probability of Holding Selected Beliefs, by Religious Tradition, Party Affiliation and Perceptions of Scientific Consensus	340
19.3	Beliefs about Homosexuality, by Education, Ideology, Religious Affiliation and Attendance	342
19.4	Details of Logistic Regression Predicting Beliefs about Evolution, Homosexuality and Climate Change	349
20.1	Views on the Moral Status of the Embryo by Religious Beliefs	358
20.2	Beliefs about the Moral Status of the Embryo by Religiosity	359
20.3	Beliefs about the Moral Status of the Embryo by Biological Literacy (%)	360
20.4	Explanatory Model of Attitudes to Biomedical Embryo Research	364
21.1	P-Engagement and SPS-Engagement for Nonscientists and Scientists by Problem	379
21.2	Correlations between P- and SPS-Engagements for Nonscientists and Scientists by Problem	380
21.3	Test of Difference between the Nonscientists' and Scientists' PEP/IS Index Correlations, by Problem	381
22.1	Scientific, Didactic, and Media Fields	389
22.2	Formal Education and Non-formal Education	391
23.1	Average of Total Items and Science Items Published over the Year 2008 and in May 2008—Mass Indicator in the Media	406
23.2	List of Materials Published on the Peak—O Estado de S.Paulo—14 May 2008	410
23.3	Index of Consumption of Scientific News in Different Media (%)	413
24.1	Items Capturing Participation and Competence in Science	421
24.2a	Conditional and Prior Probabilities for a Joint Cross-national Model of Participation in Science	423
24.2b	Conditional and Prior Probabilities for a Joint Cross-national Model of Competence in Science	424
24.3	Estimated Percentages in Each Class of Participation and Competence, by Country	428
25.1	Measure of Knowledge, Interest, Activity, and Esteem Derived from the 2001 Eurobarometer Survey (in units of the EU average)	439
25.2	Changes in Knowledge and Interest Levels for 11 EU Member States in Terms of European Averages*	441

25.3	Government Activity in the Promotion of RTD Culture and PUS	442
26.1	Distribution of Relevant Articles and the Index of Salience by Newspaper (2008)	454
26.2	Distribution of Relevant Articles among Sections of Online Daily Newspapers Excluding the Home Page (values expressed in %)	455

Acknowledgments

An edited book like this one is always a collective effort of many hands and minds, the names of whom do not appear in the final product. Here we would like to thank those hidden then colleagues at the Royal Society, Darren Bhattachary, Chloe Sheppard, Matthew Harvey, and Nicola Hern, without whose engagement this project would never have got off the ground. We also acknowledge the contributors to our initial Royal Society Meeting, 5–6 November 2007, who did not in the end contribute to this volume, but whose discussions remain part of the project. Thanks to the anonymous and not so anonymous reviewers of the chapters whose helpful comments found their way into the final product. Further thanks go to Benjamin Holtzman and Max Novick who at Routledge in New York supported this project with their continued patience through all the unmet deadlines. The English editing and the shaping up of all the chapters and artwork is due to unbeatable Susan Howard, who has done a marvellous job of bringing this project to closure.

Martin W. Bauer
Rajesh Shukla
Nick Allum

London, 16 November 2010

1 Towards Cultural Indicators of Science with Global Validity

Martin W. Bauer, Rajesh Shukla and Nick Allum

On November 5–6, 2007, a group of researchers met in the august halls of London's Royal Society to discuss recent developments on 'International Indicators of Science and the Public'. Participants arrived from twenty-one countries covering all five continents, a fact that our sponsors at the Royal Society noted with satisfaction.[1] Ian Pearson, then Minister for Science & Innovation, presented his new and short-lived vision for 'Science in Society'. Proceedings were interrupted when Queen Elizabeth II passed in full estate en route to opening the Houses of Parliament, providing fitting symbolism for the academic deliberations of which we here present the outcome.

The purpose of the meeting was **to take stock** of survey research on public understanding of science (PUS), i.e. measures of science literacy, interest, attitudes, and public engagement with science, and to initiate **step-changes for future research**.

In research on public understanding of science, large-scale national surveys have built a global momentum since the 1970s. The time has come to take a fresh look at this material and its conceptualisation to invigorate the research effort. In hindsight, the motto of the workshop might well have been '**Working better with what has been achieved and developing it further**'.[2] Some of us have been involved for some time in calling for activities such as reassessing survey-based PUS research in handbook and review papers, not least with a view to opening the agenda to cultural indicators of science (Bauer, Allum & Miller, 2007; Bauer, 2008; Allum, 2010).

We question the phenomenon under study, most generally considered the '**great societal conversation about science**'. From a comparative perspective, this societal conversation fluctuates in intensity, topics covered, engagement of the population, and focus on controversies. It is influenced by the cultural context of language, political culture, local history of science, and current levels of technological development. Societal conversation implies more than opinions expressed in survey interviews. It encompasses writings in print and news media, exhibitions, stakeholder consultations, science policy documents, informal and formal learning by young and old, to name but some. The workshop explored how to compare this societal conversation across context and time. Ultimately, what is at stake is a dynamic **process**.

Nationally representative surveys are highly developed techniques of societal self-observation, but insufficient to map societal conversations. Complementary data streams are needed to understand the symbolic environment of a typical survey respondent and surrounding scientific culture. After all, a respondent will answer questions on the basis of his or her internal (cognitive and emotional) and external resources (the semiotic environment). Questionnaires return standard answers, but do not tell us how to interpret these in different contexts. The need to contextualise survey research raised the double agenda of this meeting:

- How to improve survey research, the perception indicators of science;
- How to mobilise complementary data streams: indicators of performance and societal communication of science.

Jointly, the subjective indicators of perceptions and objective indicators of activities can map the societal conversation, ultimately bringing forth the scientific culture of a nation. The present book is the outcome of these proceedings of 2007 at the Royal Society, revised and elaborated in the light of discussions. It comprises twenty-six chapters grouped into five parts.

Introducing the problem, Benoît Godin (Montreal) offers a glance at the history of ideas of cultural indicators of science. We are encouraged by the fact that this idea harks back to earlier UNESCO initiatives of the 1950s, which were displaced by the quest for economic indicators of science and technology. The time is ripe to return to cultural indicators in a globalising world distancing itself from a 'one best way' of doing things. The idea of cultural indicators starts from the intuition that there is no strong correlation between cultural richness and economic prowess, and that one might be a lever of the other. The remainder of the book is divided into five parts, four presenting step-changes in survey-based indicator research. The fifth demonstrates the deployment of complementary data streams and argues for new conceptual developments.

PART I: LONGITUDINAL ANALYSIS

In many world regions, surveys of PUS have been conducted for some time. It is now time to consolidate existing data and assess changes and stabilities over time. Because it is the primary data of many researchers in PUS, some considering it the 'gold standard' of social research, we dedicate much space to survey data streams. The research infrastructure has made considerable progress in recent years, and data consolidation has been achieved by teams in several regions. Here is a list of databases comprising micro-integrated data for longitudinal and comparative research:

- **France** (1972, 1982, 1989, 1992, 2001, 2007; see Boy);
- **US** (1979–2006, biannually; N ~20000; see Losh);
- **EU12** (1989, 1992, 2001, 2005, N ~ 50,000; see Bauer & Shukla);
- **India** (2001, 2004, 2007, N ~ 50,000; see Shukla & Raza);
- **India-EU** (integrated India04 and EU05; N~60,000, see Shukla & Bauer);
- **China-EU** (integrated China-Hefei 2007 and EU05; see Liu & Bauer);
- **Bulgaria-UK** (1992, 1996, 2005), N ~ 4000; see Petkova & Todorov);
- **US-EU** (The literacy database, see J Miller);
- **ROSE** (database of 15-year-olds in **40 countries**, see Sjøeberg & Schreiner).

This progress means that in at least eighteen countries we are now in a position to conduct systematic longitudinal and comparative research, analysing trends in adult science literacy and public attitudes to science over two decades. The workshop further documented, shown in Table 1.1 (see Appendix 1), similar nationally representative surveys in many other contexts that have comparable elements and where micro-integration of data will bring a step-change in the analysis of material on a global scale. In all of these global contexts, it will require effort and political-academic enthusiasm to recover data files, document the data, and systematically integrate files for analysis. However, the benefits of such an undertaking are considerable: teams can move from reporting headline figures for news making to quasi-cohort and trend analysis, and to explore in their respective contexts how these indicators are useful and how they are moving.

The first part of this book documents in six chapters the progress in longitudinal analysis with data from France, US, Bulgaria, the UK, Japan, China, and across the old Europe of EU12. A common feature of these chapters is the focus on generational analysis using age cohorts rather than the age variable. This allows us to compare trends in public understanding of science along two lines: across time period and across generation. We expect that different generations move differently through time because their experiential formation is different.

Daniel Boy (Paris) presents an analysis of France from 1972 to 2007 and maps increasing ambivalence of the French public towards science and technology, showing the complex interaction between period and generational groups. Susan Losh (Florida) presents her work on the integrated NSF science indicators database, and traces changes in literacy and para-scientific beliefs through time and across generational groups. Christina Petkova and Valery Todorov (Sofia) compare Bulgarian and UK cohorts over the years 1992 and

2005, providing evidence for surprising similarities and differences between contexts that are historically so very different. In that period, Bulgaria was awakening from Stalinist hibernation. It demonstrates how such comparisons constitute explananda that call for cultural-historical insights to come to terms with them. A similar exercise is presented by Martin Bauer's (London) first analysis of the EU12 database, which spans the years 1989 to 2005 and four waves of multinational surveys. Patterns of change and stability of indicators over time and across cohorts provide evidence for the post-industrial model: the polemical deficit model according to which knowledge drives positive attitudes to science is at best a special case, to be historically defined. Kinya Shimizu and Takuya Matsuura (Hiroshima) offer a trend analysis of Japanese adult literacy from 1991 to 2001. Their detailed analysis assesses the extent to which each question item is suitable for this comparison, present an overall score based on items response theory (IRT), and examine this score in relation to various predictor variables including generational groups. The last chapter in this part is the contribution by Ke Wang and colleagues (Beijing), who describe efforts to survey the science literacy of China since the early 1990s. They compare two versions of a literacy index and its diagnostic powers for purposes of conducting comparisons within China.

PART II: CROSS-NATIONAL COMPARISONS

Examples of horizontal data integration and analysis are presented in the second part. Four chapters offer cross-national comparisons on a number of indices. Xuan Liu and colleagues (Hefei and London) analyse a newly integrated database of a large-scale regional Chinese survey of 2007 and the 2005 Eurobarometer survey. They document difficulties of micro-integration of data with items that are semantically equivalent but do not have the exact-same format, and arising informative comparisons. The analysis suggests a typology of public understandings of science profiled on literacy, interests, engagement and attitudes to science, cognitive tolerance for para-science, age, sex, education, and urban and rural life. Distributions of these types of PUS are very different across populations of China and EU27. Carmelo Polino and Yurij Castelfranchi (Buenos Aires and Belo Horizonte) report on recent efforts to coordinate PUS surveys across Latin America and offer comparisons of seven urban environments. The core is a 'science & technology information index' modelled for the different contexts. Another example of index construction is offered by Rajesh Shukla (Delhi) and Martin Bauer (London) who have integrated data for India 2004 and EU 2005. They construct and validate a composite science culture index comprising state-level STS data (GDP per capita, R&D, etc.) and individual-level PUS measures (knowledge, interest, attitudes, engagement) for twenty-three Indian states and thirty-two European national units. On this index all units can be ordered and profiled in terms of objective and subjective science culture, creating a score that can offer politically useful information comparable to the UN's Human Development Index (HDI). Finally, Svein Sjøeberg and Camilla Schreiner (Oslo) present an analysis of the forty country ROSE database, which collects data on attitudes to science among adolescents aged 15. Comparing these countries and in particular looking at the persistent gap between girls and boys on several indicators, they find strong correlations between PUS indicators and the human development index (HDI). The ROSE effort contrasts with the OECD's PISA measures of educational achievement, which in 2006 focused on science literacy. PISA does not consider youth's attitudes to science as part of its remit.

PART III: MEASUREMENT ISSUES AND THE SOPHISTICATION OF INDEX CONSTRUCTIONS

Workshop discussion also focused on measurement issues and how PUS survey data might grow in sophistication. Existing questionnaire items on literacy, interest, attitudes, and engagement with science afford further examination to determine their value for the construction of indicators for global comparison. Are items diagnostic of differences between populations? Much of this part focuses on standard literacy or knowledge items, sometimes erroneously called the 'OXFORD scale' because a team of Oxford researchers used them, but most of these items are older than the 1988 UK study. The literacy items are textbook type items; in most cases a correct or incorrect answer can be determined on good authority, but this authority is not uncontroversial.

The question was raised whether these literacy or knowledge items are useful for assessing people's understanding of science at all. For example recent Latin American efforts do not include literacy items because they are considered culturally biased and unable to pick up locally relevant practical knowledge (see Polino & Castelfranchi). Many items taken out of context have led to strange usage in public discourse on a country's 'deficient public', which most researchers who do not want to be held hostage to fortune will want to avoid. An older argument stipulates that literacy items measure little other than level of formal education and are therefore redundant (Boy in France). By contrast most efforts in Far Eastern countries, such as Japan and China, focus on literacy, and the issue of public attitudes to science is approached with hesitation. We will see how this plays out in the context of discussions of scientific culture, which clearly must consider both cognitive and affective-evaluative aspects of public attitudes to science.

Jon Miller (Chicago) offers a history of his lifelong quest for Civic Science Literacy (CSL) measures and reports **confirmatory factor analysis** of relevant items combining EU and US 2005 data, including coded open-ended questions. Items are combined and calibrated to form a 'civic scientific literacy' score on a scale of 0 to 100. Miller examines CSL both as a continuous and a threshold measure.

6   *Martin W. Bauer, Rajesh Shukla and Nick Allum*

Sally Stares (LSE) considers knowledge items as observables of an otherwise latent cognitive ability, using **latent trait models** to construct a scale that is reliable for multinational comparisons in Europe. This psychometric scale construction exercise not only considers correct and incorrect answers, which are subject to guessing, but also 'don't know' responses, which vary considerably between items. With this procedure, the best combination of diagnostic items can be determined for any comparative context.

Anil Rai and Rajesh Shukla (Delhi) consider how, what from a strictly statistical point of view is mostly qualitative information in PUS surveys, to transform such information into properly quantitative measures. They demonstrate the logic of **log-linear models** and **logistic regressions** and exemplify their usefulness on the rural subset of India 2004 science survey, creating derived indices on which Indian regions can be compared.

Gauhar Raza and Surgit Singh (Delhi) present an alternative approach, starting from intuition of a **cultural distance index** pertaining to democratisation of knowledge: how far 50% of the population is from the authoritative answer to a knowledge item, expressed in levels of education. The method builds on the innovative Indian format of literacy items, which not only gauge correct or incorrect answers and DKs, but also allow for widespread answers that might be considered supernatural and mythical in nature.

PART IV: CULTURAL ASPECTS OF SENSITIVE TOPICS

Existing survey data also allows for investigations of cultural markers. Survey items are necessarily framed with a particular agenda, but nothing prohibits us from reframing for secondary analysis. We can put a new gloss on an old problem of the field: Is tolerance of pseudoscientific beliefs a matter for the definition of 'literacy' (e.g. the literate need to reject pseudosciences) or is this definition the cultural variable in itself? Existing evidence, which incidentally has been collected to frame literacy by exclusion of pseudoscience, might cede to a more culturally sensitive analysis. The issue is one of **tolerance of different knowledge systems** within any one individual, or **'cognitive polyphasia'**. Such tolerance is the cultural variable that can play out between astrology and science, or between science and religion, or on the strength of any boundary between animals and humans. Are these competing cognitions, or are they compatible as a matter of NOMA (Gould's non-overlapping magisteria)?

Part IV demonstrates how data can be queried to address questions of variable 'individual tolerance'. Nick Allum and Paul Stoneman (Essex) explore a number of hypotheses on the prevalence of beliefs in **astrology** across Europe using **multi-level** modelling and test their compatibility with an informed understanding of science.

Apparent conflicts between **religion and science** might reveal themselves as conflicts of political ideologies. Here, comparison of scientific and political issues is essential. Scott Keeter and colleagues (Washington) demonstrate that scientific controversy over such issues as creationism/evolution cannot be analysed in isolation from other political issues in the US.

Another facet of cultural sensitivity is manifested on the issue of animal experimentation and animal cloning. Fabienne Crettaz von Roten (Lausanne) presents an analysis of German, French, and Italian linguistic communities in Switzerland and shows how systematic cultural variations are at work in defining the **human/animal boundary**. This opens up questions for further research on existing databases. Are these differences a matter of language, equally in evidence when we compare Germany, France, and Italy?

Finally, Rafael Pardo (Madrid) investigates complexities of attitude manifest in the context of embryonic stem cell research. On the basis of a multinational survey the notion of literacy is extended to **worldviews** that determine the moral status of embryos, rooted in dominant religious traditions and personal religious practice. He examines how cognitive literacy interacts with worldview on matters of stem cell research.

PART V: COMPLEMENTARY DATA STREAMS

The last part collates six papers offering new ideas for construction of indicators of science culture, also exploring data streams that differ from survey efforts explored so far. Progress in construction of science culture indicators has to open the field for data other than survey-cum-questionnaire data, consider its limitations, and complement these efforts with other types of data without throwing the baby out with the bath water. The issue is not to substitute the survey effort, but to consider it as only part of the solution, reallocating resources to other data streams.

Hak-Soo Kim (Seoul) presents a novel approach: **Public Engagement with Problems in Consideration of Science (PEP/IS)**. This approach critiques the implicit communication model of PUS as producer driven and product oriented. The alternative PEP model is process and receiver oriented within the functional cycle of **attention, cognition, problem solving, and moving on**. This pragmatist model suggests a fundamental reformatting of survey efforts towards ranking social problems (attention, cognition) and ranking perceived contribution (solutions) of science to these problems from the respondents' point of view. Social categories can be compared on both indicators.

If the culture of science manifests itself in the societal conversation of science, different **genres of communication** are key variables in that conversation. For the domain of science communication a key distinction pertains to **different systems of discourse** in laboratory science, science education, and public communication of science. The former tends towards a conceptual language converging on monosemy, establishing the denotation of terms and concepts; the latter is dominated by narrative and polysemy, connotations to key concepts and issues. Bernard Schiele (Montreal) explores how genres of formal and informal education, including museum exhibitions, lend themselves to construction of science culture indicators that represent the mix of monosemic and polysemic genres of science communication in any particular context.

An important genre of the societal conversation of science is the mass media. In particular newsprint and the Internet are readily amenable to cost-effective streaming, monitoring, and analysis with novel tools of content and semantic network analysis. In this context Carlos Vogt and colleagues (Sao Paulo and Belo Horizonte) present an **automatic system for continuous media monitoring** of science coverage based on 'artificial intelligence' (**SAPO**). The system has been tested in Brazil and Italy (see Bucchi & Neresini). A filter selects relevant 'science' materials from the total newspaper output on any one day. It is envisaged that the filter algorithm 'learns' new relevant inputs. **Intensity of news coverage** is reported daily on three indices: frequency (% of total coverage), density (relative space), and depth of coverage (relative space over relative frequency). The system was tested against manual coding, showing that automatic monitoring is sufficiently reliable to produce a daily, weekly, or monthly index. This promising development enhances potential for content analysis to rival opinion polls in their capacity of taking the pulse of public opinion. Under the heading of 'Science & Society', activities of a bewildering variety of formats have been proliferating: consensus conferences, tables rondes, hearings, national debates, scoping exercises, scientific café, etc. Such events are mostly on specific topics like nuclear waste disposal, GM foods or crops, nanotechnology, and so on. These public events are by aspiration polysemic (see above), aiming at dialogue between scientists and public. This flurry of activity calls for development of a **global score board of deliberative events**. A number of comparative dimensions are relevant: is public participation a **legal right** or an expedient concession? What formats define the local **practice**, on what **issues**? When does **deliberation enter the policy cycle**, upstream or downstream? What are their **functions**: foresight, scoping concerns, public information, or conflict management of science and technologies? On these analytic dimensions a participation scoreboard is a realistic prospect, but far from reality. A stream of studies has traced the global reception of one particular format, e.g. the Danish consensus conference, and how this reception changed the format in different contexts (Einsiedel, 2008). Clearly, proliferation of such event formats does not guarantee enthusiastic public participation and uptake of opportunities. To address this issue, Neils Mejlgaard (Aarhus) and Sally Stares (London) present an **index of 'scientific citizenship** by combining participation in

Towards Cultural Indicators of Science with Global Validity 9

public events and the competence to engage with science. Using European survey research of 2005 they develop scores of type and intensity of participation in each country where the distribution of types of citizens varies.

The new millennium saw an expanding sector of public event making in and around science: the global revival of science museums, science festivals, and a plethora of other events aimed at bringing science closer to the public, and vice versa. These **public engagement activities** have spawned a small private sector of consultancy that organise these events on behalf of sponsors. There is an increasing need to document comparatively actors and activities, posing problems of defining 'PUS activities'. Steven Miller (London) presents a recent effort to **benchmark levels of public engagement activities** across Europe. So far, little information is available on financial inputs, although in some countries this has taken the institutional format of a fixed percentage of the overall R&D budget: e.g. 3% on public engagement activities in Portugal during a particular government. In the context of public event making and general science communication, PUS surveys are important climate indicators, information about the public that needs to be taken into account. Attempts to construct indicators of science culture must assess the level of activity in terms of actors involved, activities undertaken, and financial resources, but data for this purpose is still hard to come by.

An example of **how to monitor scientific culture of a country** in the medium- and long-term is offered by Massimiano Bucchi (Trento) and Federico Neresini (Padua) on the example of OBSERVA-Italy. This Italian non-profit organisation is dedicated to observing the Italian public sphere and its scientific references through several data streams including official statistics, educational attainment, mass media monitoring, regular opinion polls, and targeted research projects. They report on challenges arising from setting up, validating, and sustaining these complementary data streams for continuous operation.

CONSIDERATIONS FOR FUTURE RESEARCH

To close this introduction, we draw attention to several general issues that have arisen in our discussions of how to develop cultural indicators of science with global validity.

Building a campaign guidance system or comparing cultural systems?

A fundamental tension leads at times to misunderstandings. One might characterise the problem as one of two implicit operational frameworks: are we constructing a campaign guidance system or attempting to compare cultural symbol systems. The pragmatic focus of these two agendas overlaps, but the agendas also pull in different directions.

If seen within a strategy of developing a **campaign guidance system**, our indicators of public engagement with, and mass media coverage of, science are monitoring vectors (input) that direct the public mind, assessed as public understanding of science (output). A metaphorical image might be that of a gunner who loads the artillery gun with ordnance aimed at a target: in order to do that effectively he will need a guidance system. Our model is then one of input, output, and outcome, and we are considering the performance of the system, which needs to be improved according to some test criteria. The system is one of strategic action.

Alternatively, if the agenda is the comparison of **cultural symbol systems**, then the focus will be on analysing different genres of science communication, informal or formal, and their interactions. Here, one is not shot at the other like a bullet to its target, but different modalities are relatively autonomous. Symbol systems are co-evolving processes of written and spoken references to science, embodied in words and visual images for that matter. The useful metaphor in this context is **'resonance'** between oscillating systems or different genres of communication; they mutually reinforce or dampen their level of activity. Here the guiding notion is the 'climate of opinion', the symbolic worldviews to be understood and appreciated in the first place. This system is not the target of intervention, but the context of strategic action.

This duality of frameworks is fundamental, remaining an issue. Tensions need not be resolved because they can coexist, not least as they suggest construction of similar data streams for empirical investigation. They do, however, diverge in interpretation.

General or Specific Indicators

Are we observing science in general or does this not always entail particular issues and controversies? Monitoring **science-in-general versus specific developments** (e.g. nuclear power, biotech, Nanotech, etc.) continues to fuel discussions. The present book is dedicated to general indicators of science, and we excluded discussions of specific issues such as water fluoridation, environmental pollution, global warming, or nanotechnology for which there is a plethora of specific research.

An argument against general indicators holds that general attitudes do not exist, because at any moment in time responses to general questions are based on the perception of specific issues; it is better to know what the specific issues are even at the cost of forfeiting time-series comparability. Also, specific indicators prove more useful as campaign guidance information, demonstrating differential public judgements, and this can be used a campaign guidance information.

The argument for general indicators goes as follows: we construct general PUS indicators for the purpose of time-series comparability. The construction of time-series requires **perennial items** that do not date. The pressing issues of the 1950s and 1960s, e.g. fluoridation or nuclear fall-out, are non-issues in the 2000s. Time-series data of knowledge of these specific issues would be poor indicators of science culture. Assessment of cognitive competence must rise above such temporal concerns; the same applies to general attitudes and interests in science.

Rather than substituting the general with accumulation of many specific issues, indicators on **specific issues and general indicators should be empirically compared**. There will be interesting variation across contexts and times in how specific issues aggregate to general literacy and attitudes, and that very relationship between specific and general could be the cultural variable.

How to Develop a Global PUS Research Base

International survey research encourages comparable sampling procedures, questionnaire formats, and interview protocol. The imperative of comparability demands semantic equivalence of question wording and response alternatives, but these are difficult to implement. Translation between different languages and varied speech practices (e.g. differences between 'no' or 'disagree' answers) are among many difficulties. The biggest obstacle is the lack of determination and support for global coordination of these efforts. The global network of researchers presenting this book is a step in this direction.

Practically, the problem arises of how to construct **functionally equivalent items** for purposes of comparison when constructing new questionnaires, but also when considering existing data for secondary analysis. How do we test functional equivalences when operationalising concepts? This is an area to discuss and empirically demonstrate, for example along the lines of differential implementation of literacy items in the US, EU, and India; Raza and Singh allow for supernatural response alternatives to do justice to local culture. Our discussions and the chapters presented here suggest points of consideration to move survey research in this area to a different level and a step-change:

- **Construct local time-series**; analyse in your own context which items are diagnostic and which ones are not, and drop the latter (item response analysis);
- If items are altered slightly from one wave to the next, consider **split-half designs** to calibrate the changes in the time-series;
- From the existing corpus of survey items, stick to a set of **core items for purposes of international comparisons**: *literacy, evaluation, interests and engagement activities*. Defining such a core set will be particularly useful. Something analogous to the FRASCATI Manual used by OECD to assess the R&D contributions of each country should be envisaged;
- **Develop new items** fit for purposes and concerns of your local context;

- **Exchange new items and new ideas** through reports and meetings;
- **Micro-integrate existing data sets.** This includes integration of data within a particular time-series and also across different countries. Such databases will allow for the sophistication of indicator construction and a step-change in the analysis of period and cohort effects that determine these indicators.

Typology Constructions

Some of the chapters in this book demonstrate the construction and usefulness of typologies of PUS (Liu et al.; Mejlgaard & Stares). On the basis of various indicators of literacy, interest, attitudes, and engagement with science, it is possible to identify different cultural milieus of science and anchor these in socio-economic variables such as age, level of education, urban and rural habitation, and others. Longitudinal and cross-sectional comparison must be undertaken not only on single indicators, but also on these typologies of cultural milieus. Are these milieus expanding or contracting? We consider these a wide field of future enquiry. A literature review of existing typologies and their underlying methodology would make a good start.

Perception and Activity Indicators of Science Culture

With our focus on 'culture', we contribute to the discussion on **science indicator systems**. One has to go back to the beginnings of such debates in national and international forums in the 1950s and 1960s to find a similar agenda: the combination of 'subjective' perceptual data and 'objective' activity and attainment data, **two facets of scientific culture, perception and action**. With the present book we work towards reintegration of PUS indicators with highly developed objective S&T indicator systems (R&D expenditure, scientific personal, high-tech balance sheet, publications, and impact measures).

In the US, the NSF Indicator Reports continue to publish chapters on both types of data, but with no attempt to bring these data into conversation with each other. Similarly, FAPESP in **Brazil** includes both types of data in their reports on the science system of the state of Sao Paulo, the largest in Brazil. In the EU these monitoring activities are compartmentalised between the Office of Statistics, which does the R&D figures, and Eurobarometer, which does the occasional perception survey. We discern no coordination and no communication between these agencies. The **Indian** Science Report of 2004 started up with holistic ambitions, but did not accomplish its own ambition; the final report does not map objective and subjective indicators together. In **China**, science literacy is seen as part of Human Resources Development, for which comprehensive input–output indicators systems are in the making. In this context, it might be worthwhile to remind ourselves of the old UNESCO agenda: to measure science communication activities as SRA (science-related activities) combined with R&D and science manpower.

APPENDIX 1: NATIONAL REPRESENTATIVE SURVEYS OF LITERACY, ATTITUDES, INTEREST AND ENGAGEMENT WITH SCIENCE IN GENERAL. THE TABLE GIVES THE YEAR AND THE NATIONAL AGENCY RESPONSIBLE FOR THE SURVEY[3]

Table 1.1 Nationally Rpresentative Survey of PUS, Science Literacy and Attitudes from across the Globe

year	EU	UK	F	E	Fin	Bulg	US	Can	Russia	AUS	NZ	Jap	Korea	Malay	India	China	Brazil	Argent	Venez	Mexico	Colum
1957							Michigan														
1970																					
1971																					
1972			SOFRES																		
1973							Harvard														
1974																					
1975																					
1976																					
1977	EB7	EB7																			
1978	EB10a	EB10a	EB10a																		
1979							NSF														
1980																					
1981																					
1982			SOFRES																		
1983							NSF														
1984																					
1985							NSF														
1986		MORI/Teleg																			

(continued)

APPENDIX 1 (CONTINUED)

year	EU	UK	F	Spain	Fin	Bulg	US	Can	Russia	AUS	NZ	Jap	Korea	Malay	India	China	Brazil	Argent	Venez	Mexico	Colum
1987																	CNPq				
1988		ESRC					NSF														
1989	EB31	EB31	SOFRES/EB	EB31				MST													
1990							NSF								NISTED						
1991												NISTEP									
1992	EB38.1	EB38.1	EB38.1	EB38.1		IS	NSF									CAST					
1993																					
1994			SOFRES													CAST					ColSci
1995							NSF		CSRS	STAP											
1996		WELL				IS			CSRS							CAST					
1997							NSF		CSRS		MST										
1998							NSF														
1999									CSRS												
2000		MORI											KSF	STIC							
2001	EB55.2	EB55.2	SOFRES/EB	55.2	FSB/EB		NSF					NISTEP			NISTED	CAST				Conacyt	
2002	EBspec			FECYT		EB							KSF				FAPESP	SeCyt			
2003							NSF		CSRS							CAST				Conacyt	

APPENDIX 1 (CONTINUED)

year	EU	UK	F	Spain	Fin	Bulg	US	Can	Russia	AUS	NZ	Jap	Korea	Malay	India	China	Brazil	Argent	Venez	Mexico	Colum
2004				FECYT	FSB								KSF		NCAER			SeCyt	MCT		ColSci
2005	EB63.1	MORI/EB	EB63.1	63.1	EB63.1	EB63.1							KSF			CAST	MCT				
2006				FECYT			NSF						KSF								
2007	(EB)	(EB)	(EB)	(EB)	FSB (EB)	(EB)							KSF		NISTED	CAST	FAPESP	SeCyt	MCT		
2008		RCUK					GSS						KSF								
2009							PEW								NCAER						
2010	EB73.1	MORI/EB	EB73.1	EB73.1	EB73.1	EB73.1	GSS			ANU			KSF			CAST	MCT				

EU-7 1977 and 1978
EU-12 for 1989 and 1992
EU-25 iinclusive former Eastern Europe in 2001 and 2002
EU-28 + CH, NO, Turkey, and Iceland in 2010

APPENDIX 2

Speakers

- Allansdottir Agnes (University of Siena)
- Allum Nick (University of Essex)
- Bauer W Martin (London School of Economics)
- Boy Daniel (CEVIPOV, Paris)
- Bucchi Massimiano (University of Trento, Italy)
- Crettaz von Roten Fabienne (University of Lausanne, Switzerland)
- Einsiedel Edna (University of Calgary)
- Hellsten, Iina (Royal Dutch Academy of Arts and Sciences)
- Ke, Wang (China Association for Science & Technology)
- Keeter, Scott (Pew Foundation, Washington)
- Kim, Hak-Soo (Sogang University, Seoul)
- Losh, Susan (Florida State University)
- Miller, Jon (Michigan State University)
- Miller, Steve (STS, University College London)
- Niels, Mejlgaard (University of Aarhus)
- Peters, Hans-Peter (Julich Research Centre, Germany)
- Petkova, Kristina (Bulgarian Academy of Science, Sofia)
- Polino, Carmelo (RICTY, Buenos Aires)
- Raza, Gauhar (NISTEDS, Delhi)
- Schiele, Bernard (University of Quebec, Montreal)
- Shukla, Rajesh (NCAER, Delhi)
- Shimizu, Kinya (University of Hiroshima, Japan)
- Sjøeberg, Svein (University of Oslo, Norway)
- Stares, Sally (London School of Economics)
- Vogt, Carlos (FAPESP, Sao Paulo Brazil)
- Williams, Wendy (Innovation, Industry & Regional Development, Australia)

In Attendance

- Castelfranchi, Yurij (University of Belo Horizonte, Brazil)
- Choi, Jin-Myung (Sogang University, Seoul)
- Damonse, Beverley (SAASTA, South Africa)
- Claessens, Michel (EU-DG Research, Brussels)
- Qiu, Chengli (Chinese Ministry of Science & Technology)
- Ren, Fujun (China Research Institute of Science Popularisation)
- Candy, Sara (Wellcome Trust, London)
- Shi, Shunke (China Research Institute of Science Popularisation)
- Townsend, Saffron (RCUK, Science & Society Unit)
- Todorov, Valery (University of Sofia, Bulgaria)
- Zhang, Chao (China Research Institute for Science Popularisation)

NOTES

1. This meeting received support of the Royal Society's Science & Society Group, on condition we avoided the phrase 'public understanding of science' in the title. The RS's public language has moved to the phrase 'public engagement' because 'public understanding' seemed to connote a deficit that

it was anxious to avoid. Bauer and Shukla have been discussing ideas for an international meeting on cultural indicators for science with the Foreign Relations Team. A note of this idea accidentally reached Darren Bhattachary in 2006, who before leaving his RS position dedicated his remaining budget to this meeting. The organisation of the event was in the competent hands of Chloe Sheppard, Matthew Harvey, and Nicola Hern. We are grateful for the support which this academic project has received.
2. On a different occasion, Cheng Donghong who at CAST sponsors much of these research activities in China referred to this idea of taking a new look at old survey materials with a characteristically metaphorical expression: Hitherto we have cooked our dish (i.e. PUS indicators) exclusively with sweet and sour sauce, but now let us consider cooking it with black bean sauce to discover its real flavour.
3. Table 1.1 lists the known-to-us surveys by year and agency conducting the survey in the particular country. The state of knowledge about these surveys remains scattered. Much information and not least the data often remain informally on somebody's personal computer. It would be desirable to conduct a survey of surveys and document their exact sources/contexts, make the questionnaires available in English, and explore the status and accessibility of the raw data, preferably as a web resource. Such a survey was initially envisaged for this book, but it was too large to handle. This documentation is a project of its own considering the resources in terms of time and effort required to complete it. But this table can serve as an initial guide for any such undertaking.

REFERENCES

Allum N (2010) Science literacy, in: S Hornig Priest (ed) Encyclopedia of Science and Technology Communication, Los Angeles, Sage, p724–727.

Bauer MW (2008) Survey research and the public understanding of science, in: M Bucchi & B Trench (eds) Handbook of Public Communication of Science and Technology, London, Routledge, p111–130.

Bauer MW, N Allum, & S Miller (2007) What can we learn from 25 years of PUS research? Liberating & widening the agenda, Public Understanding of Science, 15,1,1–17.

Easton D (1976) A re-assessment of the concept of political support, British Journal of Political Science, 5, 435–457.

Einsiedel E (2008) Public participation and dialogue, in: M Bucchi & B Trench (eds) Handbook of Public Communication of Science and Technology, London, Routledge, p173–184.

2 The Culture of Science and the Politics of Numbers

Benoît Godin

Every one of us is born into a world where science and technology is omnipresent: much of our specialized and everyday knowledge (and assumptions) is rationally based and relies on mechanical principles and models. We are accustomed to scientific thinking early on at school, we hear about discoveries and inventions on the news daily, and we use technology every day through the commodities we buy. This whole process (the "scientification" of culture) began in the early seventeenth century, and was largely completed by the early nineteenth century. The end result is often called the modern era or modernity.

However, as M. Jacob put it, science is not a foreign object in need of assimilation (Jacob, 1988: 5–6). In fact, we are already socialized or acculturated to science before we make any conscious effort to understand it or think about it. There is no separation between science on one hand, and its communication on the other. Neither are they sequential; it is not a case of science first and then its communication. Science (including its dissemination) is an integral part of Western culture.

A culture of science is a culture defined, partly or wholly, by and through science. It is a culture in which a central set of institutions and activities are concerned with science, broadly defined, including the uses and effects of the productions arising from these institutions and activities. A culture of science is more or less developed depending on the existence and strength of these institutions, their activities, their productions, and their effects on society. Only the consideration of this whole set allows one to understand a culture of science. From an analytical point of view, a culture of science is that sum of dimensions or subsystem that includes

- Institutions (research);
- Productions (graduates, knowledge, technologies);
- Diffusion, use and users (education, transfer, communication);
- Impacts (effects on society, the economy, the individual);
- Environment (laws, economic system, social values).

In this sense, "scientific literacy" and the "public understanding of science" is only one dimension of a culture of science. Most of the chapters in this book deal with this dimension. I suggest the concept "culture of science" as an alternative and more encompassing concept (Godin, 1999; Godin and Gingras, 2000). And to the five dimensions of a culture of science listed above, I would add a sixth: measurement, or the culture of numbers. Measurement is a major component of a culture of science, and measurements of science itself, and of the culture of science, are no exception (Godin, 2005a; 2005b). Measurements of science have been conducted since the mid-nineteenth century, and became even more important after World War II. Many economic and government decisions concerning science rely directly on numbers. This necessarily has effects on science and the culture of science. Measurement conveys a representation of what is measured. It is evidence of how our societies understand science, and studying measurement allows one to understand how and why scientists and institutions act in a particular way and not another, and thus to better understand what our culture of science is.

How have researchers and governments measured the culture of science? In what ways have these measurements in turn affected the culture of science? It is through measurement, statistics, and their history that I look at the culture of science in this chapter: how we measure the culture of science today and why we chose historically to measure the way we do and not some other way. The first part of the chapter documents the origins and development of measuring the culture of science, and the "biases" introduced according to the contexts in which the measurements developed. The second part discusses the cultural representation of science resulting from these measurements. The third part turns to the efforts of one organization, namely UNESCO, to offer a different conception.

THE CULTURE OF SCIENCE

Statistics on science emerged in the mid-nineteenth century.[1] At the time, the statistics came from scientists themselves, and measured the number of men of science (scientists): their demography and geography (Table 2.1). The sociopolitical context explains the kind of statistics produced. It was a time when the progress of civilization, later called culture, and the contribution of great men to it, were central to many discussions. In fact, many thought that the lower social classes and "unfits" reproduced more than men of science did, representing a danger to the human race. This led to the idea of measuring the number of men of science a nation produces (F. Galton; A. de Candolle). Soon, the idea came to some of using the numbers as an indicator of the culture of science and how a nation supports science. In fact, one of the first widespread uses of statistics would be to contribute to the advancement of science and to improve the social conditions of scientists. Scientists complained that they received too little recognition for their work: low salaries, few prizes. The number of men of science came to serve as indicator of the social condition of these scientists (J. M. Cattell) and of the culture of science: a low number of men of science in a society was seen as indicative of insufficient public support to the scientific profession, and to a low culture of science (Godin, 2007a).

From the mid-twentieth century onward, the statistics being collected changed completely. Thereafter, the main producers of statistics were governments and their statistical bureaus, and the most cherished statistics became the money spent on research and development (R&D) activities. Again, the socioeconomic context explains the situation. First, efficiency has been on the agenda of most organizations since the beginning of the century, so controlling expenses has become a priority. It was precisely in this context that a national budget for research would be constructed. This would soon be followed by statistics on production from investments in research activities. Second, and in a more positive sense, governments sought to contribute to the development of science, and needed quantifiable objectives. A ratio on the amount of GDP devoted to R&D, among others, would serve this end (Godin, 2007b).

The end result of this historical development is a set of statistics, which I analyze here according to the dimensions listed above,[2] and which define a culture of science. I concentrate on official (i.e., government) statistics. In point of fact, it is official numbers that define our measurement of a culture of science. The financial resources governments invest in collecting statistics, as well as the recurring production of these statistics, help to ensure that the official statistics (rather than those of academics) are the statistics spontaneously used to discuss science and the culture of science. Governments automatically bring legitimacy to the numbers they produce.

Table 2.1 Historical Development of Statistics on Science (19th–20th Centuries)

Stages	Source	Main statistics
Emergence (1869–circa 1930)	Scientists (Galton, Candolle, Cattell)	Number of scientists
Institutionalization (1920–circa 1970)	Governments and national statistical offices (pioneering role of the United States)	Monetary expenditures
Internationalization (1960 and after)	International organizations (UNESCO, OECD, European Commission)	Indicators International comparisons

Institutions

A culture of science rests first of all on institutions devoted to the production of scientific knowledge (for a specific nation, this is only partly true, as one could theoretically buy or absorb all its scientific knowledge from outside sources). Accordingly, the identification, listing, and counting of these institutions was one of the first tasks to which statisticians devoted themselves decades ago, and collecting such data was considered, in the 1950s and 1960s, to be the first step toward a sound science policy. Cattell's work in the early twentieth century began by producing lists of individuals, universities, and scientific societies, and was continued by the US National Research Council from the 1920s onward, then by the US National Science Foundation beginning in the 1950s. Early in their existence, UNESCO and the OECD also collected and published lists of international organizations, starting in the 1960s.

With time, institutions have become a mere unit for surveying. Today, it is virtually impossible to obtain a complete list of scientific institutions for a given country, or even just the number of such institutions. In the case of industrial laboratories, for example, data are kept secret for confidentiality reasons. It has become customary instead to classify institutions according to the economic sector to which they belong, and to produce only aggregate numbers by sector: industry, government, university, nonprofit (and foreign).

Such a measurement developed for policy purposes: construction of aggregates and simple statistics useful to policy-makers. That "useful" statistics is the "national science budget", or Gross Domestic Expenditures on Research and Development (GERD). It is the sum of the monetary expenditures devoted to research activities in all of the above economic sectors. GERD is a measure of what has come to be known as input: the resources invested in research activities. It is often accompanied by numbers of personnel involved in research activities.

Without doubt, GERD has become the measure par excellence of a culture of science: the more resources are invested in research activities, the more a nation is said to be scientifically oriented. Until the 1950s it was the number of men of science that measured a civilization or a culture; now it is GERD. The GERD/GDP ratio is used to compare nations with regard to their efforts toward developing a culture of science. The ratio serves as an objective for policy. For example, a ratio of 3% of a nation's economic resources devoted to research activities is said to be the optimal level of national investments in science, and has been the goal pursued by governments since the early 1960s.

Production

One does not assess or evaluate the state of a culture of science simply by counting the institutions and the resources devoted to research activities. Institutions lead to productions or outputs, so says the traditional model, and it is these outputs that ultimately motivate the establishment and funding of institutions. What are the outputs that were measured and that have come to define a culture of science?

Input → Research activities → Output

Figure 2.1 The traditional model for measuring science.

The institutions devoted to science produce different kinds of output. Some is purely scientific (knowledge); others are technological (inventions); still others are human (graduates). Over time, technological output has taken central place over scientific output in the measurements. Certainly, since the beginning of the twentieth century, bibliometrics has devoted itself to measuring scientific knowledge (the number of papers), and a whole "industry" of researchers has become active in this type of study. But to most "statisticians", above all official statisticians, what matters most is useful knowledge or knowledge put to use, namely technology. A look at the annual series of indicators produced by governments and international organizations, and at the methodological manuals available to national statisticians for measuring science, bears

Table 2.2 OECD Methodological Documents (Year = first edition)

Manuals
The Measurement of Scientific and Technical Activities: Proposed Standard Practice for Surveys of Research and Development (Frascati manual) (1962).
Proposed Standard Practice for the Collection and Interpretation of Data on the Technological Balance of Payments (1990).
Proposed Guidelines for Collecting and Interpreting Technological Innovation Data (Oslo manual) (1992).
Data on Patents and Their Utilization as Science and Technology Indicators (1994).
Manual on the Measurement of Human Resources in Science and Technology (Canberra manual) (1995).
Measuring Productivity (2001).

Handbooks
OECD Handbook on Economic Globalisation Indicators (2005).

Guides
Guide to Measuring the Information Society (2005).

Frameworks
A Framework for Biotechnology Statistics (2005).

Others
Bibliometric Indicators and Analysis of Research Systems: Methods and Examples (1997).

witness to this understanding: the culture of science is a culture of technology (Table 2.2).

How do we explain this? Behind the culture of science and its measurement is economics, which in governments' hands has become a doctrine. One measures the quantifiable, the tangible, and what is economically efficient: money spent on R&D, patents, technological balance of payments, trade in high-technology products, technological innovation, and productivity—all economic measures.

Despite this purely economic focus, one different set of statistics has remained central to official statistics for over a century: the number of graduates. To some, this is a measure of a "culture" of a society, or knowledge. However, to others graduates are "human capital", and they are measured because they contribute to economic growth. Whatever the meaning of the statistics, one thing is sure: the reasons for producing the statistics have changed over time and reflect the context of its uses.

Diffusion and Uses

In such a context, it is surprising that few if any of the official measurements are devoted to the diffusion and use of science. If science is to be useful, it has to be diffused and used through society and the economy. Productions themselves are not enough. One must ask, "is science used, and to what extent?" However, even UNESCO, to which the transfer of scientific knowledge has always been central to policy, has never developed indicators on the diffusion of science. With regard to the OECD, measurements of diffusion and use are quite recent, and are limited to information and communication technologies (ICT).

The reasons for this situation are multiple. Certainly, a major reason is ideological. For decades, policy on science has relied on a rhetoric of promises: fruits follow automatically from basic research and its funding. Similarly for technology: once technologies are there, they will automatically be used and lead to progress. One forgets that nothing is automatic. There are costs, time lags, and resistance. Everything requires effort.

However, there exist some measures (or proxies) for statistics on diffusion among official statisticians. The first series is that on education. In fact, the school is the basic "medium" for diffusing a culture of science. For decades, nations have thus measured the literacy of their populations in terms of enrollment and diplomas. Many have now gotten involved in measuring student achievement (OECD's PISA surveys). Unfortunately, statistics on education are usually not part of series on science statistics. In government departments, separate divisions are responsible for education and science, so separate documents and statistical series are produced.

Another proxy of official measurement on diffusion is scientific literacy (knowledge of scientific facts). The US National Science Foundation and the European Commission have included a chapter on "scientific culture" in their regular series of indicators on science since the 1990s. However, it is to academic researchers that we owe the most serious of such measurements. The literature on the Public Understanding of Science has produced a series of studies and measurements, often commissioned by governments, on many dimensions relating to the communication of science to the public, and the relation between the public and scientists (Bauer et al., 2007). The literature has also dealt with an important aspect of the culture of science—an aspect too often relegated to environment or context—the social values held by individuals concerning science (i.e., interests and attitudes).

Impacts

As are diffusion and use, the impacts or effects of science are central to a culture of science. Unused, science remains isolated from society; similarly a science with no impact is "wasted". However, with regard to the diversity of productions arising from science and to their uses, the impacts are poorly measured.

The range of impacts of science on society is very broad: economic, social, cultural, political, environmental, health, etc. This wide spectrum gave rise to an early conceptual framework for study and policy. From the 1920s onward, and for thirty years, US sociologist William F. Ogburn studied technology and measured its effects on society. Ogburn was interested in explaining what he called cultural lags, or maladjustments, between technology and society.

Ogburn had few followers. The measurement of the culture of science soon turned economic. Of all the impacts of science on society, the only ones for which there are ongoing series of measurements are economic growth and productivity. Econometric studies abound on linking science (R&D) to economic growth and productivity. Policy documents use the results of these studies regularly, and public organizations put statistics on science side by side with those on productivity on scoreboards of indicators.

The reasons for this state of affairs are many. Methodological limitations are certainly important. There are problems of causality and attribution in the measurement of impacts. However, this is not the whole story. Don't we have (imperfect) measures of economic impact (productivity)? Aren't the limitations as important in measuring the impact of science on productivity as in measuring the impacts of science on the family? We come back to the economic doctrine then. The obsession with economics has truly defined our measurements: our culture of science is economically oriented.

A CULTURAL REPRESENTATION

The relationship between the culture of science and its measurement is a dialectical one. We have just discussed how particular ideas about science and contexts affect the numbers produced. In turn, the numbers affect our conception of science, and this in turn affects the culture of science. Numbers carry a social or "cultural" representation of the world, control over which is an issue of national political significance.

Our current representation of science is the result of a conceptual and statistical construction over the twentieth century. This construction culminated in the OECD's methodological work, particularly the Frascati manual (OECD, 1962), now in its sixth edition, a methodological manual for conducting surveys on R&D. Four elements characterized the official definition of science, and therefore the culture of science, over the twentieth century (Godin, 2009). First, science has been defined (and measured) based on the concept of "research". This is a purely social construction, as science could also be defined otherwise than as activity or research. Scientists and philosophers have long defined science by its output (knowledge) and method, economists have defined it as information, and sociologists have defined it by its institutions and practices. Early officials' definitions also varied. Until recently, the USSR and the communist countries, for example, used a broad definition, in which science included more than just research, that is, covered areas excluded from the OECD definition of research. UNESCO, for its part, developed the concept of scientific and technological activities, which included not only research, but also education and scientific and technological services.

Defining science as research came about due to the institutionalization of research as a major phenomenon of the twentieth century. By the 1960s, most large organizations had recognized research as a contributor to economic growth, performance, and innovation, and many organizations were devoting an increasing share of their budget to these activities. Hence the need for a better understanding of what was happening, and thus for measuring the efforts expended on research.

However, the definition of science as research is also due to a second factor, namely accounting and its methodology. There are activities that are easily measurable and others that are not. There are activities for which numbers are available, and others for which they are not. There are activities that can be identified and distinguished easily, and some that in practice are difficult to separate. Officials chose to concentrate on the more easily measurable activities (research), for methodological reasons having to do with accounting (costs) and its measurement. One measures research activities rather than research outputs (or knowledge), research activities rather than (research plus) scientific and technological services, research and development rather than research exclusively, and purely systematic research rather than (systematic and) ad hoc. Let's look at these choices.

The second characteristic of science as defined by governments and their statistics is R&D. Research is defined essentially as R&D, where "D", for development, corresponds to over two-thirds of expenses. Development is composed of several activities like scale activities, pilot plants, and design. It is an important category of taxonomies on research. Since the survey on industrial research by accountant R. N. Anthony from Harvard University, conducted for the US Department of Defense in the early 1950s, research has been defined as composed of three categories: basic research, applied research, and development. Development got into R&D for many reasons, among them because of its importance in industrial (and military) research, and because of the difficulty of separating (and budgeting) development from other activities, such as research proper. It also owes its presence in the definition to the priority that technological development had on the science policy agenda.

The third characteristic of the official definition of research is the idea of "systematicness". Industrial research underwent expansion after World War I. Most big firms became convinced of the necessity to invest in research, and began building laboratories for the purpose of conducting research. It was considered that research had to be "organized and systematized". The issue of "systematically" organizing industrial research was on every manager's lips. This is the rationale behind the official definition of research. Research is organized research, i.e., laboratory research.

The definition of research arising out of the industrial survey has had influence on the whole methodology of questionnaires, including questionnaires for surveying government and university research that used the same definition. The main link here is, again, US accountant Anthony. In the survey he conducted for the Department of Defense, Anthony showed that firm size was one of the main variables explaining R&D investment. Consequently, he suggested concentrating on larger firms. Hence, research was thereafter equated with systematized research, namely large organizations with dedicated laboratories. This rationale soon came to be related to another one: the costs of conducting a survey. Because there are tens of thousands of firms in a country, units surveyed have to be limited to manageable proportions. This was done by introducing a bias in industrial surveys: the survey identified all major R&D performers, that is, big firms with laboratories (or "organized" research) and surveyed them all, but selected only a sample of smaller performers, when they selected any. This decision was also supported by the fact that only big firms had precise bookkeeping practices on R&D, because the activity could be located in a distinct and formal entity, the laboratory.

The fourth and last aspect of the official concept of science as research is the exclusion of a certain type of activities, namely scientific and technological services. The choice made was to separate research from other (routine) activities, however indispensable they may be to research: planning and administration, expansion of R&D plant, data collection, dissemination of scientific information, training, and testing and standardization. In fact, firms had accounting practices that did not allow these activities to be easily separated.

The four characteristics of science just identified and the representation of science emerging from the numbers have not gone unchallenged. The decision to concentrate on research, or R&D, had its opponents. In fact, at the same time as the now-official definition was being constructed, others were suggesting different approaches. The issue was partly a matter of whose vision would define science.

ALTERNATIVE REPRESENTATIONS

We owe to UNESCO the development of a more inclusive definition of science. The fact that UNESCO was devoted to educational and cultural development as much as economic development explains its interest in a broader concept. Also, the fact that the organization was dominated by scientists, not economists as was the case at OECD, was an influential factor in defining science differently. Certainly very early on the OECD was developing projects to broaden its measurements, but the organization never put these ideas into practice.

Scientific and Technological Potential

The first early UNESCO attempt to broaden the definition of science, and therefore to offer a different representation of the culture of science, was the concept of "scientific and technological potential" accompanied by a manual on methodology for surveying such potential. According to the organization, "the national scientific and technological potential (STP) comprises the whole of the organized resources a country has at its sovereign disposal for the purposes of discovery, invention and technological innovation, and for the study of national and international problems that science and its applications involve" (UNESCO, 1969a: 20; 30–32), namely:

- human resources;
- financial resources;
- physical resources;
- information centers and services;
- research programs;
- decision centers.

In UNESCO's view, a survey of scientific and technological potential should cover the social and human sciences as well as the natural sciences, and include a survey of the natural resources of the country. Briefly stated, UNESCO's view of a culture of science and its measurement was not limited to R&D activities, but included a larger set of activities. To UNESCO, "such activities play an essential part in the scientific and technological development of a nation. Their omission from the survey corresponds to a too restricted view of the scientific and technological potential and would constitute an obstacle to the pursuance of a systematic policy of applying science and technology to

28 Benoît Godin

development" (UNESCO, 1969a: 21). The obstacle was perceived to be bigger in the case of developing countries because of their reliance on knowledge produced elsewhere, that is, on knowledge transfer:

> What would be the use of transfer of technology or knowledge derived from R&D if the countries to which they were passed lacked the infrastructure necessary to make them operational? (Bochet, 1977: 5)

> Programmes of R&D in the developing countries are not sufficient to guarantee a rise in the scientific and technological activities of a country. In addition to those important activities it has been found necessary to create an infrastructure of scientific and technological services which, on the one hand, support and aid R&D proper, and on the other hand, serve to bring the results of R&D into the service of the economy and the society as a whole. (Bochet, 1974: 1)

Related Scientific Activities

The manual on scientific and technological potential, updated in 1980, came from a Policy Division. The scope of such a survey was very broad, probably too broad. So UNESCO's Division of Statistics turned to a more limited set of activities: scientific and technological services, or "related scientific activities" as they came to be called.

UNESCO and its consultants regularly challenged the definition of science centered on R&D, insisting on adding related scientific activities. The official argument they offered in document after document was the contribution of these activities to science:

> The priority given to R&D in data collection is only a matter of expediency, and does not mean that the importance of an integrated approach to R&D seen within a full context of educational and other services is underestimated. One may even argue that it is only in close conjunction with these services that R&D can be meaningfully measured—because they are *indispensable* for research efficiency (...) and should precede rather than follow the emergence of R&D in a country. (Gostkowski, 1986: 2)

Interest in related scientific activities was the consequence of UNESCO's basic goal of extending standardization beyond industrialized (i.e., OECD) countries. The first step in that program, initiated in 1967, was Eastern Europe. As early as 1969, UNESCO published a paper titled *The Measurement of Scientific and Technical Activities*, written by C. Freeman. The document was concerned with the standardization of data between Western and Eastern Europe, and with the necessity of measuring related scientific activities: R&D is "only part of the spectrum of scientific and technological activities (...). It is considered essential at the outset to visualize the whole and to begin to build the necessary framework for establishing a viable data collection system covering the whole field" (UNESCO, 1969b: i). The document led to a guide and a manual on science and technology statistics, both published in 1984.

What was peculiar to eastern countries at the time was the fact that R&D was not designated as such. The USSR, for example, put all its statistics on science and technology under the heading "science". Moreover, government science, for example, included training, design, and museums. UNESCO thus had to choose between two options for standardization: follow the OECD and concentrate on R&D, or measure, as in Eastern Europe, both R&D and related scientific activities. To UNESCO, related scientific activities were defined as:

- Activities which, whilst not being actually innovative in character, form the *infrastructure* necessary for the effectiveness of R&D;
- Activities which, within the framework of science and technology, maintain the continuity of the routine *competence* necessary for R&D activity, although not playing a direct part in it;
- Activities which, whilst not being innovative in character, have, in varying degrees, *connections* with R&D activities, created according to circumstances, either internally or externally to R&D. (Bochet, 1974)

As such, related scientific activities included information and documentation, standardization, museums, topography, prospecting, etc. From UNESCO's efforts to measure related scientific activities came a guide on (but limited to) scientific and technical information and documentation, drafted in 1982, tested in seven countries, and published in a provisional version in 1984. The guide defined scientific and technical information and documentation as "the collection, processing, storage and analysis of quantitative data concerning information activities (...)" (UNESCO, 1984: 5). To UNESCO, the principal items to be measured were the institutions and individuals performing these activities, the amount of financial resources and physical facilities available, and the quantity of users.

In the end, the program to include Eastern Europe failed. Equally, UNESCO never collected data on related scientific activities, including information and documentation. Why? The reasons are many. First, UNESCO itself came to concentrate on R&D. The activity was said to be easier to locate and to measure, and had the virtue of being an "exceptional" contribution to science and technology. R&D was perceived as a higher order of activity. No argument was needed to convince people of this hierarchy. It was taken for granted by almost everybody that "soft" activities like market studies or design, for example, were not part of science. This was the general understanding of the time. The little interest that did exist in counting related scientific activities among countries was generally motivated by political considerations, such as the need to present a higher science and technology performance. Hence, while

UNESCO pushed for the concept of related scientific activities, it simultaneously argued for the centrality of R&D. Here is one example, among many, of the rhetoric used:

> Because of the *unique* (*"exceptionnel"* in the French version) contributions that R&D activities make to knowledge, technology, and economic development, the human and financial resources devoted to R&D, which might be called the *core* of science and technology, are usually studied in greater detail. (UNESCO, 1986: 6)

The second reason that UNESCO never pursued work on related scientific activities was linked to the fact that, in the end, few countries were interested in these activities. A meeting of experts on the methodology of collecting data on scientific and technical information and documentation activities was held in 1985 to assess the lessons learned from the pilot surveys. It was reported that the activities were not deemed all that important or urgent, that the purpose for measuring them was not obvious, and that there were difficulties in interpreting the definition.

But the main reason that UNESCO failed in its efforts to measure related scientific activities was that the United States left the organization in 1984, accusing UNESCO of ideological biases. The decision had a considerable impact on the UNESCO Division of Statistics in terms of financial and human resources. It led to the decline, and almost the disappearance, of UNESCO in the measurement of science.

Scientific and Technological Activities

In its efforts to extend science measurement, UNESCO faced two challenges, corresponding to two groups of countries: "The methodology so developed [OECD] must be adapted for use by Member States at widely varying levels of development and with diverse forms of socio-economic organizations", UNESCO explained (UNESCO, 1966b: 3). The first group [developing countries] had almost no experience in the field of science and technology statistics, whereas the second [Eastern European countries] had an economic system that required important adaptations to fit OECD standards:

> A statistical methodology developed in a country with 40,000 scientists and 200,000 engineers in all fields of science and technology may be of little use in a country with only 50 scientists and 200 engineers; a questionnaire suitable for use in a country with a highly developed statistical organization may be impractical in a country where few professional statisticians are struggling to gather the most basic demographic and economic data essential to planning. (UNESCO, 1966a: 3)

The task was enormous: "The Secretariat does not underestimate the formidable problems which are involved in such an undertaking, but is confident that, with the help of Member States having experience in this field of statistics, much progress can be made toward this goal" (UNESCO, 1966a: 4). "Worldwide" standards were consequently suggested as early as 1969 (UNESCO, 1969b). The UNESCO manual dealt with the necessity of measuring related scientific activities, as discussed above, but also with another concept, that of "scientific and technological activities".

The concept of "scientific and technological activities" was the third and last effort of UNESCO to broaden the definition and measurement of a culture of science, and would become the basis of UNESCO's philosophy of science measurement:

> Broadening of the scope of science statistics is particularly appropriate to the conditions of most of the developing countries which are normally engaged in more general scientific and technological activities, rather than R&D solely (OECD, 1969c: 9). In developing countries proportionally more resources are devoted to scientific activities related to the transfer of technology and the utilization of known techniques than to R&D per se. (UNESCO, 1972: 14)

According to the UNESCO recommendation, adopted by member countries in 1978, scientific and technological activities were composed of three broad types of activities: R&D, scientific and technical education and training, and scientific and technological services (or related scientific activities) (Figure 2.2) (UNESCO, 1978). The UNESCO recommendation was short-lived. In 1986, the director of the UNESCO division of statistics on science and technology concluded that "Due to considerable costs and organizational difficulties, the establishment of a system of data collection covering at once the full scope of scientific and technological services and S&T education and training in a country has been considered not practicable".

A few years after the UNESCO recommendation, the OECD appropriated the concept of scientific and technical activities in a new chapter added to the 1981 edition of the Frascati manual. Certainly, the concept of "scientific activities" had already been present in the manual since 1962, and that of scientific and technical activities in the title of the manual. But now it appeared in an introductory chapter "addressed principally to non-experts and (...) designed to put them in the picture" (OECD, 1981: 13). The purpose was not to measure scientific and technical activities but "to distinguish R&D, which is being measured, from S&T education and training and scientific and technical services which are not" (OECD, 1981: 15). It had correspondingly few consequences on the standard definition of science and its measurement.

However, one decision has had a huge impact. The same edition of the Frascati manual that introduced the concept of scientific and technical activities also introduced that of innovation. Of all non-R&D activities and related scientific activities, innovation is the only one in the history of OECD statistics on science that was given a certain autonomy, and a status equivalent to R&D: in 1992, the OECD member countries adopted a manual devoted specifically to the measurement of innovation—the Oslo manual (OECD, 1992). Since then, innovation is the real benchmark for assessing a culture of science.

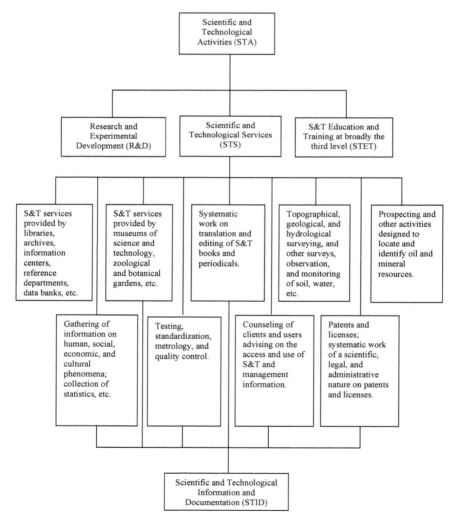

Figure 2.2 S&T activities (UNESCO).

CONCLUSION

The story of this paper attests to the politics of numbers. Over the twentieth century, the development of statistics on science and the culture of science evolved from issues on civilization, culture, and men of science to issues on money, efficiency, and technological innovation. At the international level, there were also issues about who would control the field and impose its vision. In attempting to accommodate its member countries, UNESCO's efforts were guided as much by the desire to generate a larger range of standardization than the OECD did as by political reasons. The OECD surprised UNESCO when, in 1963, it adopted a standard methodology for conducting R&D surveys, a manual that, according to the OECD, "attracted considerable interest in other international organizations and in member countries (. . .), [and was] one of the most important [items] in the Committee's program" (OECD, 1964: 11). As a matter of fact, as early as 1960 UNESCO was trying to assess resources devoted to science and technology in developing countries. Was it not UNESCO's role, then, to deal with international standards? By 1958, UNESCO had already produced standards for education and was working on further standards for periodicals (1964) and libraries (1970).

Given the OECD Frascati manual, if UNESCO wanted to get into the field of science measurement, it needed to distinguish itself. It did so by taking related scientific activities more seriously than did the OECD. But it was, in the end, only slightly more interested in these activities themselves than was the OECD. UNESCO had to find a niche where it could become a credible player in the methodology of science and technology statistics. Moreover, UNESCO simply followed Eastern Europe's experience, since that was the easiest way to standardize statistics outside of OECD countries.

The OECD member countries refused to follow, which would have meant departing from their practices, because, as reported by the OECD Secretariat in its responses to an ad hoc review group on statistics, "les pays de l'OCDE *perdraient le contrôle complet* qu'ils détiennent actuellement sur leurs normes et méthodes" (OECD, 1977: 16):

> The time is not ripe for "world-wide" science standards and (. . .) the official adoption of the current draft of the UNESCO Manual in a fit of *empty internationalism* would be unlikely to bring any practical benefits. (. . .) The current draft is, in our view, rather too ambitious and insufficiently based on practical experience to play this role. (OECD, 1977: 18)

Over the twentieth century, what has been the place of statistics on scientific literacy, public understanding of science, and dialogue between science and society in official statistics? This is part of the third dimension of a culture of science, which we have called diffusion and use. If one were to use a restricted definition of a culture of science (limited to scientific literacy and/or public understanding of science) one could reasonably argue that the culture of science is badly measured: statistics on this dimension are few among official measurements. Nevertheless, there have been many official measurements on other dimensions of the culture of science for over a century. In fact, statistics on the culture of science abound, for better and for worse.

Statistics carry a given representation of what a culture of science is. The official statistics and indicators are witnesses to a moral economy or political hierarchy in public representations of science—a hard core and a periphery: R&D and other activities; the natural and experimental sciences and other sciences; high-technology and other technologies; and the scientist and the public. Certainly, from time to time governments and international organizations have got into the measurement of the "public". The Eurobarometer is one of the very few instruments to this end. However, in general the field is left to academics, as is bibliometrics. In the end, numbers definitively attest to our social, political, and cultural values.

NOTES

1. Science is used here as a shorthand for science, technology and innovation.
2. This paper does not deal with the last dimension (environment).

REFERENCES

Anthony, R. N. and J. S. Day (1952), *Management Controls in Industrial Research Organizations*, Boston: Harvard University.
Bauer, M. W., N. Allum and S. Miller (2007), What Can We Learn from 25 Years of PUS Survey Research? *Public Understanding of Science*, 16 (1): 79–95.
Bochet, J.-C. (1974), The Quantitative Measurement of Scientific and Technological Activities Related to R&D Development, CSR-S-2, Paris: UNESCO.
Godin, B. (1999), *Les usages sociaux de la culture scientifique*, Québec: Presses de l'Université Laval.
Godin, B. (2005a), *La science sous observation: cent ans de mesures sur les scientifiques, 1906–2006*, Quebec City: Presses de l'Université Laval.
Godin, B. (2005b), *Measurement and Statistics on Science and Technology: 1920 to the Present*, London: Routledge.
Godin, B. (2007a), From Eugenics to Scientometrics: Galton, Cattell and Men of Science, *Social Studies of Science*, 37 (5): 691–728.
Godin, B. (2007b), Science, Accounting and Statistics: The Input–Output Framework, *Research Policy*, 36 (9): 1388–1403.
Godin, B. (2009), What is Science: Defining Science by the Numbers, 1920–2000, Two Parts, *Foresight*, Forthcoming.
Godin, B. and Y. Gingras (2000), What is Scientific and Technological Culture and How is it Measured? A Multidimensional Model, *Public Understanding of Science*, 9: 43–58.

Gostkowski, Z. (1986), *Integrated Approach to Indicators for Science and Technology*, CSR-S-21, Paris: UNESCO.
Jacob, M. C. (1988), The Cultural Meaning of the Scientific Revolution, New York: Alfred A. Knopf.
OECD (1962), *The Measurement of Scientific and Technical Activities: Proposed Standard Practice for Surveys of Research and Development*, DAS/PD/62.47.
OECD (1964), *Committee for Scientific Research: Minutes of the 11th Session*, SR/M (64) 3.
OECD (1977), *Response by the Secretariat to the Questions of the Ad Hoc Group*, DSTI/SPR/77.52.
OECD (1981), The Measurement of Scientific and Technical Activities: Proposed Standard Practice for Surveys of Research and Experimental Development, Paris: OECD.
OECD (1992), *Proposed Guidelines for Collecting and Interpreting Technological Innovation Data*, DSTI/STII/IND/STP (91) 3.
UNESCO (1966a), *Problems Encountered in the Development of a Standard International Methodology of Science Statistics*, UNESCO/CS/0666.SS-80/5.
UNESCO (1966b), *Science Statistics in UNESCO*, UNESCO/CS/0666.SS-80/3.
UNESCO (1969a), Manual for Surveying National Scientific and Technological Potential, Series on Science Policy Studies and Documents, No. 15.
UNESCO (1969b), *The Measurement of Scientific and Technical Activities*, Paris: UNESCO.
UNESCO (1969c), *Science Statistics in Relation to General Economic Statistics: Current Status and Future Directions*, UNESCO/COM/CONF.22/2.
UNESCO (1972), *Considerations on the International Standardization of Science Statistics*, COM-72/CONF.15/4.
UNESCO (1978), *Recommendation Concerning the International Standardization of Statistics on Science and Technology*, Paris: UNESCO.
UNESCO (1984), *Guide to Statistics on Scientific and Technological Information and Documentation (STID)*, ST-84/WS/18, Paris.
UNESCO (1986), *Provisional Guide to the Collection of Science Statistics*, COM/MD/3, Paris.

Part I
Longitudinal Analysis

3 Attitudes toward Science in France 1972–2005

Daniel Boy

In 1972, officials at the French *Délégation Générale à la Recherche Scientifique et Technique* commissioned a study from the *Fondation Nationale des Sciences Politiques* on public attitudes toward scientific research in France. This request was made as a result of certain concerns prevalent among those who managed public research. They were anxious to avoid public protest against directions taken by investment in scientific research, which was beginning to be criticized elsewhere and particularly in the United States. In 1971, the OECD commissioned a report from Dean Harvey Brooks of Harvard University on changing science policies. The report criticized directions taken by scientific and technological development, and, in particular, it underscored a rising trend in opinion that still holds true today: "There is today a certain emotional reaction against science and technology and some serious criticism of them. It is realized that the immense social benefits that have flowed from science and technology are sometimes accompanied by social drawbacks. Thus, policies concerned with science and technology in the next decade will have to take into account, much more explicitly than in the past, the benefits and disadvantages, real and potential, that may result from the application of science or the deployment of technology." The study on French attitudes toward research was entrusted to a research team at the *Centre d'Etude de la Vie Politique Française* (CEVIPOF) and the field survey was conducted by the SOFRES survey organization in 1972.

Since that initial study, five other substantial surveys have been carried out: in 1982, 1989, 1994, 2001 and 2007.[1] Some of the survey questions were repeated word for word from one survey to the next, other questions were asked only in some of these surveys.

An integrated database was then built using elements in common to the six surveys. This database provides an opportunity to measure changes in French attitudes toward science over a span of thirty-five years. Moreover, this database allows the concept of "generation" to be used: the youngest generation (18 to 35) questioned in the first survey (1972) is now aged 53 to 60 years old. To what extent has this generation (and other generations, by comparison) maintained or changed their initial attitudes toward scientific issues? To address this question, and facilitate the comparison of French data with other data, I will identify the different generations according to Martin Bauer's empirical definition:[2]

- **New Order, born after 1977**: this is the youngest cohort of respondents. They grew up after the end of the Cold War and woke up to the rhetoric of the 'new world order' and the final victory of the capitalist economics. They have lived through the rhetoric of the IT and biotech 'revolutions' of the late twentieth century. This is the generation of the PC and Internet euphoria of 1995–2000.
- **Gen Xers** were born between **1963 and 1976**. They are the outcome of the birth control 'revolution'. They grew up during the oil crisis of the 1970s, the nuclear issues of the 1980s, antinuclear protest, nuclear armament debates, and the Star Wars initiative.
- **Baby Boomers** were born between **1950 and 1962**. They grew up in the postwar period, one of optimism and drive to modernize. They witnessed the longest period of economic prosperity in history. During this period, Western societies became 'affluent' and free of material concerns. This is the protest generation of the 1970s. Their worldviews are idealistic and they are more skeptical with regard to progress and its links to science and technology.
- **The War and Crisis** generation were born between **1930 and 1949**. This generation witnessed WWII and was the immediate postwar generation that entered the Cold War. This generation is also marked by the 'nuclear enthusiasm' of the 1950s, which promised a scientific revolution and 'energy too cheap to meter' in the atomic society.
- **The Roaring '20s** produced the generation born **before 1930**. They grew up through the buzzing period of the 1920s that ended in the crash of 1929 and the economic crisis that followed.

In France, as in other industrial countries, it is often argued that the "idea of progress" has been declining in the public mind over the last thirty years. The various so-called scientific or technological crises of the '80s and the beginning of the '90s (contaminated blood, the BSE crisis, fear of GMOs) offer a possible explanation for this increasing lack of trust in science and technology. However, before attempting to find an explanation for this supposed change in attitude, an analysis of our database should be undertaken to ascertain whether traces of it can be found there. In particular, there is a need to determine whether public attitudes toward science and technology are more and more hostile or at the very least ambiguous. If this is the case, then it needs to be ascertained whether this is true whatever the domain taken into account or only for certain specific fields. Finally, the question of whether older generations remain the keenest supporters of scientific progress whereas younger ones tend to condemn it needs to be verified.

SCIENTIFIC RESEARCH: AN INCREASING STATE PRESENCE?

In France, the state plays a far greater role than in other comparable countries. Moreover, it is often argued that the vast majority of French citizens, whatever their ideological preferences, tend to be in favor of maintaining a large public sector. As far as scientific research is concerned, one might assume that this pro-state conviction should result in widespread support for state funding of scientific research rather than private sector funding. In three of our six surveys (1989, 1994, 2001) two questions were asked on this theme: "Do you believe that research in science and technology in France is financed mostly by the state, mostly by private firms or by the state and private firms in equal shares? And "In the future, is it preferable that research in science and technology be financed mostly by the state, mostly by private firms or by the state and private firms in equal shares?" The results show that a fairly large majority of French people believe that the burden is equally shared, only one-fourth of them believing that such funding is mostly ensured by the state. In reality, in 1971, the state took care of 63.3% of research and development expenditures (private firms 36.7%). In 1990, it funded 53.1% of research, whereas in 2001 the share of state funds in research financing had dropped to 43.7%. It is evident therefore that the public has not perceived this relative withdrawal of the state. What's more, a growing proportion of French people would like to see most research financed by the state—24% in 1989, 28% in 1994, 31% in 2001—whereas, in actual fact, its contribution has continually decreased.

One question, asked only in the last three surveys, shows significant change on the issue of conducting basic research, that is to say, scientific research mostly financed by the state: in 1994, 39% of the people interviewed agreed that research should be pursued "even when it is unsure whether it will have practical applications"[3] compared to 52% in 2001 and 59% in 2007. What does this trend mean? There is a good deal to suggest that the scientific crises of the '90s convinced a large proportion of the public to call for a new direction in scientific research: asking for more basic research, for more state intervention, is also a way of asking for more precaution. The scientific crises, (BSE, GMOs, asbestos etc.), were viewed by the French public as the consequence of a lack of precaution in the management of science and technology. And, for many people, the main reason for this lack of precaution was the unjustifiable pursuit of profit. As long as science and technology are considered to be increasingly concerned with making a profit, the public tends to forget the idea of progress made in these areas as being in the general interest. The demand for more basic research and more state intervention noticeable in these surveys may provide an answer to this situation.

THE IMAGE OF RESEARCHERS

Science is not only perceived by the public through its achievements but also very often in a more concrete manner through the emblematic figures of "scientific researchers." The surveys show that the figure of the "researcher" or "scientist" materializes in the social imagination under two apparently contradictory aspects that continue to coexist: on one hand there is the image of the good scientist, embodied in France by historic figures such as Louis Pasteur or Pierre and Marie Curie. On the other hand, there is the figure of the evil scientist illustrated by the myth of Frankenstein invented by Mary Shelley in the nineteenth century, countless versions of which can be found in contemporary fiction (comic strips, the movies and television). The following two statements represent two illustrations of these social beliefs: "Research scientists are devoted people working for the good of humanity" *and* "Due to their knowledge, research scientists have power that can make them dangerous." These two images have always dominated public opinion. However, a shift has been noted since 1972 indicating a slight deterioration in the positive aspect and a slight reinforcement of the negative aspect (when "strongly agree" answers are examined).

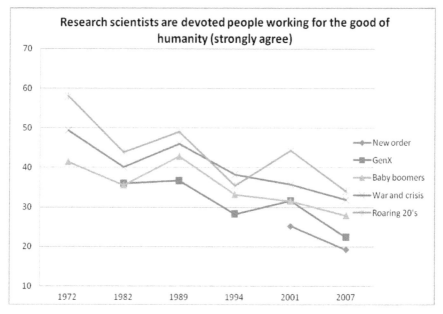

Figure 3.1a Opinion toward scientists by generation.

Attitudes toward Science in France 43

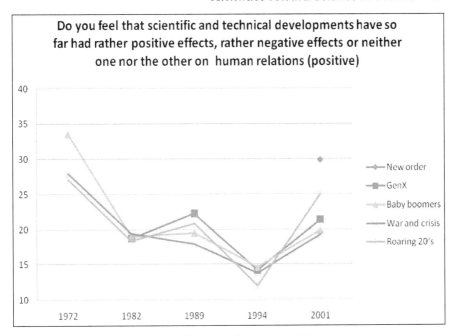

Figure 3.1b Effect of science on human relations: positive.

Looking at changes in these opinions among the different generations, it appears first that, the "myth of science", i.e., the image of the "good scientist", has always been more embedded among the older generations (the "Roaring 20s" and "War and Crisis" generations) than among the younger ones' ("Baby Boomers", "Gen X"). But the main point is that, whatever generation is concerned, the old stereotypes are tending to crack little by little, even if these images are still dominant among the public.[4]

SCIENTIFIC AND TECHNICAL DEVELOPMENT: POSITIVE OR NEGATIVE CONSEQUENCES

The Effects in Various Areas

Does the development of science and technology have positive effects or not? The answers to this question naturally depend on the field examined, but they have not changed significantly since the first survey, at least when relatively vast domains are taken into account. As regards, for example, "standard of living" and "working conditions", the perceived consequences are highly positive and assessments have not varied very much over the period.[5] In terms of "health," the percentages of positive assessments have tended to increase, particularly in the last two surveys.

The more sensitive questions of a "sense of morality" and "human relations" have always been perceived in a more negative light. Respondents probably understand the term "sense of morality" to mean the human propensity to abide by ordinary moral standards in everyday behavior. It is hardly surprising that attitudes here have always been very pessimistic: the idea that "modern" societies (thus, those in which science has developed) have not made "moral" progress has probably been part of the dominant beliefs since Jean-Jacques Rousseau. This interpretation is borne out by the results, always negative and relatively unvarying, pertaining to the question as to whether scientific developments make "people better".

The idea that "human relations" have deteriorated also constitutes one of the features of this nostalgia for a golden age that is so common in our societies.

If the generational effect is considered (Figure 3.1b), a clear decline in positive attitudes can be observed, especially in the 1994 survey, followed by a small increase in the 2001 survey.[6] But, by and large, the differences between the generations are relatively limited.[7]

Two Specific Fields: The Environment and Ethics

The environment and ethics are the two areas in which the consequences of scientific and technical development have been the most often investigated in recent years. The following two questions, asked regularly since 1972, cover these issues.

The first one[8] shows that, as regards the environment, the fear of "chemicals" (in this case "chemical fertilizers") is longstanding, unwavering and on the rise in recent years: in 1972, 72% of respondents approved the idea of eliminating chemical fertilizers, in 2001, 90% did.

Looking at the generational effect (Figure 3.2a) there have always been significant differences in attitudes. New generations ("Gen X", "Baby Boomers") tend to be less opposed to the use of fertilizers. And this difference is a constant throughout the period during which the questions were asked.

The results of the second question[9] are more difficult to interpret: the question alludes to research that raises "ethical" issues with the example of artificial procreation. Attitudes have varied in this area, although a clear trend has not emerged. It may be that the answer to the statement is determined by the context in which the question is asked: at the start of the period, "test-tube babies" (still very theoretical) called up the image of the motherless child. In the 1980s, in-vitro fertilization became a fairly common solution for couples having difficulty conceiving, thereby legitimating the procedure and reinforcing the benefits of scientific development. In the 1990s, it may have been the issue of cloning that provoked negative perceptions of this type of application. But in more recent years,

Attitudes toward Science in France 45

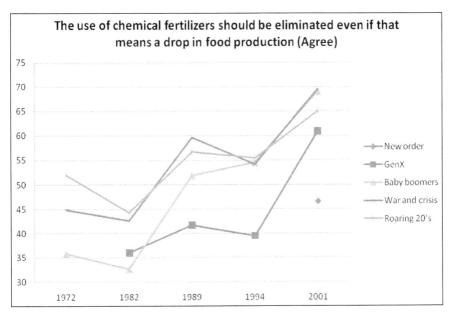

Figure 3.2a Attitudes toward chemical fertilizers.

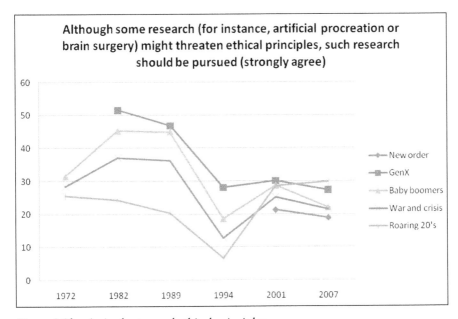

Figure 3.2b Attitudes toward ethical principles.

clear legislation (laws on bioethics passed in 1994 and revised in 2002) probably reassured the public on the ethical limits placed on scientific development.

Once again, it appears that the different generations have made different judgments on this issue (Figure 3.2b) especially at the beginning of the period we are examining: before the '90s, younger generations were far more in favor of scientific research even if it threatened ethical principles. This support declined in all generations in the 1994 survey. In the last surveys (2001, 2007), it seems that there was general agreement among the generations on this issue.

The Overall Assessment

Beyond sector-based judgments, the survey elicited an overall assessment through the following question: "Do you feel that science does humanity more good than harm, more harm than good, or about as much good as harm." The major changes observed since 1972 are clear:

- negative answers (more harm than good) have remained very low,
- the positive answer (more good than harm) has clearly decreased,
- the median answer (as much good as harm) has clearly increased.

On the whole therefore, science does not prompt more negative opinions but more *balanced or ambiguous judgments*.

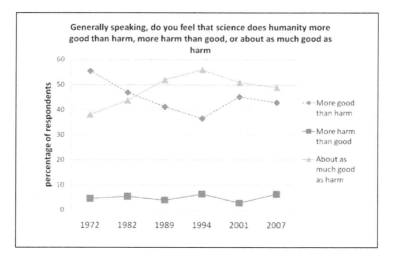

Figure 3.3a Generally speaking, do you feel that science does humanity more good than harm, more harm than good, or about as much good as harm?

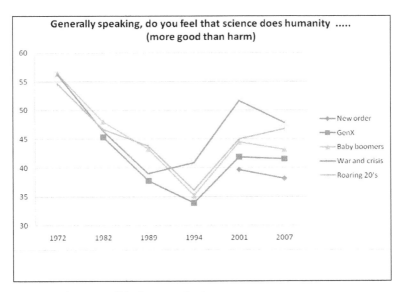

Figure 3.3b Overall judgment of science.

Looking at these changes among the different generations, it appears that the decrease of the most positive opinion (more good than harm) until 1994, followed by a new increase in 2001 and 2007, is confirmed for all generations. However, while during the '70s and '80s there were few differences between the different generations on this subject, since the '90s, the feeling that "science does humanity more good than harm" has become more widespread in older generations than in younger ones.

Changes in this overall assessment of the effects of science and technology reinforce a finding frequently noticed in this research: during the last thirteen years, attitudes toward science in France have been characterized by an initial period of optimism (during the '70s and '80s), followed by a strong decline in the '90s, and then by a new increase in optimism in the last two surveys. How can this pattern be explained? One possible explanation might be the context of our surveys and especially the emergence of so-called scientific or technological crises throughout the '80s and the beginning of the '90s. In France, the blood contamination scandal was probably viewed by the public as the first major failure of scientific or technological developments. At almost the same time, the BSE crisis reinforced the public in their feeling—right or wrong—that science and technology were overly influenced by profit and that risk management was poorly organized in France. By the beginning of the new century, these fears had not totally vanished. However, confidence in science and technology seems to have been reinforced by a series of political responses. The creation of state agencies[10] in charge of public expertise on scientific or technological controversies has probably contributed to this change.

A RISE IN THE IRRATIONAL?

Periodically, science administrators and mediators become concerned about the rise in irrationality in our societies. All manners of parascience (astrology, telepathy, table-turning, etc.) seem to be steadily gaining in appeal. This might partly explain the lack of interest in genuine science. The public is thought to be turning away from real science because more attractive substitutes are being offered. Adequate survey data on parascience consumption is not available (i.e., the number of volumes of parascientific literature, including books, magazines, and today, websites), nor are there turnover figures for those who make a living from these industries. On the other hand, since 1982, surveys on French attitudes toward science have measured degrees of belief in so-called paranormal phenomena.

To illustrate developments in these beliefs during this twenty-five year period, and check the generational effect, an index was built using five "parasciences":

- explanation of personality by astrological signs,
- bewitchment or sorcery,
- prediction by astrological signs, horoscopes,
- table-turning, spiritualism,
- long-distance communication of thought.

Figure 3.4a shows the percentage of people declaring they believe in at least three of these phenomena (of the five presented):

It has frequently been assumed that the decline of confidence in science must be linked with a rise of irrationality in society. Clearly this hypothesis is not borne out by the facts: though we have established that in France there has been a relatively strong decline in support for science during the '90s, there is no corresponding rise in beliefs in parascience during the same period. On the contrary, it appears that, generally speaking, the level of belief in parasciences was relatively stable from 1982 to 2007.

But it is also clear that these beliefs are much more widespread among the younger generations (Baby Boomers, Gen X, New Order) than among older ones (Roaring 20's, War and Crisis). Contrary to possible expectations, believing in "irrational" phenomena is not a vestige of less educated older generations, but a trait of younger and educated generations.

Attitudes toward Science in France 49

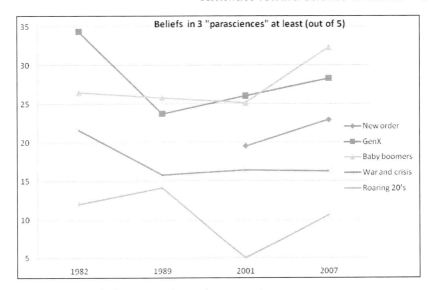

Figure 3.4a Beliefs in parasciences by generations.

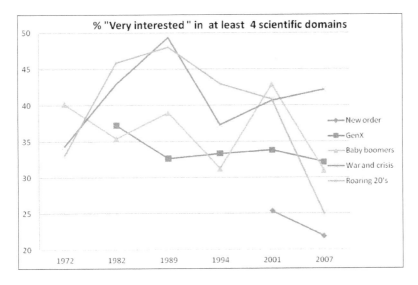

Figure 3.4b Index of interest for scientific culture.

SCIENTIFIC CULTURE

Interest in scientific and technological culture has been taken into account through the following question: "For each of the following domains could you tell me if you are very interested, somewhat interested, or not at all interested in getting information on:

> space exploration
> medical research
> research on matter
> research on life
> major technological achievements
> major explorations."

An index was drawn up to check overall change in interest for scientific culture. Respondents who declared themselves to be "very interested" in at least four of the six scientific domains quoted above were considered to be interested in scientific culture. This general index reached a peak at the end of the '80s (43% in 1989, as opposed to 34% in 1972) and then tended to decline, reaching 32% in the 2007 survey.

However, the main findings of this insight into scientific culture is the fact that, as measured by our index of interest in six domains, scientific culture has always been preferred by the older generations (Roaring 20s, War and Crisis) more than by the younger ones (Generation X, New Order).

CONCLUSION

What has happened to science in France during the last thirty years? Clearly, there has been a decline in traditional links between science and society which can be diagnosed on the basis of four indications:

- The traditional image of scientists as "devoted people working for the good of humanity" has always been more present among older generations and has declined slightly over time.
- There has been a decrease in positive attitudes toward science and technology in several domains. This is the case in particular for the supposed effect of science and technology on "human relations", the problem of the use of chemical fertilizers, the development of certain research threatening ethical principles, and the overall effect of science on humanity. However, in most of these cases the decline halted from the beginning of the 2000s.
- Interest in scientific culture is also slightly decreasing.

How can these different variations be explained? Several hypotheses might be suggested. First the idea that in France, maybe more than in other comparable countries, certain catastrophic events, like the contaminated blood scandal or the BSE crisis, have had a dramatic impact on public opinion. France is a country where confidence in science and technology was traditionally very strong: the images of Louis Pasteur or Pierre and Marie Curie were the main elements of the myth of science, that is to say, the core of the old "ideology of progress". Scientific and/or technological crises have been seen by public opinion as the end of this golden age. The idea that the search for profit was the main reason for these crises has been largely developed through the mass media. Therefore, the only way to remedy this situation, in the French cultural and political context, is to reinforce public supervision of science and technology, especially by means of public agencies mandated to supervise risk as a result of technological development.

Another possible explanation is that the old ideology of progress became less prevalent because it characterized older generations whereas new generations tend to be characterized by a more critical view of science and technology.

To choose among these different explanations, a series of models of regression were imagined, taking the different attitudes toward science (images of scientists, effect of science and technology on "human relations", use of chemical fertilizers, development of certain kinds of research threatening ethical principles and the overall effect of science on humanity and interest in scientific culture) as dependent variables and gender, date of the survey, level of education, and generation as explanatory variables.

Looking at the table that sums up these results for the most significant change—the overall judgment of science—(Table 3.1), it appears that all selected explanatory variables are significantly linked to the dependent variable. However, the most significant variable is still the level of education: people with high levels of education (especially when they have degrees in science subjects) tend to judge the overall effect of science on humanity more positively. The second factor to explain changes in attitudes toward science is the time of the survey, in other words, the effect of the general context. This result confirms that scientific crises in the '80s and '90s have had major consequences for the public image of science.

Last, the generation one belongs to also affects the overall judgment of science: generally speaking, older generations are more deferential toward science than younger ones. This is quite paradoxical because, on average, older generations tended to have a lower level of education. In other words, attitudes to science were more favorable among older generations, including the less well-educated part of the population, whereas science has partly lost its prestige among less educated younger generations.

Table 3.1 Logistic Regression of the Overall Judgment of Science

"Do you feel that science does humanity more good than harm, more harm than good, or about as much good as harm?" The "more good than harm" answer is opposed to "more harm than good", or "about as much good as harm".

	B	E.S.	Wald	ddl	Signif.	Exp(B)
SEX (reference: male)						
Female	0,331	0,046	51,855	1	0,000	0,718
LEVEL OF EDUCATION (Reference: primary)			189,759	4	0,000	
Intermediate	0,152	0,061	6,183	1	0,013	1,164
Secondary	0,439	0,073	35,926	1	0,000	1,551
University (not scientific)	0,796	0,081	96,589	1	0,000	2,217
University (scientific)	0,999	0,091	120,341	1	0,000	2,716
DATE OF SURVEY (Reference: 1972)			86,974	5	0,000	
1982	0,310	0,081	14,682	1	0,000	0,733
1989	0,499	0,082	36,720	1	0,000	0,607
1994	0,758	0,084	80,704	1	0,000	0,468
2001	0,380	0,087	19,254	1	0,000	0,684
2007	0,459	0,098	22,090	1	0,000	0,632
GENERATION (Reference: new order)			43,411	4	0,000	
Gen Xers	0,202	0,121	2,781	1	0,095	1,224
Baby boomers	0,406	0,121	11,184	1	0,001	1,500
War and crisis	0,575	0,121	22,748	1	0,000	1,777
Roaring 20s	0,609	0,128	22,599	1	0,000	1,839
Constant	0,327	0,140	5,435	1	0,020	0,721

Cox & Snell's R2	Nagelkerke's R2
,046	,062

NOTES

1. Technical note about the surveys.
2. Bauer, chapter 6 of this book.
3. The other possible answer was "Scientific research should only be developed when it is believed that it will have practical applications."

4. At least, when the total percentages of "Agree", and not only "Strongly agree", are considered.
5. However, these series of question were not asked in the most recent survey (2007).
6. This question was not asked in the 2007 survey.
7. The significance of the generational effect on this opinion, as measured by the Cramer V coefficient is only 0.6.
8. Do you agree or disagree with the following statement: "The use of chemical fertilizers should be eliminated even if that means a drop in food production"?
9. *"Although some research (for instance, artificial procreation or brain surgery) might threaten ethical principles, such research should be pursued. Do you strongly agree, partly agree, partly disagree, or strongly disagree?"*
10. AFSSA (The French Food Safety Agency), created in 1999, is a public independent body that contributes to the protection and improvement of public health, animal health and welfare, vegetal and environmental health through monitoring, research and research instigation. AFSSET (The French Agency for Environmental and Occupational Health Safety) was created in 2001 and is a public body reporting to the French Ministers for ecology, for health and for employment.

BIBLIOGRAPHY

Bauer M., Allum N., Miller S. (2007), "What Can We Learn From 25 Years of PUS Research? Liberating and Widening the Agenda", *Public Understanding of Science*, No 2, Vol. 11, pp. 79–95.

Bauer M., Durant J., Evans G. (1994), "European Public Perceptions of Science", *International Journal of Public Opinion Research*, No 1, Vol. 16, pp. 163–186.

Boy D. (1999a), "Politiques de la science et démocratie scientifique", *Revue Internationale de Politique Comparée*, No 3, Vol. 6, pp. 613–625.

Boy D. (1999b), "Les biotechnologies et l'opinion publique européenne", *Futuribles*, No 238, pp. 37–56.

Boy D. (2006), "Science et société : de la culture à la démocratie" in J.-Ph. Leresche, M. Benninghoff, F. Crettaz von Roten, M. Merz (eds.), *La Fabrique des sciences. Des institutions aux pratiques*, PPUR, Lausanne, pp. 261–281.

Crettaz von Roten F. (2006), "Do We Need a Public Understanding of Statistics ?" *Public Understanding of Science*, No 2, Vol. 15, pp. 243–249.

Evans G., Durant J. (1995), "The Relationship between Knowledge and Attitudes in the Public Understanding of Science in Britain", *Public Understanding of Science*, No 1, Vol. 4, pp. 57–74.

Hornig Priest S., Bonfadelli H., Rusanen M. (2003), "The 'Trust Gap' Hypothesis: Predicting Support for Biotechnology across National Cultures as a Function of Trust in Actors", *Risk Analysis*, No 4, Vol. 23, pp. 751–766.

Miller J.D. (1983), "Scientific Literacy: A Conceptual and Empirical Review", *Daedalus*, No 2, Vol. 112, pp. 29–48.

Miller J.D. (2004), "Public Understanding of, and Attitudes toward, Scientific Research: What We Know and What We Need to Know", *Public Understanding of Science*, No 3, Vol. 13, pp. 273–294.

Novotny H., Scott P., Gibbons M. (2003), *Repenser la science. Savoir et société à l'ère de l'incertitude*, Paris, Belin.

Pardo R., Calvo F. (2002), "Attitudes toward Science among the European Public: A Methodological Analysis", *Public Understanding of Science*, No 2, Vol. 11, pp. 155–195.

Pardo R., Calvo F. (2006), "Mapping Perceptions of Science in End-of-Century Europe", *Science Communication*, No 1, Vol. 28, pp. 3–46.

Peters H.P. (2000), "From Information to Attitudes? Thoughts on the Relationship between Knowledge about Science and Technology and Attitudes toward Technologies" in M. Dierkes, C. von Grote (eds.), *Between Understanding and Trust: The Public, Science and Technology*, Harwood Academic Publishers, Amsterdam, pp. 265–286.

4 American Public Understanding of Science 1979–2006

Susan Carol Losh

This study examines how generation and age, net of gender, high school science and math, and college science, affected American adult civic science literacy (CSL) using the National Science Foundation Surveys of Public Understanding of Science and Technology 1979 to 2006. In 2007 alone, US science and technology research and development expenditures exceeded *350 billion dollar*s (US Bureau of the Census: 2009, Table 769). Yet many industrial leaders, educators, scientists, and politicians believe our youth are unprepared for college, that a "brain drain" occurs from science to other educational areas, and that adults cannot discuss science at the level of a major newspaper (Burris, 2006; Gates 2005; Miller, 2000, Lemonick, et al., 2006; Seymour, 2006). These criticisms ensue despite sizable changes in science education.

Although US adults express considerable science interest, some issues, e.g., genetic engineering, require basic knowledge that many lack (Kumar & Chubin, 2000). Given investments in public discourse, education, and research and development, it is vital to understand continuities from formal education to adult CSL. Adults raise children, educate youth, and formulate actions from school board decisions to international policy. Adult CSL can be critical for intelligent policy discussions and a supportive research climate (Allum, Sturgis, Tabourazi, & Brunton-Smith, 2008).

A different challenge emanates from pseudoscience purveyors, e.g., psychics or "creationists," who vie for public legitimacy, political clout, and consumer dollars. The "deficit model" often tacitly assumes that those lacking science literacy cannot distinguish "real" from ersatz science, thus risking exploitation from pseudoscience practitioners.

ISSUES IN SCIENCE KNOWLEDGE AND PSEUDOSCIENCE BELIEF

In international science comparisons, American students appear mediocre (Schmidt, McKnight & Raizen, 1997; US Department of Education, 2007). Yet Kadlec, Friedman, and Ott (2007) found that Midwest parents and secondary students were sanguine about science instruction and student preparation. Students depicted their science courses as dull and "largely irrelevant". Kadlec, et al. describe an "urgency gap" between "leaders" and "experts" versus parents and students about science education. Perhaps complacency occurs because students *do* elect more math and science than before (US Department of Education, 2007). Simultaneously, recent generations are more technologically fluent (Losh, 2010; Pew Research Center, 2007).

Pseudoscience belief among Americans (including school teachers) also causes concern; its levels may be stable or even increasing (Eve & Dunn, 1989; 1990; Goode, 2000; Pew, 2009b). Pseudoscience beliefs are "cognitions about material phenomena that claim to be 'science,' yet use non-scientific evidentiary processes [e.g.,] authoritative assertion...anecdote... or unelaborated 'natural' causes" (Losh, Tavani, Njoroge, Wilke, & McAuley, 2003). Many adults believe that astrology is scientific, and "extraterrestrials" visit a "6000-year-old Earth" (Davis & Smith, 2009; Gallup News Service, 2001). Although reading a horoscope can be fun, pseudoscience is rife with untested "cures" or unsubstantiated reports of "flying saucers".

Although some scholars find that studying pseudoscience helps explore basic science literacy (Goode, 2000; 2002; Martin, 1994), educators typically avoid teaching about it (Lilienfeld, Lohr, & Morier, 2001). Beliefs in Biblical creation or ghosts can provide a "back door" scaffold to understand layperson conceptualizations about science, by identifying evidence they find compelling and causal mechanisms they believe operate in the material world. Despite its popularity and its costs, very little research assesses American adult pseudoscience support (for exceptions, see Pew, 2009a; 2009b).

WHAT'S HAPPENED WITH SCIENCE EDUCATION?

Science education, and ultimately science literacy, is assumed to stimulate pseudoscience rejection. However, educational *level* inconsistently predicts these beliefs, depending on the domain addressed (Taylor, Eve, & Harrold, 1995). Moreover, many influences on pseudoscience attributed to degree level instead result from other causes either correlated with educational attainment, or products of it (e.g., science factual knowledge, Losh, et al., 2003).

As noted earlier, modern Americans *are* better educated, and thus have more formal science exposure (US Department of Education, 2007). Only 25% of 1940 adults completed high school, and 5% had at least a baccalaureate; by 2004, 85% had graduated high school and 29% had a BA (National Center for Educational Statistics, 2004: Table 8; U.S Department of Education, 2007). However, since most Americans still do not graduate college, recent changes in elementary and secondary school science education remain important.

By the millennium, many US educators were de-emphasizing factual memorization, focusing more on science inquiry, hands-on experience, and the context of science and technology (AAAS, 1993; Gess-Newsome, 2002; Schiebiner, 1999; Sunal & Sunal, 2003). Although evidence is mixed that these approaches increase performance at different grade levels (Burkam, Lee, & Smerdon, 1997; Gess-Newsome, 2002; Lee & Burkam, 1996; Lilienfeld, et al., 2001; Moss, Abrams, & Kull, 1998; Scanlon, 2000), at least some changes may "trickle through" to understanding science among *more recent generations* of adults who have had such exposure.

In this research, I track six generations over 11 to 28 year periods, disaggregating cohort and age effects on adult CSL dimensions, net of gender and educational variables. Although "chronological age" is often a CSL predictor, it is still treated cavalierly, typically as a vague demographic, whereas cohort remains mostly ignored. "Young adults" at almost any time consistently appear more interested or knowledgeable in science than seniors (National Science Board, 2008).

Findings here will show that the generational construct provides valuable information about CSL and that earlier research using only the variable "age" can mislead. Thus I juxtapose assumed but unexamined "adult development issues" (i.e., "age") versus cohort experiences. Simply studying CSL over time tells us our past, age effects in a single period photograph the present, but generation can inform us about future trends as recent cohorts age and replace earlier ones.

Issues of Age and Generation

My emphasis on generation is not just semantics. Social changes occur in many ways; it is important to disentangle them. For example, perhaps *older adults* find it harder to assimilate scientific advances due to slower working memory, memory decay, or even midlife presbyopia, which makes reading small print difficult (Boyd & Bee, 2009; Woolfolk, 2007). Thus, media accommodations for seniors could boost science literacy. Second, in *cultural transformations* or *"period effects"*, for example, more positive fictitious treatments of scientists or wider dissemination of scientific discoveries could motivate an entire society, irrespective of age or generation, to become more knowledgeable over time.

In *cohort replacement on concomitant variables*, generations differ on specific attributes, which in turn directly predict CSL. "Baby Boomers" are better educated than earlier generations and education raises adult CSL; as "Boomers" replace earlier cohorts, overall CSL should rise through enhanced schooling among this large cohort. Here, cohort analysis directs our attention to educational rather than aging variables. Finally in *direct cohort effects*, a specific generation experiences relatively unique events, predisposing its members to adopt particular behaviors or attitudes. I take such an approach. For example, adults born before 1900 experienced early electronic communication and air travel; "Millennials" born in the 1980s matured using computers and the Web at school (US Bureau of the Census,

2008: Table 253); thus ease of Internet science information access could foster CSL among Millennial adults.

Generation also can relate to *what kind* of science education a student received. Although changes in science education methods began roughly 30 years ago, the initial impact was largely limited to major research universities. Allowing at least 10 years for these perspectives to diffuse to other postsecondary institutions means that any cohort effects on CSL from such educational changes would begin with "Generation X" and be more pronounced among Millennials.

"One-shot" cross-sectional analyses of a general population survey inevitably confound age and cohort. Simultaneous estimates of age, cohort, *and* period effects present logical and statistical problems because any of these three variables depends upon the other two. However, with several sequential studies, cohort and age effects can be partially disentangled. Although there are statistical attempts to handle multivariate estimation of all three variables (e.g., Glenn, 2005; Mason, Winsborough, Mason & Poole 1973; Mason & Brown, 1975), in this study I build and track synthetic cohorts, contrasting generational with age effects, leaving period changes largely reflected in the joint age and cohort patterns.

If American science education innovations contribute to adult CSL, then not only should knowledge rise among recent cohorts, *but cohort effects on understanding science inquiry should be particularly pronounced*. Similar effects may occur for how generation affects pseudoscience belief. Because more recent cohorts have more formal schooling, including greater science and math exposure (Carlson, 2008), I control several educational variables.

RESEARCH QUESTIONS

How does generation, compared with age and net of gender and educational variables, affect understanding science inquiry and knowing basic science facts in American adults?

How does generation, compared with age and net of gender, educational and knowledge variables, affect pseudoscience support among American adults?

METHODS

The National Science Foundation Surveys

The most comprehensive set of US adult "science literacy" studies is the NSF Surveys of Public Understanding of Science and Technology 1979–2006 (directed by Jon Miller, 1979–1999; ORC-MACRO, 2001; and Davis & Smith, 2009). The *total archive*, monitoring several science literacy dimensions, comprises 23,906 unweighted (23,994 weighted) interviews in 12 probability samples (1979; 1981; 1983; 1985; 1988; 1990; 1992; 1995; 1997; 1999; 2001; 2006). The series also has considerable detail on adult educational achievements. Although the 1979 and 2006 surveys were conducted in person, those in between used Random Digit Dialing.[1]

Measures: Basic Science Knowledge

I measured knowledge three ways. Cognitive processes underlying factual memorization differ from those for science inquiry. For example, facts can show memory decay, especially if they are mostly irrelevant to adult daily lives. However, inquiry processes apply to many science areas; thus understanding inquiry should "age better" than factual memorabilia. I first created an index from either eight (1988 and 1990) or nine closed format items (1992–2006; see Table 4.1 for all questions used) about basic science facts taught in primary school and reviewed in middle school (Cain, 2002; Sunal & Sunal, 2003), e.g., a true-false item asks whether "antibiotics kill viruses and bacteria." Balanced response items ask whether the earth goes around the sun or vice versa. These have been called the "**Oxford items**" (Allum, et al., 2008). The index score is the percent correct of the total. The remaining two measures speak more to science inquiry.

Second is an **applied probability score** of the percent correct for either four (1988–2001) or two items[2] (2006) about a couple planning a family who carry genes for a hereditary illness. Third is a **methods** question (1995–2006): whether researchers should test a new medication by (1) administering it to 1000 patients or (2) giving 500 patients the new drug with a 500-patient control group.

Measures: Pseudoscience Support

Three **pseudoscience** belief items include: (1) whether astrology was rated very, somewhat, or not at all scientific (asked 1979 to 2006; very and somewhat responses were combined); (2) a true/false item suggesting some UFOs are alien spacecraft (1985, 1988, 1990 and 2001); and an evolution *support* item (dichotomized as "true" versus "false/other"; 1985 to 2006).

I treat these items as separate domains, first because they represent distinct dimensions: traditional pseudoscience; a modern "sci-fi" item; and straightforward endorsement of evolution. Second, prior research indicates these items have different predictors; e.g., inerrant religiosity negatively correlates with evolution support, but there is no *a priori* reason to expect it to predict UFO responses. Finally, these items do not empirically cluster (r, astrology and evolution = 0.06; r, astrology and the UFO item = 0.10; r, evolution and UFO items = 0.12).

Table 4.1 Questions Used to Measure Adult Public Understanding of Science

A. Factual Questions (asked 1988–2006)

1. The center of the Earth is very hot. Is that true or false?
2. All radioactivity is manmade. Is that true or false?
3. It is the father's gene that decides whether the baby is a boy or a girl. Is that true or false?(Available after 1990.)
4. Lasers focus sound waves. Is that true or false?
5. Electrons are smaller than atoms. Is that true or false?
6. Antibiotics kill viruses as well as bacteria. Is that true or false?
7. The continents on which we live have been moving their location for millions of years and will continue to move in the future. Is that true or false?
8. Does the Earth go around the Sun, or does the Sun go around the Earth?
9. How long does it take for the Earth to go around the Sun: one day, one month, or one year?(Only asked to those responding Earth goes around the Sun.)

B. Inquiry Items (asked 1988–2006)

I. Now, think about this situation. A doctor tells a couple that their genetic makeup means that they've got one in four chances of having a child with an inherited illness.
   1. Does this mean that if their first three children are healthy, the fourth will have the illness? (Not asked in 2006.)
   2. Does this mean that if their first child has the illness, the next three will not?
   3. Does this mean that each of the couple's children will have the same risk of suffering from the illness?
   4. Does this mean that if they have only three children, none will have the illness? (Not asked in 2006.)
II. Now, please think about this situation. Two scientists want to know if a certain drug is effective against high blood pressure. The first scientist wants to give the drug to 1000 people with high blood pressure and see how many of them experience lower blood pressure levels. The second scientist wants to give the drug to 500 people with high blood pressure, and not give the drug to another 500 people with high blood pressure, and see how many in both groups experience lower blood pressure levels. Which is the better way to test this drug? (Asked from 1995 forward.)

C. Pseudoscience items (asked 1979–2006)

Would you say that astrology is very scientific, sort of scientific, or not at all scientific?

Some of the unidentified flying objects that have been reported are really space vehicles from other civilizations. (Asked 1985–2001.)

Human beings, as we know them today, developed from earlier species of animals. Is that true or false? (Asked from 1985 forward.)

Source: The National Science Foundation Surveys of Public Understanding of Science and Technology.

Constructing Birth Cohort and Age Categories

There is sizeable debate over when cohorts begin or end (Carlson, 2008; Glenn, 2005, Pew, 2007). Rather than one constant interval (e.g., 20 years), cohorts are often created using time duration *and* significant events occurring when individuals could experience them. For example, most Millennials would not remember "the Reagan White House astrologer" (Quigley, 1990; Regan, 1988) featured in considerable late night TV talk show humor in the late 1980s.

One example of these debates is the "Baby Boom", which scholars agree *began* in 1946. Some end it in 1957, when *birth rates* peaked, others in 1961, when the *number of births* peaked, still others in 1964, when completed fertility dropped below three children. Since "Generation X" is generally agreed to begin in the early 1960s, I ended "the boom" in 1961. I created six cohorts: "**Gen Y**," often called "**Millennials**", born 1979–1988; "**Generation X**" (1962–1978); "**Baby Boom**" (1946–1961); "**The Lucky Few**"[3] (1930–1945); **Post-WWI** (1918–1929); and **WWI** (1891–1917). I coded 86 respondents born before 1891 to "missing" both because of their scarcity and because dementia increases after age 80.

For cross-tabulations and analysis of variance, five age categories approximately correspond to US government usage: 18–24; 25–34; 35–44; 45–64; and age 65 and over. Although age group and cohort correlate ($r = 0.65$) because older adults in the years of the NSF surveys tend to be from earlier cohorts, some independence still exists between the two variables.

Gender and Educational Predictors

Gender is an important predictor of science interest, careers, and knowledge (Aldrich, 1978; Burkam, et al., 1997; Fox & Firebaugh, 1992; National Science Board, 2008). *Degree level* was coded: high school or less; two-year college degree; baccalaureate; or advanced degree. Exposure to high school biology, physics, or chemistry, and level of high school math, was available 1990–2001. High school math was coded 1 (none; general or business math); 2 (algebra I or geometry); or 3 (algebra II, calculus, precalculus, or statistics). I summed high school science courses (0–3).[4] The number of *college* science courses ranges from 0 to 10 or more.

RESULTS

Cohort and Education

The NSF data agree with other US educational attainment reports: more recent cohorts had more education: 87% of the WWI cohort had at most high school completion compared with 67% of Gen X adults (many Millennials had not finished schooling). Nine percent of the WWI cohort had at least a BA compared with 18% of Gen Xers and 24% of Baby Boomers ($X^2_{(15)}$ = 907.50 Cramer's v = 0.11, p < .001). Similarly, 51% of the earliest cohort had at most "general" or "business" high school math, decreasing to 8% among Millennials. Forty-one % of Millennials had advanced math, compared with only 8% of the WWI cohort ($X^2_{(10)}$ = 952.88 v = 0.19, p < .001).

Formal exposure to science also rose. Over twice as many Millennials as the WWI cohort had high school biology (87% vs. 36%, $X^2_{(5)}$ = 995.03 Φ = 0.27, p < .001); triple the percentage had chemistry (69% vs. 20%, $X^2_{(5)}$ = 531.41 Φ = 0.20, p < .001), and about double had physics (39% versus 22%, $X^2_{(5)}$ = 114.60 Φ = 0.09, p < .001).[5] Millennials had over double the number of high school science courses as the WWI cohort (1.95 vs. 0.77, $F_{5, 13200}$ = 163.90, p < .001 η = 0.24.) Because recent generations more often attended college, they also elected more college science classes. The WWI cohort averaged 0.41 courses compared with Baby Boomers (1.64), Gen Xers (1.58), or even Millennials (who, at 1.09 classes, were still completing their formal education, cohort $F_{5, 20708}$ = 169.63 p < .001, η = 0.20).

Cohort, Age, Education, and Basic Civic Science Literacy (CSL)

Table 4.2 shows the three CSL measures by time and (separately) how cohort affected basic science knowledge. The "Oxford Items" and drug test item clearly rose over time (Oxford Index: $F_{7, 15354}$ = 28.01 p < .001 η = 0.11; drug item: $F_{4, 9275}$ = 17.87 p < .001 η = 0.09). The ANOVA for the probability score was statistically significant ($F_{7, 15347}$ = 7.92 p < .001 η = 0.06).

Table 4.2 presents *science knowledge* variables by cohort and age. Given that later cohorts took more science and math courses, their CSL should rise for that reason alone. But, *will cohort effects hold net of educational variables?* To illustrate how cohort affected CSL, net of gender, age, high school and college math and science (and for pseudoscience items, the Oxford Index and probability score),[6] I used a presentation program often linked to analysis of variance: Multiple Classification Analysis. MCA adjusts for other predictors in ANOVA, to produce, for example, "net cohort effects". Typically in MCA adjusted generation differences shrink compared with one-way ANOVAs, which do not control for other predictors of the dependent variables. Figure 4.1 shows MCA *adjusted cohort effects* on the Oxford Index, applied probability, experimental drug item, astrology, extraterrestrials, and evolution items.

Controlling gender and educational variables, younger adults from later generations knew more basic science facts (mean correct on the Oxford Index was 60.5%). In the age and cohort ANOVA net of controls, age had modest effects, but cohort sizable effects, and the entire model had considerable predictive utility (age $F_{4, 13188}$ = 7.57 p < .001; cohort $F_{5, 13188}$ = 81.65 p < .001. Total η = 0.56).

How age and cohort affected applied probability is less clear. More recent cohorts tended to score higher; age effects were small (ȳ= 78.8%, age $F_{4, 13188}$

Table 4.2 Civic Science Literacy Variables by Time, Age Group and Cohort

Civic Science Literacy Variables over Time								
Civic Literacy Variables	Year →	1988	1990	1992	1995	1997	1999	2001	2006
Oxford Qs Mean %		55.9	56.0	59.0	60.0	60.0	61.3	63.6	64.0
Probability Score %		82.2	77.5	79.6	77.2	77.3	78.7	80.3	80.8
Drug Question % Correct					69	73	73	79	79
Minimum n		2041	2033	2000	2006	1999	1881	1574	1818

"Oxford Index" Mean % Correct by Cohort and Age Group 1988–2006						
Age	Cohort → WW I	Post WW I	Lucky Few	Baby Boom	Gen X	Millennials
18–24	—	—	—	—	61.3	65.8
25–34	—	—	—	61.1	63.5	68.0
35–44	—	—	67.5	65.0	64.8	—
45–64	—	51.4	56.4	65.3	—	—
65+	42.9	46.9	50.9	—	—	—

Applied Probability Mean % Correct by Cohort and Age Group 1990–2006						
Age	WW I	Post WW I	Lucky Few	Baby Boom	Gen X	Millennials
18–24	—	—	—	—	80.9	79.8
25–34	—	—	—	83.8	84.3	82.0
35–44	—	—	89.8	82.9	82.0	—
45–64	—	75.0	79.1	81.9	—	—
65+	60.0	65.5	69.6	—	—	—

Experimental Question Mean % Correct by Cohort and Age Group 1995–2006						
Age	WW I	Post WW I	Lucky Few	Baby Boom	Gen X	Millennials
18–24	—	—	—	—	82	83
25–34	—	—	—	69	79	88
35–44	—	—	—	76	81	—
45–64	—	—	67	76	—	—
65+	43	62	63	—	—	—

Please see text for tests of statistical significance.

= 12.35 p < .001; cohort $F_{5, 13188}$ = 36.39 p < .001. Total η = 0.34). Cohort affected the drug test item but age had no net effect. Compared with 43% of the WWI cohort and 62% of the post-WWI cohort, 84% of Gen Y, and 80% of Millennials chose the correct control groups response. ($\bar{y}$= 74%, age $F_{4, 9154}$ = 0.93 p = 0.45; cohort $F_{5, 9154}$ = 35.15 p < .001. Total η = 0.25).

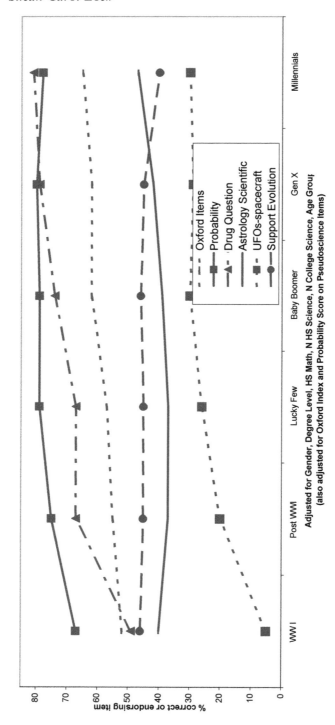

Figure 4.1 Adjusted cohort effects on science and pseudoscience measures.

Cohort, Age, and Pseudoscience Support

Data in Table 4.3 are more discouraging. Given that recent cohorts have more education, all else equal, they should more often answer, "true" to the evolution question and less often support astrology or extraterrestrial UFOs. If so, again, do *cohort effects hold net of education, other background variables, plus an adult's level of basic science knowledge?*

Table 4.3 Support for Pseudoscience by Time, Age Group and Cohort

% Pseudoscience Support Over Time											
Year → Responses	1979	1983	1985	1988	1990	1992	1995	1997	1999	2001	2006
Astrology Scientific	50.0	49.0	42.6	40.5	39.7	38.2	40.0	40.9	41.0	44.2	35.1
Evolution True			45.2	45.9	44.8	45.1	43.8	43.7	45.3	53.3	42.8
Aliens-UFOs			42.8	24.9	24.3				29.6		
Minimum n	1635	1645	2018	2041	2033	2000	2006	1999	1881	1574	1818

% Astrology "Very" or "Somewhat" Scientific by Cohort and Age Group 1979–2006						
Age	Cohort → WWI	Post WWI	Lucky Few	Baby Boom	Gen X	Millennials
18–24	—	—	—	51	49	52
25–34	—	—	62	42	41	38
35–44	—	—	43	38	37	—
45–64	66	46	37	33	—	—
65+	48	42	42	—	—	—

% Agreeing Some UFOs Really Space Ships by Cohort and Age Group 1995–2001						
18–24	—	—	—	61	31	25
25–34	—	—	—	38	27	—
35–44	—	—	47	31	25	—
45–64	—	26	31	33	—	—
65+	19	24	29	—	—	—

% Agree, "Human beings...developed from earlier species" by Cohort and Age Group 1985–2006						
18–24	—	—	—	49	53	51
25–34	—	—	—	50	49	39
35–44	—	—	48	49	46	—
45–64	—	37	40	45	—	—
65+	37	37	39	—	—	—

Please see text for tests of statistical significance.

Instead, pseudoscience seemed to appeal *more* to recent generations. The top panel of Table 4.3 shows support over time. The apparent reduction in astrology support is illusory; young Millennials later in Table 4.3 *most often* said astrology was very or somewhat scientific (52%) followed by the WWI cohort (49%) with other cohorts in between ($F_{5, 20553}$ = 14.99 p < .001). Fifty percent of adults aged 18 to 24 endorsed astrology as did 44% of those at least 65, with age groups in between less supportive ($F_{4, 20553}$ = 35.10 p < .001; Total η = 0.11). Interestingly, Table 4.3 indicates that astrology belief fell with age for every sample cohort, including Millennials.

By 2006, the adjusted percent endorsing evolution was at its lowest ebb in the 21 years the NSF had asked the question. Initially age and cohort each (younger adults and recent cohorts) appeared to predict evolution support: 52% of 18 to 24 year olds endorsed it, compared with 37% among those 65 and older (age $F_{4, 17290}$ = 47.55 p < .001). Forty-nine percent of Millennials answered "true" compared with 37% of the WW I cohort (cohort $F_{5, 17290}$ = 3.78 p = .002; total η = 0.11).

However the MCA graph in Figure 4.1 shows that controls for age, gender, and educational variables shrank the difference between cohort extremes; *in the adjusted parameters Millennials were the least likely in terms of net cohort effects to say evolution is "true"*. In the regressions presented later I added a polynomial cohort-squared term to capture the slight nonlinear effects. Table 4.3 also shows evolution support tending to drop with age for each cohort.

Answers to the UFO item showed a different curvilinear pattern. The percent answering "true" or don't know peaked among Baby Boomers at 35%, dropping among cohorts on either side ($F_{5, 7613}$ = 11.09 p < .001). About one-third of those aged 25 to 64 endorsed "UFO-ology" compared with 22% of those over 64 (age $F_{4, 7613}$ = 14.43 p < .001 Total η = 0.12). The most striking support was found among *young* Baby Boomers: 61% answered "true"—a number that swiftly dropped as this cohort aged.

Overall *adjusted* cohort values thus differed substantially from unadjusted figures. Young Millennials were the most credulous about astrology, comparable to Baby Boomers and Gen X in endorsing extraterrestrials, and supported evolution the least. In other words, the most recent American adult generation, *net of educational achievements*, had more basic science knowledge *and* greater pseudoscience susceptibility. The apparently greater evolution support among Millennials probably reflects the increased education among this generation.

Multivariate Analyses

Table 4.4 first presents standardized beta weights from regressions on the Oxford Index, applied probability score, and experimental question. It shows additions to the R^2s (ΔR^2) in order from (1) gender and degree level, (2) *n high school* science and math classes, (3) *n college* science classes, and (4) age and birth cohort. The latter half of Table 4.4 presents beta weights

Table 4.4 Standardized Multiple Regression Effects on Basic Science Knowledge

Predictors Dependent Variable→	Oxford Index	Probability Score	Drug Question
Gender (Male = 1)	0.14***	0.01	-0.02*
Degree Level	0.13***	0.08***	0.04**
High School Math	0.14***	0.14***	0.10***
Number High School Science Courses	0.13***	0.05***	0.03*
Number College Science Courses	0.21***	0.07***	0.05***
Age	0.07***	0.04*	0.02
Generation	0.18***	0.16***	0.15***
R² Gender and Degree Level	0.183***	0.042***	0.016***
Δ R² High School Math and Number High School Science Courses	0.084***	0.041***	0.023***
Δ R² Number College Science Courses	0.027***	0.003***	0.002***
Δ R² Generation and Age Categories	0.014***	0.014***	0.017***
Total R²	0.309***	0.101***	0.058***
R	0.556	0.318	0.241
N	13,205	13,205	9,171

Standardized Multiple Regression Effects on Pseudoscience Support

Predictors Dependent Variable→	Astrology Scientific	Evolution True	UFO-Alien Spaceship
Gender (Male = 1)	-0.04***	0.06***	-0.04*
Degree Level	-0.05***	0.09***	-0.07**
High School Math	-0.06***	0.00	-0.07***
Number High School Science Courses	0.00	0.03**	-0.02
Number College Science Courses	-0.04***	0.05***	-0.03
"Oxford Index"	-0.15***	0.13***	0.07***
Probability Score	-0.07***	0.00	0.02
Age	-0.09***	-0.09***	0.07***
Generation	-0.23***	0.01**	-0.30***
Generation Squared Term	0.21***	0.04	-0.41***
R² Gender and Degree Level	0.037***	0.041***	0.006***
Δ R² High School Math and Number High School Science Courses	0.007***	0.008***	0.003***
Δ R2 Number College Science Courses	0.004***	0.003***	0.000***
Δ R² Knowledge and Probability Scores	0.021***	0.012***	0.006***
Δ R² Generation and Age Categories	0.011***	0.004***	0.011***
Total R2	0.080***	0.068***	0.023***
R	0.283	0.261	0.160
n	13,205	13,205	3,569

* p < .05   ** p < .01   *** p < .001

from the regressions on *pseudoscience* beliefs. A cohort-squared polynomial was added to capture nonlinear effects on the pseudoscience dependent variables. I again show ΔR^2 as each group of predictors is added to the equation (gender and degree; high school math and science; college science; two science knowledge measures; and finally, age and cohort.)

Generation is a robust predictor of adult CSL. Controlling education and gender, cohort trumped age effects on all three science knowledge variables. As initially suspected its effects were strongest on the applied probability and drug items. High school math, and high school and college science exposure, also predicted CSL variables. Men had higher Oxford Index scores, women slightly more often answered the drug question correctly, and no sex difference occurred on the applied probability score.

Gender influenced all three pseudoscience beliefs. Women endorsed astrology or linked UFOs to space aliens slightly more, and supported evolution slightly less. Better-educated adults more often rejected pseudoscience or endorsed evolution, although these effects were weak. Adults with more high school math more often rejected astrology or UFOs; high school science exposure positively (but weakly) predicted accepting evolution. Taking more college science also weakly affected rejecting astrology or endorsing evolution.

Higher Oxford Index scores predicted rejecting astrology, accepting evolution—but *also* endorsing extraterrestrials. Net of controls, older adults more often rejected astrology *and* evolution. Cohort effects (including curvilinear effects on astrology or UFO beliefs) contributed more weakly to the explained variance on pseudoscience beliefs than they did to science knowledge. Generation had no net multivariate effect on the evolution item.

DISCUSSION AND CONCLUSIONS

Quite simply, it is insufficient to present adult responses over time alone to assess changes in the public understanding of science. When I disaggregated age and cohort in this study, the results differed from gross change over time, and *adjusted* cohort effects differed further still. Recent American generations have more formal schooling, more science and math, and, net of these, greater science knowledge. Unfortunately, this does not translate into pseudoscience rejection. Indeed, the Millennials cohort often seemed the most credulous.

The Cohort Factor

Pundits and scientists often malign US science achievement, concerned that educational quality and citizen basic CSL are declining. *Yet, my analyses, explicitly including generation, find the opposite.* Partly because the average American stays in school longer, they elect more science and math.

Between the World War I and Millennial cohorts, advanced secondary math enrollment quintupled; high school biology or chemistry enrollment tripled. Controlling degree level and college science, high school science exposure still improved adult science literacy.

Cohort especially influenced understanding inquiry: its beta weights for the probability and experimental drug measures were stronger than those for educational effects. In contrast, adult age had virtually no effects on science knowledge. In probably the most important finding from these analyses, it appears *age effects on CSL in a single cross-sectional survey almost certainly reflect generational rather than aging processes.*

Unfortunately, education and knowledge among recent cohorts provide little resistance to pseudoscience susceptibility. Net of controls, Millennials most often endorsed astrology. Perhaps they were too young to remember jokes about Nancy Reagan's astrologer, who advised the Reagans when to have surgery and cast the horoscopes of Cold War adversaries (Quigley, 1990). More recent cohorts also more often believed UFOs were extraterrestrials. Although cohort had no net effect on evolution support, Millennials had the lowest *adjusted* percent supporting evolution. And those with higher Oxford Index scores more often rejected astrology or endorsed evolution—but slightly more often *agreed* with the UFO-aliens item.[7]

Net of education and cohort variables, older adults had slightly *higher* Oxford Index and applied probability scores, arguing against memory or other cognitive "decay" as Americans age. Indeed, within each cohort, seniors endorsed astrology or "UFO-ology" less; perhaps adults learn through experience that many folk tales, e.g., zodiac compatibility, don't "deliver" and that extraterrestrial visits are unlikely. On the other hand, evolution beliefs may reflect religious attitudes not measured here (e.g., inerrant religiosity) that may increase with age. In addition, "creationist" and "Intelligent Design" challenges to American science education escalated during the later 1900s, perhaps creating a true period effect for adults from all cohorts.

Education Factors

Increased exposure to math and science consistently fostered adult science knowledge. High school exposure rivaled the influence of college science; wisely so, since even among Gen X, only one-third had any postsecondary degree. For most, high school may be the last time educators have to *formally* influence science literacy.

Educational variables affected evolution responses more than age or generation. Although evolution is not addressed in *every* high school or *every* college science course, [rejecting] evolution was nonetheless the only pseudoscience-related topic here likely to be studied in a formal education setting. K–12 curricula ignore most pseudoscience beliefs, including "sci-fi" fantasies. Perhaps educators are too embarrassed to address such topics. School administrators may fear even raising these issues partially legitimates them. And, as Eve and Dunn (1989; 1990) note, many educators personally *hold* pseudoscience beliefs, thus seeing nothing that needs modification. This pedagogical vacuum leaves students free to construct beliefs from alternative sources without any formal correction or authoritative skepticism.

Historically, the well educated glean scientific and technological information earlier than others. However, early information can be false or misleading, which may be reflected in the space fantasies often apparently unsquelched by a college education or basic science knowledge. If an individual lacks specialized particulars explaining new technological or scientific advances, one marvel (e.g., time travel) may seem as plausible as another (e.g., genetic splicing). In an era when science becomes highly specialized and technology seemingly miraculous, beliefs about visitation from outer space or "alternative medicine" may continue unless the educational system explicitly tackles them.

Men knew more science facts than women, and held fewer pseudoscience beliefs (but gender effects were small). I have noted (Losh, 2001; Losh, et al., 2003) that some pseudoscience belief does not necessarily support a "science deficit" model (Bauer, Allum, & Miller, 2007). For example, these analyses and others report women support astrology more, *net of education variables and science knowledge*. This may occur partly because most women still rely on male financial support. If much of women's economic well-being depends on their male partners, by claiming to help select "the right mate," astrology may provide its adherents with a superficial sense of control. If formal education ignores such beliefs, they can easily continue, reinforced by ubiquitous daily horoscopes and popular television programs.

Altogether, the impact of age, generation, education, and science knowledge on different CSL facets make sense. When classrooms address a topic, e.g., inquiry understanding or evolution, formal education variables influence adult CSL. When formal education neglects a particular pseudoscience sector, age and cohort assume more importance. Diminished pseudoscience beliefs as adults age, e.g., astrology, may reflect life experiences. On the other hand, cohort effects (e.g., extraterrestrials) draw our attention to the influence media can wield when formal education abdicates the topic.

Why Cohort Change?

There are two likely possibilities for the simultaneous increase among recent generations in science knowledge—and some pseudoscience support. The first addresses science education changes, which have diffused to K–12 educators, emphasizing more science inquiry, contextual issues, and "how scientists think". This explanatory inference is indirect at best, because we only know *which* high school science and math courses respondents took *and not how* these were taught. Unfortunately, if academic jargon (e.g., "emphasis on inquiry") were included in general public surveys it is doubtful that adults would recognize it, or remember how their high school teachers addressed science topics to be able to tell us more. Nevertheless, these analyses suggest that more recent graduates have received a more fruitful US science education than in the past.

More recent generations also more often attended college; there, they learn appreciation of scientific methods from *several* disciplines. *Nearly every basic college text in the social, educational, or behavioral sciences now includes at least one chapter on systematic methods of study*, thereby reinforcing student experiences in "other" science courses.

Regrettably, by educators ignoring pseudoscience topics, mass media rush to fill the gaps, invoking a second possible explanation of cohort effects on adult CSL. Media can ridicule pseudoscience phenomena, e.g., the "White House astrologer". On the other hand, it's unsurprising that young Baby Boomers, who matured on *Twilight Zone* and *Star Trek*, linked UFOs to alien spacecraft. In other research, I have noted that boys, especially, fed a diet of *Power Rangers*, infuse their drawings of scientists with superhuman powers (Losh, Wilke, & Pop, 2008). More recent generations more often access satellite or cable television and the Internet, both of which are brimming with uncorrected sci-fi fantasies.

The results from this study suggest recent changes in science education may have boosted science knowledge among young adults in particular. The findings support those who wish to see such innovations continue and extended to US *college* science courses, hopefully to forestall the "brain drain" among talented undergraduates away from science careers (Burris, 2006). At a time when there are potential spending cuts on science education, it is important to highlight possible positive consequences, indirect though these may be, of pedagogical changes.

On the other hand, it is sobering to see the disarrayed assertions among American industrial, government, and educational leaders about science education. It is relatively easy to sketch an "idealized" science curriculum—especially if proponents don't know about *already occurring* changes in science education and science and math advances among our students. Parents and teachers who recognize these improvements (regardless of further desired developments) may resent what they see as jabs at student achievements, and overemphases on math and science education. None of these parties seem aware of advances in cognitive science that assert both domain knowledge (e.g., science facts) and critical thinking skills are required to educate new generations of Americans. Clearly, increased communication across these spheres is critical to bridge the sizable gaps among them.

ACKNOWLEDGMENTS

Thanks to the American Educational Research Association. Their research grant, through the National Science Foundation's REC-0310268, made it possible to complete construction of the major database and to generate these analyses. Thanks also to Ryan Wilke, Brandon Nzekwe, Alice Robbin, Ray Eve, Ken Feder, Martin Bauer, Nick Allum, Bob Bell, and Jeri Mulrow for insight and assistance. The responsibility for errors and interpretation is, of course, my own.

NOTES

1. Only 2006 respondents with landlines or cell phones—95% of the sample—are analyzed to maximize comparisons with the 1981–2001 RDD surveys.
2. The percentage score is used for the Oxford items and the applied probability score to standardize for the different numbers of items asked over time. Coefficient Alpha for the Oxford Index is 0.68 with each item contributing about equally to the coefficient. Indices from earlier studies were shortened for the 2006 NSF Surveys, based on IRT and other measurement analyses reported in Bann & Schwerin, 2004.
3. Carlson (2008) adopts this term because this relatively small generation matured during a period of affluence following World War II, thus enjoying considerable educational and occupational opportunities.
4. This sum is a conservative estimate because courses omitted from the NSF questionnaire (e.g., ecology) could not be included.
5. It is unclear whether the oldest cohort meant physics, or perhaps a physical science, e.g., "earth science".
6. Experimental drug responses were excluded here as a predictor because this item was only asked after 1992. Its inclusion would cause severe data loss on all other variables and restrict the UFO-alien analysis to 2001 alone.
7. The lack of high school science and math effects in the final equations for pseudoscience support largely occur because these variables predict the Oxford Index. A structural equation model could disentangle these educational effects, *but is tangential to these analyses*, which focus on the net effects of birth cohort or generation.

REFERENCES

Aldrich, M. (1978). Women in science. *Signs, 4* (1): 126–135.
Allum, N., P. Sturgis, D. Tabourazi, & I. Brunton-Smith (2008). Science knowledge and attitudes across cultures: A meta-analysis. *Public Understanding of Science, 17* (1): 35–54.
American Association for the Advancement of Science (1993). *Benchmarks for Scientific Literacy*. New York: Oxford University Press: http://www.project 2061.org/tools/benchol/bolinto.htm
Bann, C.M. & M.J. Schwerin (2004). Public knowledge and attitudes scale construction: Development of short forms (Contract No. GS10F-0097L). Research Triangle Park, NC: RTI International.
Bauer, M.W., N. Allum, & S. Miller (2007). What can we learn from 25 years of PUS survey research? Liberating and expanding the agenda. *Public Understanding of Science, 16* (1): 79–95.
Boyd, D. & H. Bee (2009). *Lifespan Development (5th edition)*. Boston: Pearson.

Burkam, D.T., V.E. Lee, & B.A. Smerdon (1997). Gender and science learning early in high school: Subject matter and laboratory experiences. *American Educational Research Journal, 34* (2): 297–331.

Burris, J. (2006). Testimony Offered to the Research Subcommittee of the Committee on Science of the US House of Representatives Hearing on Undergraduate Science, Math & Engineering Education: What's Working? March 15.

Cain, S.E. (2002). *Sciencing.* Saddle River, N.J.: Merrill Prentice Hall.

Carlson, E. (2008). *The Lucky Few: Between the Greatest Generation and the Baby Boom.* New York: Springer.

Davis, J.A. & T.W. Smith (2009). *General Social Surveys, 1972–2008.* Chicago: National Opinion Research Center and Storrs, CT: The Roper Center for Public Opinion Research, University of Connecticut (distributor).

Eve, R.A. & D. Dunn (1989). High school biology teachers and pseudo scientific belief: Passing it on? *Skeptical Inquirer, 13*: 260–263.

Eve, R.A. & D. Dunn (1990). Psychic powers, astrology & creationism in the classroom? Evidence of pseudoscientific beliefs among high school biology & life science teachers. *The Biology Teacher, 52* (1, January): 10–21.

Fox, M. & G. Firebaugh (1992). Confidence in science: The gender gap. *Social Science Quarterly, 73* (1): 101–113.

Gallup News Service (2001). Did men really land on the moon? Poll analyses. February 15. Available at http://gallup.com/poll/releases/pr010215f.asp. (Retrieved from source.)

Gates, B. (2005). *High Schools for the New Millennium: Imagine the Possibilities.* Retrieved from: http://www.gatesfoundation.org/nr/Downloads/ed/EdWhitePaper.pdf on September 18, 2007.

Gess-Newsome, J. (2002). The use and impact of explicit instruction about the nature of science and science inquiry in an elementary science methods course. *Science & Education, 11*: 55–67.

Glenn, N.D. (2005). *Cohort Analysis (2nd edition).* Thousand Oaks, CA: Sage Publications.

Goode, E. (2000). *Paranormal Beliefs: A Sociological Introduction.* Prospect Heights, IL: Waveland Press.

Goode, E. (2002). Education, scientific knowledge, and belief in the paranormal. *Skeptical Inquirer, 26* (No. 1, January/February): 24–27.

Kadlec, A., W. Friedman, & A. Ott (2007). *Important but not for me: Parents and students in Kansas and Missouri talk about math, science and technology education.* New York: Public Agenda, released September 18, 2007 (retrieved from source): http://www.publicagenda.org/importantbutnotforme/pdfs/important_but_not_for_me.pdf

Kumar, D.D. & D.E. Chubin (eds.) (2000). *Science, Technology, and Society: A Sourcebook on Research and Practice.* New York: Kluwer Academic/Plenum Publishers.

Lee, V.E. & D.T. Burkam (1996). Gender differences in middle-grade science achievement: Subject domain, ability level, and course emphasis. *Science Education, 80* (6): 613–650.

Lemonick, M., R.W. Keegan, & I. Ybarra (2006). Is America flunking science? *Time*, February 13, 23–33.

Lilienfeld, S.O., J.M. Lohr, & D. Morier (2001). The teaching of courses in the science and pseudoscience of psychology: Useful resources. *Teaching of Psychology, 28* (3): 182–191.

Losh, S.C. (2001). Science and pseudoscience. *Public Perspective, 12* (5): 24–26.

S.C. Losh (2010). Generation, education, gender and ethnicity in American digital divides. Chapter 11 (refereed) in E. Ferro, Y.K. Dwivedi, J.R. Gil-Garcia, and M.D. Williams, (eds.) *Handbook of Research on Overcoming Digital Divides:*

Constructing an Equitable and Competitive Information Society. Hershey, PA: IGI Global, pp. 196–222.

Losh, S. C., C.M. Tavani, R. Njoroge, R. Wilke, & M. McAuley (2003). What does education really do? Educational dimensions and pseudoscience support in the American general public, 1979–2001. *The Skeptical Inquirer, 27,* (September/October): 30–35.

Losh, S.C., R. Wilke, & M. Pop (2008). Some methodological issues with 'Draw a Scientist Tests' among young children. *International Journal of Science Education, 30* (6): 773–792.

Martin, M. (1994). Pseudoscience, the paranormal, and science education. *Science and Education, 3:* 357–371.

Mason, K.O., H.H. Winsborough, W.M. Mason, & W.K. Poole (1973). Some methodological issues in cohort analysis of archival data. *American Sociological Review, 38:* 242–258.

Mason, R. & W.G. Brown. 1975. Multicollinearity problems and ridge regression in sociological models. *Social Science Research, 4* (2): 135–149.

Miller, J.D. (2000). The development of civic scientific literacy in the United States. In D.D. Kumar & D.E. Chubin (eds.) *Science, Technology, and Society: A Sourcebook on Research and Practice.* New York: Kluwer Academic/Plenum Publishers: 21–47.

Miller, Jon D., Linda Kimmel, ORC Macro and NORC. National Science Foundation Surveys of Public Attitudes Toward And Understanding of Science And Technology, 1979–2006 [Computer file]. 3rd Roper Center version. Tallahassee, FL: Susan Carol Losh, Florida State University, Department of Educational Psychology & Learning Systems/Arlington, VA: National Science Foundation, Division of Science Resources Statistics/Arlington, VA: American Statistical Association [producers], 2009. Storrs, CT: Roper Center for Public Opinion Research [distributor], 2009.

Moss, D.M., E.D. Abrams, & J.A. Kull (1998). Can we be scientists too? Secondary students' perceptions of scientific research from a project-based classroom. *Journal of Science Education and Technology, 7* (2): 149–161.

National Center for Educational Statistics (2004). *Digest of Educational Statistics.* Washington, D.C.: US Government Printing Office.

National Science Board (2008). *Science & Engineering Indicators 2008.* Arlington, VA: National Science Foundation (NSB-08–01).

Pew Research Center for the People and the Press Pew (2009a). A survey conducted in collaboration with the American Association for the Advancement of Science. http://people-press.org/report/528/ (Accessed 10 July 2009).

Pew Forum on Religion and Public Life (Pew 2009b). Many Americans mix multiple faiths: Eastern, New Age beliefs widespread. Washington, D.C.: Pew Research Center. http://pewforum.org/Other-Beliefs-and-Practices/Many-Americans-Mix-Multiple-Faiths.aspx (Accessed 5 March 2011).

Pew Research Center (2007). *How Young People View their Lives, Futures and Politics: A Portrait of "Generation Next."* Washington, D.C.: Pew Research Center for the People and the Press.

Quigley, J. *What Does Joan Say: My Seven Years as White House Astrologer to Nancy and Ronald Reagan.* New York: Birch Lane Press, 1990.

Regan, D.T. (1988). *For the Record.* New York: Harcourt Brace Jovanovich, 1988.

Scanlon, E. (2000). How gender influences learners working collaboratively with science simulations. *Learning and Instruction, 10:* 463–481.

Schiebiner, L. (1999). Gender studies of STS: A look toward the future. *Science, Technology & Society, 4* (1): 95–106.

Schmidt, W.H., C.C. McKnight, & S.A. Raizen (1997). *A Splintered Vision: An Investigation of US Science & Mathematics Education.* Boston: Kluwer Academic Press.

Seymour, E. (2006). Testimony offered to the Research Subcommittee of the Committee on Science of the US House of Representatives Hearing on Undergraduate Science, Math & Engineering Education: What's Working? March 15.
Sunal, D.W. & C.Z. Sunal (2003). *Science in the Elementary and Middle School.* Upper Saddle River, NJ: Merrill Prentice Hall.
Taylor, J., R.A. Eve, & F.B. Harrold (1995). Why creationists don't go to psychic fairs. *Skeptical Inquirer, 19* (6): 23–28.
US Bureau of the Census (2008). *The Statistical Abstract of the United States Table 253.* Washington, D.C.: US Government Printing Office.
US Department of Education, National Center for Education Statistics (2007). *The Condition of Education, 2007 (NCES 2007–064).* Washington, D.C.: US Government Printing Office.
Woolfolk, A. (2007). *Educational Psychology (10th edition).* Boston: Allyn & Bacon.

5 The Image of Science in Bulgaria and UK, 1992–2005
Does Generation Matter?
Kristina Petkova and Valery Todorov

THE CONTEXT: KNOWLEDGE SOCIETY AND PUBLIC UNDERSTANDING OF SCIENCE

In societies of late modernity knowledge and the institutions of knowledge become of crucial importance. Unlike previous societies these no longer depend on agriculture and the manufacturing industry. In the last two decades (and more) the idea of a knowledge-based society has been promoted. Its economy is a 'learning economy' (Lisbon European Council, March 2000), and the entire society creates, shares and uses knowledge for the prosperity and well-being of its people (European Commission, 2000, p. 5; see also *Handbook of Knowledge Society Foresight*, 2002). Being the 'leading actors of knowledge societies' (op. cit., p. 7) people are expected to engage in a constant quest for new knowledge or, in other words, to become learning citizens (Martin, 2003).

In this context, the public understanding of science becomes an indispensable part of the massive effort on the part of the European Commission to bring science closer to society and to demythologise the image of science and the scientist.

It is thus not surprising that during this period there has been a growing interest in measuring the public understanding of science. Good examples in this respect are representative national surveys carried out within the framework of the Eurobarometer (for a review see Bauer, Allum & Miller, 2007). As a result we are now in the position to conduct systematic longitudinal analysis of trends in adult scientific literacy, attitudes to science, and interest in science.

WHY COMPARE BULGARIA AND UK?

Bulgaria and the UK represent two different cases regarding both science development and the public understanding of science. The UK is a classic example of a country with long traditions in science development and firmly established scientific institutions.

From the point of view of the scientific community, public understanding of science (PUS) can be defined as a policy of the community that aims to spread (popularise) science and scientific activities among lay people. The necessity of such a policy becomes clear when public support for science appears to be questioned either by the state or by the wider society or by both. Perhaps the most dramatic challenge to the relationship between science and the wider society was witnessed (at least for the Western countries) in 1986 when the Chernobyl disaster happened. In order to act, however, the scientific community needs autonomy and to have the appropriate units of actors to implement its policies. The UK has had longstanding independent scientific institutions capable of initiating and promoting their own policies. Thus, when power changed hands in the mid 1980s and the state withdrew its unconditional support for science in the UK, the Royal Society took action and in order to gain the support of the wider public launched a long-term campaign for enhancing public understanding of science (Royal Society, 1985).

The historical background of Bulgaria is completely different. Bulgarian statehood was restored at the end of 19th century after 500 years of Ottoman domination. At that time promoting strongly developed Bulgarian science and education was at the core of Bulgarian liberation and independence. Support for science and education became a priority for the newly established Bulgarian state. A Bulgarian scientific community gradually came about, striving for its autonomy, but the establishment of a communist dictatorship in 1944 was a turning point in its development. Science and education were subjected to the ideological doctrine of the Communist party, and the scientific community was strictly controlled by the party-state. There was no chance of even mentioning the issue of autonomy. In 1989 the Communist party officially stepped down from the totalitarian governing of the state. The extremely hard economic transition that took place afterwards led to a dramatic cut in the financing of science. Added to this was the heavy and ineffective organisational structure of scientific institutions. Thus, Bulgaria entered a situation in which, like in the UK, though for different reasons, the state questioned its support for science. The scientific community, having been completely controlled by the state, fell into a deep crisis.

What was needed was a new approach to the relationship between science and society. For almost twenty years, however, scientific institutions continued, by force of old habits, entirely dependent on state support. As a result, these institutions gradually started fading away with almost no recruitment of young people. Coming to the verge of its very existence in the last couple of years the Bulgarian scientific community for the first time showed concern for the public understanding of science. Finally, it took two important steps. First the General Assembly of the Academy of Sciences, which is the biggest science institution in Bulgaria, ordered a special survey on lay people's attitudes to science and particularly to the Academy. The second decision was to establish a new administrative unit for public relations at the Academy, and in addition, to create a Council for public relations at the General assembly, which comprises both scientists and journalists.

THE PRESENT STUDY

With this study we are continuing a track of comparative work on public knowledge and attitudes to science (Bauer, Petkova, Boyadjieva, 2000) and long-term trends in public representations of science (Bauer et al., 2006). The findings showed some striking differences in the long-term trends in attitudes to science and in science reportage across Bulgarian and UK contexts. The question that emerges is: How will generations in the two countries relate to science in these different contexts?

The study examines the impact of birth cohort on three main public understanding of science indicators—science literacy, interest in science, and attitudes towards science in two European countries, the UK and Bulgaria, for the period 1992–2005.

The expectations of the study are that the trends of these three indicators will be cohort dependent and could be interpreted in terms of the place of science within the two types of society.

EMPIRICAL BASIS

The database of our study comprises integrated longitudinal data collected in five representative surveys on public understanding of science carried out in the UK (1992 and 2005, Eurobarometer) and in Bulgaria (1992, 1996, and Eurobarometer 224/2005). The overall number of participants in these surveys is 5543. Sample size for Bulgaria varied between 878 and 1008 respondents. Sample size for UK varied from 1307 to 1374 respondents.

The variables with similar connotations that were identified in these surveys were recoded to a common base with identical scales of measurement.

Variables in the Study

The first variable we identified was interest in science. From the battery of questions related to interest in science and technology we included in the analysis the question about interest in scientific discoveries.

Answers are given on a 3-point scale from very interested to not at all interested.

The Attitude Variable

Where applicable, we attempted to develop an attitude scale. However, the items measuring attitudes towards science did not show consistent correlation (only two pairs of variables showed significant correlation that did not exceed .r = .33). This did not allow us to construct a homogeneous scale; therefore, we treated the items as facets of attitude to science:

- support for government funding
- distrust in scientists
- daily relevance of science
- belief that benefits of science are greater than harms
- belief that science makes life healthy, easy, and comfortable
- science versus faith

The Science Literacy Variable

We calculated a knowledge index on the basis of the five knowledge items that appeared in all three surveys: 'The centre of Earth is very hot'; 'Lasers work by focusing sound waves'; 'It is the father's gene which decides whether the baby is a boy or a girl' (the change in wording of the question in Eurobarometer 2005 was taken into consideration in the recoding); and 'Does the Earth go around the Sun or the Sun around the Earth?'.

The Cohort Variable

In cohort analysis literature there has been a debate over the time duration of a cohort. Rather than using a constant interval, authors (e.g., Losh, 2007) have associated the beginning and the end of a cohort not only with duration but also with distinctions of historical periods and significant events occurring within them. Thus we followed Bauer and Shukla (2011) and defined five birth cohorts.

Born after 1977

UK: The cohort of respondents that grew up after the Cold War and witnessed the final victory of the capitalist style of economy.

Bulgaria: The generation that lived through the collapse of communism and the beginning of the transition to democracy and market economy.

Born 1963–1976

UK: The generation that grew up during antinuclear protest, nuclear armament, debates, and the Star Wars initiative.

Bulgaria: The generation indoctrinated to believe that it was living in the period of 'elevated and all around expression' of 'socialist democracy'. Labeled 'developed socialism' for the duration of this period, science was supposed to be 'widely implanted in the production process, and become a direct production force' (*Economic Encyclopaedia,* 1984, p. 363).

Born 1950–1962

UK: The generation that grew up in a period of modernisation and economic prosperity. This is the protest generation of the 1970s with an idealistic worldview and scepticism regarding the role of science and technology.

Bulgaria: The generation that, to use again communist rhetoric, lived in a time characterised by the 'fight between dying capitalism and new born socialism' (*Economic Encyclopaedia,* 1984, p. 365).

Born 1930–1949

UK: The generation that witnessed the WWII and entered the Cold War, characterised by nuclear optimism and belief in the scientific revolution.

Bulgaria: The generation that took the consequences of the 'revolutionary ardour' of the proletarian dictatorship.

Born before 1930

UK: The generation that grew up through the exciting period of the 1920s and survived the dramatic economic crisis of 1929.

Bulgaria: The generation that through the years of totalitarian socialism still reminisced about the 'bourgeois' past.

Further, we defined the age variable in four categories. We did this because in one of the surveys (Bulgaria 1992), instead of listing the exact age, respondents had to choose one of the following categories; 18–29 years of age; 30–44; 45–59, or more than 60.

With regard to the education variable, after a series of transformations we defined four levels of education: none, primary, secondary, and tertiary.

REPORTING RESULTS: SCIENCE THROUGH THE EYES OF BULGARIANS AND BRITISH PEOPLE

As has been pointed out (See Bradley & Elms, 1999, for discussion) cohort analysis pertains to the influence of both age and the concrete period of time and synthesises the effects of both. This creates a logical and statistical identification problem because if the three factors are studied simultaneously their impact cannot be differentiated (Losh, 2007; Yang & Land, 2006; Bradley & Elms, 1999). That is why it is recommended that only two variables be studied at a time. In our study we had the option to explore the cohort effect either controlling for age or for period. Therefore we first ran a correlation analysis of the variables. Due to the high correlation ($r=.0.89$, $p<.001$) between age and cohort we decided to control for period. Therefore we decided not to use the age variable. In addition, due to the fact that as longevity increases, there is a tendency through generations for education to take more time and the level of obtained educational degree to be higher. Therefore, in order to see the net effect of cohort we needed to control for education as well.

To explore tendencies in the lay community's relation to science in the UK and Bulgaria through cohorts, we built several factorial ANOVA models. Before running the analysis, we standardised the attitude variables as well as the knowledge variable. In the ANOVA models, dependent variables were the different facets of attitude to science, science literacy, and interest in scientific discoveries. The independent variable was always the cohort. As mentioned above we also included control variables—period, sex, and education. To distinguish between the two countries we included the variable 'country' and tested its effect.

The Effect of the 'Country' Variable, the Interaction Effect of Country/Cohort Variables, and Cohort Effect for the UK and Bulgaria

First we carried out factorial ANOVA analyses including the 'country' variable. The effect of country and the interaction effect for country and cohort were significant for all ANOVA models at $p < .05$; with the exception of 'Science makes life healthier, easier, and more comfortable' (see Table 5.1).

However, this model does not allow us to examine the significance of the cohort effect within each one of the two countries. To do so, we carried out additional ANOVA analyses for the UK and Bulgaria separately. The estimated marginal means of the attitude variables for each of the countries are presented in Figure 5.1a to 5.1g. Table 5.2 presents effects of cohort on attitudes to science for each of the countries.

Science Literacy

As can be seen in Figure 5.1, Bulgarians are generally less knowledgeable in science than the British.

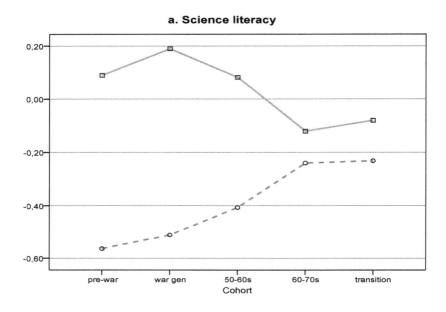

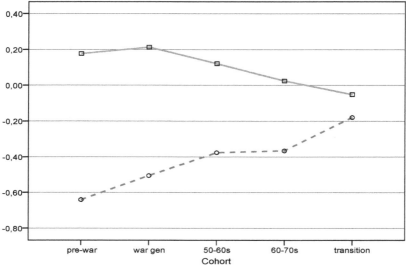

Figure 5.1a and 5.1b The estimated marginal means of attitude variables (dashed line = Bulgaria, continuous line = UK).

c. Benefits greater than harm

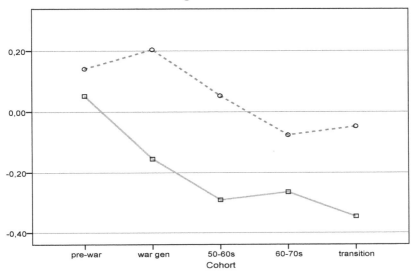

d. Not important for daily life

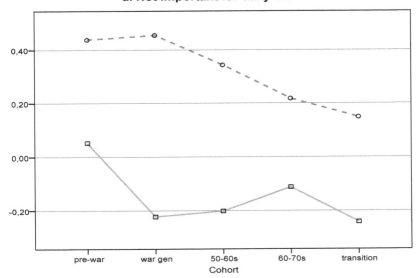

Figure 5.1c and 5.1d The estimated marginal means of attitude variables (dashed line = Bulgaria, continuous line = UK).

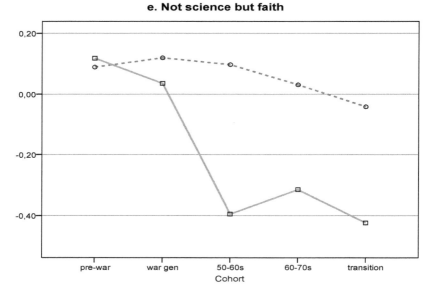

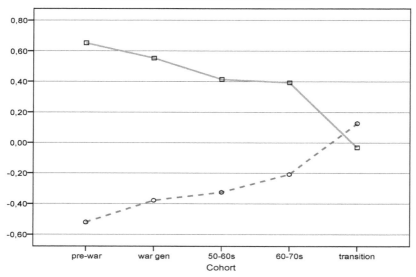

Figure 5.1e and 5.1f The estimated marginal means of attitude variables. (dashed line = Bulgaria, continuous line = UK)

g. Support for scientific research

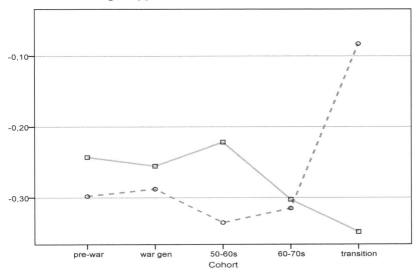

Figure 5.1g The estimated marginal means of attitude variables. (dashed line = Bulgaria, continuous line = UK)

The greatest gap between the countries is in the prewar generation. The Bulgarian generation of 1950–1962 and the following one seem to be catching up to the transition generation. For the UK we witness a steady decrease in knowledge that starts with the war generation, the generation of the '70s almost reaching the level of Bulgarians.

Interest in Science

Bulgarian cohorts consistently demonstrate lower interest in science than the British.

For the UK the trend is on the verge of significance (p = .06). As can be seen there has been gradual decrease in interest in science that starts, again, with the war generation. As opposed to this, for Bulgaria we witness a significant increase in interest in science through cohorts.

Benefits Greater Than Harms

For this facet of attitude to science, each cohort of Bulgarians is consistently more optimistic than the British and seem to believe that science is something more beneficial than harmful.

The cohort effect for the two countries separately is also significant. Both trends are downwards, showing an increased concern with the risks related to scientific development.

Daily Relevance of Science

For this variable all effects are significant.

In both countries, the tendency is towards a growing perception of relevance of science in everyday life. For the UK, however, this trend towards relevance of science seems to slow down with the WWII generation.

Science versus Faith

Rather unexpectedly, bearing in mind their socialist past, Bulgarians in general seem to be more prone to resort to religion than the British.

Within the countries the effect of cohort is significant only for the UK. Data show that the relationship between science and faith has undergone dramatic changes. Starting with a kind of plateau between the prewar and war generation, faith loses its importance rapidly.

Distrust in the Scientists

The interaction of country and cohort effect is significant. Bulgarians put more trust in science than the British.

Within the countries the trend is significant only for the UK and shows gradual increase in trust in science and scientists.

Support for Government Funding

As a whole British people are more inclined to support increased government financing of science than Bulgarians.

Within the countries the trend is significant only for Bulgaria. After more or less unchanged willingness for governmental support for science, the transition generation demonstrates an increase in support.

DISCUSSION: THE SOCIAL EMBEDDEDNESS OF PUBLIC UNDERSTANDING OF SCIENCE

We reported different trends of changes through cohorts in Bulgaria and the UK regarding science literacy, interest in science, and attitudes towards science. We end with some speculations on these trend differences, hypothesising about their relatedness to the types of societies existing in Bulgaria and the UK.

For almost half a century the Bulgarian social context was a communist regime. In the doctrine of communism, science is part of the fundament of society. Truth was considered to be an exclusive privilege of science. The cognitive power of scientific activity was thought to be unlimited, and science was viewed as the main producer of progress. Socialism was supposed to be society built on scientific ground. In communist ideology the term 'scientific' legitimised each and every political intention and/or decision. It is indicative that in all Bulgarian universities a mandatory discipline was 'scientific communism'. Science was bestowed with the full financial support of the totalitarian state. This resulted in a hypertrophied rationalism, approximating at its extreme a religious dogma, in overestimation of objectivism, determinism, and control, and in disregard of the potential risks of science development in modern life.

Hence the profession of scientist was rated extremely high, second in prestige only to a career in the Communist party. However, during totalitarian socialism science was subjected to severe ideological control. On the one hand science was considered almighty and on the other it had to comply with the requirements of the planned totalitarian society. Objectives were set by the Communist party: the scientific community was stripped of any autonomy. The only tool in its quest for knowledge was supposed to be Marxist-Leninist ideology. What needed to be 'shared' with lay people (and how and when) was decided by the Communist party with political and ideological considerations. As stated in the beginning of this article, PUS as an independent policy of the scientific community virtually did not exist (for a more detailed analysis, see Boyadjieva et al, 1994).

By contrast, in the years following WWII, the UK, together with other advanced capitalist societies, has been undergoing a long-term secular transition for which various names have been used: post-modern society (Bauman), postindustrial society (Bell), risk society (Beck) reflexive modernity (Giddens). During this period, for the first time in history, belief in inexhaustible natural resources begins to be undermined (Meadows & Meadows, 1972), and the idea of interdependence between society and nature comes to the fore. Social attention is drawn to the risk elements, which loom large and unpredictable. This process leads to the gradual evolvement of a more 'egalitarian' relationship between science and the wider society, which no longer views science as something exceptional. Science loses its privileged position with regard to state support and depends to a great extent on private initiative. Scientists start to be regarded as just one professional group among others, and the public increasingly insists on taking part in the debate on the 'internal affairs' of science. There is no longer an indispensable core of popular scientific knowledge (Bauer 2006). This triggers the autonomous scientific community to initiate a campaign for the promotion of science among lay people.

Coming back to and summarising the results of the present study, we see some striking differences as well as common tendencies that could be viewed within the context of the outlined characteristics of the UK and Bulgarian societies. Thus, referring to these differences we can interpret the decrease in science literacy in the UK through the fading away of the requirement for a 'unified canon' of science literacy (Bauer, 2006, p. 3). The decreased interest in science can likewise be explained through the tendency of disenchantment with science, which starts to be 'taken for granted' (Bauer, 2006, p.3). The much more optimistic view of the benefits of science demonstrated by Bulgarian people and their confidence in the infallibility of scientists corresponds to the key role of science for socialism and the exhilarated attitude promoted by the totalitarian state. The opening of the country after the fall of the Berlin Wall might have contributed to the growing scepticism of Bulgarians. In market-oriented society, which Bulgaria has been striving for since 1989, science is taken down from the pedestal of 'main productive force'.

The opposite tendency, towards an increase of trust in scientists in the UK, might be the result of opening science to the public, and of the activities initiated by the scientific community to promote science in the wider society, discussed in the beginning of the article. Both of these developments bring science closer to society and enhance the feeling of lay people that they will have a say in crucial scientific matters.

At first glance, the findings that Bulgarians are less willing to support increased government financing of science and are more inclined to resort to religion than to science seem inconsistent with the characteristics of the totalitarian state. However there are two things to be considered. First the question administered to the respondents was phrased 'Do you support increase in **government** spending for science'. Whereas for the UK governmental financing is one among other possible sources of support for science, in Bulgaria governmental financing of science was the only option during totalitarianism. Therefore their position is understandable. When interpreting the Bulgarian data it should also be taken into account that the 1996 survey was carried out during a period of hyperinflation: to consider additional spending was out of the question. At that time the generation of the 50–60s was in its prime and was responsible for the 'survival' of their families. This might have had an impact on their reluctance to support an increase in **government** spending for science as opposed to the younger 'transition generation'. Turning to the opposition 'science versus faith' we see an expected overall trend towards secularism for the UK and a plateau for Bulgaria—Bulgarians tending to consider the point of view of faith more than the British. When interpreting this controversial finding we should not forget that the totalitarian state had raised rationalism to the point of religious dogma, and a reaction against this could be predicted.

CONCLUSION: A FURTHER OUTLOOK

We investigated the impact of cohort on public understanding of science in two countries representing two different social systems within which science has had a different social role and prestige. We believe that it would be interesting to carry out such study in societies with similar characteristics— for example two former socialist countries, or two developed Western societies. This would allow further deepening of the analysis. Thus if different tendencies are revealed this will draw the attention to other explanatory factors—for example, cultural factors. Further, it might allow us to better outline the impact of both the specificity of the relations between science and the public, and the initiative of scientists themselves with respect to public understanding of science.

APPENDIX 1

Table 5.1 ANOVA. Effects of Country, Cohort, and interaction between Country and cohort on Science literacy, Interest in scientific discoveries, Healthy easy, and comfortable, Not science but faith, Benefits greater than harms, Support for scientific research, Daily relevance of science, and Distrust in scientists. Control variables: Year, Sex, and Education.

Dependent Variable	Independent Variable	F	df (between, within)	p (Exact)	Partial Eta Squared
Science literacy	Country	202.00	1, 5482	.000	.036
	Cohort	.81	4, 5482	.518	.001
	Country*Cohort	21.24	4, 5482	.000	.015
Interest in science and scientific discoveries	Country	287.65	1, 4613	.000	.059
	Cohort	1.30	4, 4613	.267	.001
	Country*Cohort	14.19	4, 4613	.000	.12
Healthy easy, and comfortable	Country	3.00	1, 4343	.083	.001
	Cohort	.91	4, 4343	.456	.001
	Country*Cohort	.52	4, 4343	.718	.001
Not science but faith	Country	47.75	1, 4339	.000	.011
	Cohort	12.52	4, 4339	.000	.011
	Country*Cohort	10.04	4, 4339	.000	.011
Benefits greater than harm	Country	47.64	1, 4343	.000	.011
	Cohort	8.72	4, 4343	.000	.008
	Country*Cohort	2.50	4, 4343	.000	.002
Not important for daily life	Country	169.06	1, 4233	.000	.037
	Cohort	4.77	4, 4233	.001	.004
	Country*Cohort	5.17	4, 4233	.000	.005
Support for scientific research	Country	.10	1, 4339	.755	.000
	Cohort	.68	4, 4339	.606	.001
	Country*Cohort	3.21	4, 4339	.012	.003
Distrust in science and scientists	Country	374.10	1, 3474	.000	.097
	Cohort	.46	4, 3474	.769	.001
	Country*Cohort)	30.16	4, 3474	.000	.034

Table 5.2 ANOVA. Effects of Cohort for Bulgaria and UK separately on Science literacy, Interest in scientific discoveries, Support for scientific research, Benefits greater than harms, Daily relevance of science, and Distrust in scientists. Control variables: Year, Sex, and Education.

Dependent Variable	Bulgaria				UK			
	F	df (between, within)	p (Exact)	Partial Eta Squared	F	df (between, within)	p (Exact)	Partial Eta Squared
Science literacy	7.12	4, 2824	.000	.010	4.04	4, 2654	.003	.006
Interest in science and scientific discoveries	6.24	4, 1955	.000	.013	2.27	4, 2654	.059	.003
Healthy easy, and comfortable	2.57	4, 2342	.036	.004	.75	4, 1997	.56	.001
Not science but faith	1.56	4, 2338	.182	.003	15.20	4, 1997	.000	.030
Benefits greater than harm	6.73	4, 2317	.000	.011	5.92	4, 2022	.000	.011
Not important for daily life	6.97	4, 2315	.000	.012	5.55	4, 2022	.000	.011
Support for scientific research	2.63	4, 2338	.026	.005	1.24	4, 1997	.293	.002
Distrust in science and scientists	2.25	4, 1448	.062	.006	2.95	4, 2022	.019	.006

REFERENCES

Bauer, M. (2006) Towards Post-industrial Public Engagement with Science: Revisiting 10-year-old Hypotheses for Europe 2005, 9th International Conference on Public Communication of Science and Technology, Seoul, 17–20 May 2006.

Bauer, M. (2011) The Changing Culture of Science across Old Europe 1989 to 2005. In Martin W Bauer, Rajesh Shukla and Nick Allum (eds.) (2011) *The Culture of Science—How the Public Relates to Science Across the Globe*. New York: Routledge.

Bauer, M. W. and Allum, N. and Miller, S. (2007). What can we learn from 25 years of PUS survey research? Liberating and expanding the agenda. *Public understanding of science*, 16 (1). pp. 79–95.

Bauer, M., Petkova, K., & Boyadjieva, P. (2000) Public Knowledge and Attitudes to Science: Alternative Measures That May End the 'Science War', *Science Technology and Human Values*, 25, 1 pp. 30–52.

Bauer, M., Petkova, K., Boyadjeva, P., & Gornev, G. (2006) Long Term Trends in the Public Representation of Science Across the 'Iron Curtain' (1946–1995), *Social Studies of Science*, **36**, 1 pp. 99–131.

Bauman, Z. (2002). A Sociological Theory of Postmodernity. In: C. Calhoum, J. Gerteis, J. Moody, St. Pfaff, and I. Virk (eds.) *Contemporary Sociological Theory*, London: Blackwell Publishing, pp. 429–440.

Beck, U. (1992). *Risk Society: Towards a New Modernity*. London: SAGE.

Beck, U., A. Giddens and S. S. Lash (1994). *Reflexive Modernization. Politics, Tradition and Aesthetics in the Modern Social Order*. Cambridge and Oxford: Polity Press and Blackwell Publishers.

Bell, Daniel (1973). *The Coming of Post-Industrial Society: A Venture in Social Forecasting*. New York: Basic Books.

Boyadjieva, P., Tchalakov, I., Petkova, K. (1994) *Science: Life outside the Laboratory*, Bulgarian Academy of Science Publishing House.

Bradley, H. & Elms L. (1999) Age-Period-Cohort Analysis with Noisy, Lumpy Data, paper presented for the annual meeting of Political Methodology Group of American Political Science Association, College Station, Texas.

Economic Encyclopaedia (1984) Partizdat, Sofia.

European Commission (2005) *Eurobarometer*, technical report, no 224, Brussels, April 2005.

European Commission (2000) *A Memorandum on Lifelong Learning*. Available at: http://ec.europa.eu/education/policies/lll/life/memoen.pdf

European Commission (1993) *Eurobarometer*, technical report, no 224, Brussels, April 2005.

Handbook of Knowledge Society Foresight (2002) Prepared by PREST and FRC for the European Foundation for Improvement of Living Conditions. PREST authors Ian Miles and Michael Keenan; FERC authors Jari Kaio-Oia.

The Lisbon Special European Council (2000) Towards a Europe of Innovation and Knowledge.

Losh, S. (2007) Generational and Educational Effects on Basic U.S. Adult Civic Science Literacy, paper presented at International Indicators of Science and Public Workshop, The Royal Society, London, November 5–6.

Martin, I. (2003) Adult Education, Lifelong Learning and Citizenship: Some Ifs and Buts, *International Journal of Lifelong Education*, **22**, 6, pp. 566–579.

Meadows, D., Meadows, D., Zahn, E., Miling, P. (1972) *The Limits of Growth*, New York: Universe Books.

Royal Society (1985) *Public Understanding of Science*, London: The Royal Society.

Yang, Y. & Land, K. (2006) Age-Period-Cohort Analysis of Repeated Cross-Section Surveys: Fixed or Random Effects? *Sociological Methods and Research*, **36**, pp. 297–326.

6 The Changing Culture of Science across Old Europe
1989 to 2005

Martin W. Bauer

The Eurobarometer (EB) survey, the survey instrument of the European Commission in Brussels, has been asking questions pertaining to public sentiments on science and technology since the 1970s. A series of four strongly related EB surveys have recently been integrated (N>50,000, 60 variables) into a longitudinal database to study changes and continuities in the culture of science across Europe between 1989 and 2005 across twelve EU countries (EU12). Comparable questions cover knowledge, interest, and various attitude facets in relation to level of education, religiosity, age, and gender. The chapter will compare changes over time on these indicators and their relationship (Bauer, Shukla, & Kakkar, 2008).

The chapter has four sections. In the first section we will look at the temporal trends in science literacy, interest and attitudes from 1989 through to 2005. The second section defines the age cohort groups and compares scientific knowledge, interests and attitudes across these groups to map the generational trend. In the third section, we take a closer look at gender and education gaps in the light of the postindustrial hypothesis, according to which different phases of societal development give rise to different structures of public understanding of science. The final section presents a model to explore various correlates of knowledge, attitude, and interest ceteris paribus.

TEMPORAL TRENDS ACROSS EUROPE

We have at our disposition an integrated data set of comparable questions and answers from four waves of investigations of 1989, 1992, 2001, and 2005. The items are by now well known, pertaining to scientific knowledge, interests, and attitudes. For recent reviews of using such measures see (Allum, 2010; Bauer, Allum & Miller, 2007).

The **knowledge** indicator consists of 13 quiz items, where the respondent has to indicate whether the statement of fact is true or false. For statements like 'electrons are smaller than atoms,' the correct response is 'true.' For others like 'antibiotics kill viruses as well as bacteria' the correct answer is 'false' (see appendix). The number of correct responses is taken as an index of knowledge. On average people get 7 out of 13 items correct (mean= 7.07; SD=2.40; N=47,000). Technical discussions of these indicators point to a guessing element in the responses and a possible multidimensional structure. For our purposes, we consider these indicators sufficiently reliable (Cronbach Alpha = 0. 57). For discussions of reliability related to these items for comparative purposes, see Pardo & Calvo (2004) and also Stares (in this volume) and Shimizu & Matsura (in this volume).

We focus our analysis on two **attitude** facets: *science and technology are making our lives healthier, easier and more comfortable* (quality of life), *and science makes our way of life change too fast* (pace of life). To express a positive attitude to science, respondents have to agree with the first and disagree with the second item. These items have a long history of use going back to Withey (1958).[1] People respond by indicating whether they strongly agree, tend to agree, tend to disagree, or strongly disagree with the statement, or neither/nor or don't know (DK). Rescaling the responses from -2 to +2 gives an index where positive values indicate attitudes towards science, e.g. expectations for quality of life (M= 0.77, SD=0.94; n= 41,700).

Finally, **interest** is indicated by a single item where respondents declare themselves either moderately or very interested in 'new scientific discoveries'. We take the response 'very interested' as an indicator, which is 33 percent of the population. Another third are moderately interested, and a final third shows no interest at all. Our indicator is a dummy variable 0/1 (M=0.33; SD=0.47; N=50245).

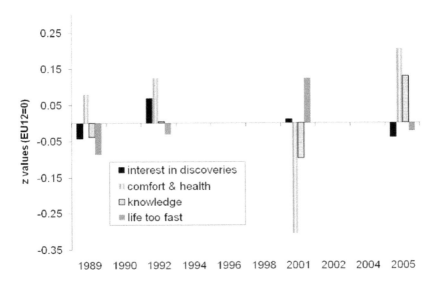

Figure 6.1 Overall trends in knowledge, attitudes and interest presented as standard z-values.

On these indicators, knowledge, interest and three facets of attitudes, we can chart the temporal trend from 1989 to 2005 (see Figure 6.1). To compare more easily, we standardised the variables to a common scale (z-values; M=0, SD=1). Note that the oscillation in the attitude 'quality of life' might be an artefact of response alternatives. In 2001, contrary to the other years, respondents did not scale their attitude, but responded with agreement or disagreement. This is likely to have polarised responses towards more disagreement, hence the index is more negative in that year.

Figure 6.1 shows the temporal trends for knowledge, interest and the two attitude facets, quality of life and life pace. Knowledge is on the increase; the attitude 'quality of life' remains stable overall. Remarkably, interest in new discoveries and 'pace of life' attitude first increases and then declines again. At the beginning of the 21st century, Old Europe is more scientifically literate, positive attitudes remain stable, and interests in science is declining.

A scientifically literate population is widely seen as a human resource that brings competitive advantage in the world economy. Where the OCED's PISA initiative assesses the performance of school children (PISA, 2003), our study offers an index of adult scientific literacy for the population of 15 years and older.

The trend in scientific literacy varies across EU12. The Netherlands, Denmark, and Luxembourg lead EU12 overall, and Portugal marks the end of the scale. The overall trend towards higher literacy is present in most countries, with the exceptions of Italy, Portugal, Luxembourg, and Ireland, where there is little progress over the period.

We compare the countries of Old Europe relative to each other in Table 6.1. In ranking of knowledge Denmark has improved its position from rank 7 in 1989 to 1 in 2005. Similarly Belgium and Germany (West) improve their position. The Germans in particular have seen a 'modernisation' drive and the end of the 'fun society' (Spassgesellschaft) since its unification with the East in 1990. In any ranking, when some gain, others fall back as Luxembourg, France, and Italy did. The UK holds the mid-field, and the 'PIGS' Portugal, Ireland, Greece, and Spain mark the tail of the adult literacy field.

GENERATIONAL TRENDS: A QUASI-COHORT ANALYSIS

The analysis of age cohorts allows us to map another type of trend: the changes across different generations within the population. Cohort analysis recognises the importance of growing up in a particular context and sharing a common experience of history, which impresses itself on a particular generation (see Rogler, 2002). I define five generational groups for our database; each cohort is assumed to have developed a particular take on science

Table 6.1 Shows the Ranking of EU12 Countries on Knowledge (know13) by Years and by Generational Cohort

YEARS	Total	2005	2001	1992	1989
NL	1	3	1	7.5	3
DK	2	1	2	3	7
LUX	3	5	4	1	1
F	4	6	5	3	2
D	5	2	6	3	5
IT	6	8	3	5	4
UK	7	7	7	6	6
B	8	4	9	7.5	8
E	9	9	8	9	11
EIRE	10	11	11	10	9
GR	11	10	10	11	10
PORT	12	12	12	12	12
K_mean	7.84	8.36	7.70	7.81	7.48
K_CV	0.35	0.31	0.34	0.34	0.39
COHORT	>1977 New Order	1963-76 Gen X	1950-62 Babyboom	1930-49 Crisis&War	<1930 Roaring 20s
F	5	5	2	5	1
LUX	4	1	6	2	2
NL	2	3	3	2	3
D-West	3	6	5	3	4
DK	1	2	1	4	5
UK	9	8	7	6	6
B	10	9	8	8	7
IT	8	4	4	7	8
EIRE	12	11	11	9	9
GR	7	10	10	11	10
E	6	7	9	10	11
PT	11	12	12	12	12
K_mean	8.64	8.41	8.20	7.50	6.31
K_CV	0.28	0.29	0.31	0.36	0.47
Int_mean	0.38	0.36	0.35	0.31	0.26
int_CV	1.29	1.34	1.36	1.49	1.69

Knowledge is based on 13 items; interest in 'new scientific discoveries' (see appendix). CV stands for Coefficient of Variability, i.e. standard deviation divided by the mean; it is a measure of variability across countries.

in terms of attaining a level of literacy, taking certain attitudes, and being interested in scientific developments.

- **Born > 1977**, the youngest cohort of respondents, grew up after the end of the Cold War and woke to the rhetoric of the 'new world order' and the final victory of the capitalist style of economy, and lived through the

rhetoric of the IT and biotech 'revolutions' of the late 20th century. This is the generation of the PC and Internet euphoria of 1995–2000: 8.8 percent (ranging from 5.3 percent to 13.20 percent across EU12).
- **Generation X** is the generation born between 1963 and 1976. They are the outcome of the birth control 'revolution' and grew up during the oil crisis of the 1970s, and the nuclear issues of the 1980s, the antinuclear protest, nuclear armament debates, and the Star Wars initiative: 26.4 percent (ranging from 23.6 percent to 29.6 percent).
- **Baby Boomers** were born between 1950 and 1962. They grew up in the optimism and modernisation drive of the postwar period. They witness the longest period of economic prosperity in history. During this period Western societies become 'affluent' and free of material concerns. This generation is the protest generation of the 1970s, with idealistic worldviews. They are more sceptical with regard to progress and its link with science and technology: 23.3 percent (ranging from 19.6 percent to 29.7 percent).
- **Crisis and War Generation** was born between 1930 and 1949. This generation witnessed WWII and formed the immediate postwar generation that then entered the Cold War. This generation also carried the 'nuclear enthusiasm' of the 1950s, which promised a scientific revolution and 'energy too cheap to meter' in the atomic society: 28 percent (ranging from 24.9 percent to 30.7 percent).
- **The Roaring '20s** is the generation born before 1930, growing up after WWI or through the buzzing period of the 1920s, which ended in the big crash of 1929 and the economic crisis that followed: 13.6 percent (ranging from 9.9 percent to 16.3 percent).

With these cohort groups we can now conduct an analysis of our observations on knowledge, interest, and attitudes towards science.

One of the key problems of such an analysis is to disentangle the confounded effects of any time (period), of belonging to a particular generational group (cohort), and of simply growing older and more mature (age). Only two of these influences can be statistically controlled at any one time (see Mason & Wolfinger, 2001). For the moment we will ignore the age effect and focus on the relative importance of period and cohort effects. It is unlikely that scientific knowledge, interest, and attitudes to science are a matter of maturity, though it could be a matter of changing concerns over the life cycle. Science might fascinate young kids, but such interest is lost when they begin looking for careers. It might become relevant again via health issues arising with childbearing or with getting old and frail; or star gazing, 'de-siderare,' is particularly appealing during the romantic phase of life when desire rules. Be that as it may, in the following I will put age effects aside and compare the relative importance of generational cohorts against that of period of study (see Losh, in this volume).

Overall we find scientific literacy increases continuously from the older cohorts to the younger ones. However, this generational trend is not uniform across Europe. In France and the UK, literacy stagnates across the younger three groups, whereas Italy sees a decline in the youngest 'new order' age group.

Table 6.1 shows the ranking of the countries on science knowledge for each of the generational groups. The variability between the countries decreases as knowledge overall increases. Young Europeans are crowding on a higher level of average science knowledge. The relative position of the countries varies considerably across the cohorts. In Denmark, the younger cohorts are better positioned than the older ones, whereas in the UK the situation is reversed: the younger cohorts are in a less favourable position than the older ones. Italy's position is better for Generation X and the Baby Boomers than it is for their youngest generation.

Interest in science equally increases across the five cohorts: the older generations are less interested than the youngest ones. But again, the national patterns vary. France shows that the prewar generation are the most interested, the youngest cohort the least. In Italy, the New Order shows less interest in science than Generation X. In Germany (West) and Portugal, interest in science is shifting from one cohort to the next, in line with the EU average. Overall we observe a convergence towards a European mean: just as for knowledge the variability of interest is smaller among the New Order than among the Roaring '20s generation.

Figure 6.2 shows knowledge and three attitude facets in relation to age cohorts, life quality, more interesting work, and pace of life (see appendix

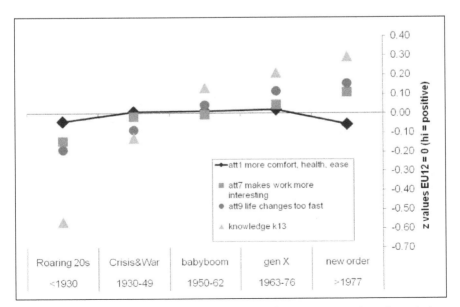

Figure 6.2 Knowledge of and attitudes to science by generation cohort: knowledge, quality of life, interesting work and pace of life shown as z-values.

for item wording). Each indicator is standardized (z-values; M=0, SD=1; a high value means 'positive attitude to science'). The three facets of attitude show different trajectories across the age cohorts. For pace of life and more interesting work, the process is linear: the younger the cohort, the more positive the expectations from science; science makes work more interesting and does not upset the pace of life.

Let us consider the expectation that science and technology improve our quality of life in Figure 6.2. This indicator shows a non-linear trajectory across the age cohorts. The oldest and the youngest are more sceptical with regard to the effects of science on life quality than the middle generations.

Again the story varies across countries. Germany follows the EU12 trend, and remains the most positive in all age cohorts. France witnesses a decline of positive expectations among the Generation X and the New Order cohort. If France was one of the most positive countries among the older cohorts, among the younger ones it has become the most sceptical (see Boy, this volume). The UK shows a dent in attitudes among the two younger age groups. Across the Channel, the Baby Boomers are the torch bearers of high expectations of science. In Portugal and Italy, positive attitudes to science increase from generation to generation. In Portugal the 'new order' experiences a dent in attitudes.

For the other two indicators, the story is more in line with expectations: the younger generations expect working life to get better due to science and technology, and do not worry about the pace of life accelerating. The older generations are more sceptical on both counts. Overall, we might conclude that knowledge increases across all five generational cohorts, and if knowledge of science increases, attitudes, in particular expectations of improvements to the quality of life, do not necessarily follow suit; different facets of attitude show a different generational path as facets of attitudes to science differentiate in public opinion. It seems that the notion of progress is not universally attributed to advances in science and technology, and this neither by the younger nor the older generations. Increased familiarity with science dents the high expectations people might otherwise entertain.

THE GENDER AND EDUCATION GAPS ACROSS COHORTS

Let us now consider the data in the light of two major concerns arising in public and science policy circles: the gaps in knowledge, attitudes, and interest across levels of education and gender. How do these gaps close or widen across the generational cohorts; in other words, are education and gender gaps a matter of the older generations, but not the younger ones? To answer this question we model the data with three dependent variables—knowledge, interest, and attitude—and cohort, education, and gender as factors, while at the same time controlling for year within each cohort (a MANOVA linear model; see appendix).

The Changing Culture of Science across Old Europe 99

Overall, all variables have an impact on knowledge, attitudes, and interest, but in slightly different order of magnitude. Education, sex, and age cohort impact on knowledge in that order; year, sex, and education impact on attitudes in that order; education and sex impact on interest in that order.

More remarkable is the interaction effect between cohort and education: across the age cohort, formal education positions respondents differently on these indicators, and this is most prominent for the differentials in interest in science. Knowledge across levels of education: primary, secondary, and tertiary education, controlling for the other variables; knowledge, interest, and attitude (quality of life) show an interaction effect. People with more or less formal education position themselves differently on these indicators depending on their generational group. In particular, people with primary education have a very different trajectory across the generational cohorts compared to those with secondary and tertiary education, controlling for effects that period and sex add to this picture.

On scientific knowledge (see Figure 6.3a), those with higher education are also more knowledgeable in science, so far no surprise. However, those with primary education gain most as they move from the older to the younger generations; whereas those with secondary or tertiary education remain on the roughly the same level across cohorts (interaction effect: $F=107$; $df=8$; $p<0.001$). The added value of primary education to science knowledge is remarkable, to the effect that, among the New Orders, those with only primary education know more about science than those those with secondary education. Students with primary education show increasing familiarity with science. This fades away among those with secondary education in the younger generations. Overall, though, the gap between levels of education is narrowing.

The picture of interest (see Figure 6.3c) is very similar to that of knowledge: the highly educated show most interest across all age cohorts. Interest is declining among secondary educated people, but among primary educated people interest in science is steadily increasing from the older to the younger generation. This leads to the effect that among New Orders and Generation X, the primary educated are more interested in science than those with secondary education (interaction effect: $F=33.1$; $df=8$; $p<0.001$), again disregarding fluctuations across the period or differences of the sexes.

On attitudes we observe the striking trend (see Figure 6.3b) that, for all levels of education, the younger generations are more sceptical as to the positive effects of science and technology on the quality of life. But for Generation X with primary education, this trend has stalled; they in particular hold higher expectations. The generation born in the 1960s carries the Internet 'revolution', incorporating computer and Internet use in their everyday life, which the younger generations take for granted. However, for the New Orders, higher education makes again all the difference when enthusiasm for science is concerned (interaction effect: $F=10.4$; $df=8$; $p<0.001$).

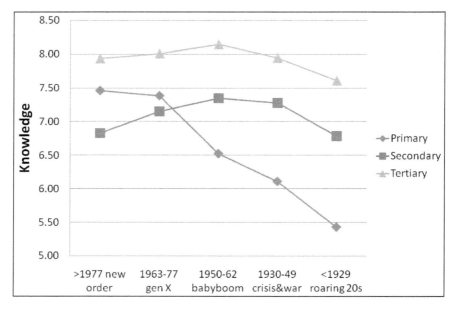

a.

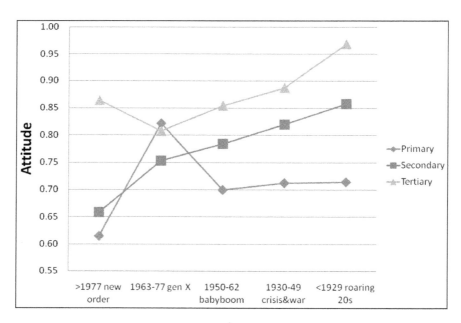

b.

Figure 6.3a and 6.3b Knowledge and attitude by education and age cohort (controlled for other variables).

The Changing Culture of Science across Old Europe 101

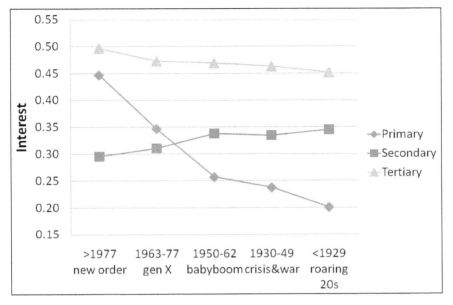

c.

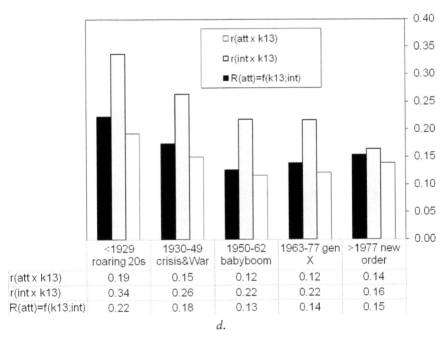

d.

Figure 6.3c and 6.3d Interest by c) education and age cohort (controlled for other variables). d) shows the correlations (coefficient r) between the three indicators and the regression of attitude on knowledge and interest (coefficient R) within each age cohort.

In contrast to these moving education differentials, the **gender gap** persists on knowledge, attitudes, and interest across the all age cohorts (not shown in the table). There is no interaction effect between sex and cohorts for any of these indicators. This means that women are consistently less knowledgeable (in terms of textbook knowledge of science), less interested in new discoveries, and more sceptical regarding the impacts of science on quality of life across all age cohorts. The overall trends are the same for both men and women: knowledge is increasing, attitudes increase and decrease again, and interest is decreasing across the generational cohorts.

The Postindustrial hypothesis of PUS

The postindustrial hypothesis of public understanding of science (Bauer, Durant, & Evans, 1994) postulated that the interrelation between knowledge, attitudes, and interests is itself variable, depending on the societal context. This formulation renders relative the predictions of the 'deficit model' according to which knowledge, interest, and attitudes are universally positively correlated. Rather than 'the more you know, the more you like it,' a different context might engender the notion of 'familiarity breeds contempt,' or if not contempt then a more sober attitude. The relevant context might be given in the socio-economic transition from an industrial to a knowledge-based society. In Europe such a transition is both a factual diagnosis as well as the aspiration of high politics (see Rohrbach, 2007). We consider our generational cohorts as an index of this transition. We assume that the older generation carries a mentality more akin with industrial society, whereas the younger generations are more in line with a knowledge society, and this is reflected in a gradient of the interrelations between our three indicators.

Figure 6.3d shows that for all generational cohorts the correlation between attitudes and knowledge is positive; the correlation is low and with very little difference ($0.11 < R < 0.20$). Here lies little evidence for a contextual dependency of these correlations. High knowledge and positive attitudes to science are associated across all age groups. The gradient is more manifest in the interrelation between knowledge and interest in science ($0.15 < r < 0.36$); for the older generations knowledge is more of a driver of interest in science than for the younger ones. This is consistent with our hypothesis; we have also observed above that the generational trend for interest is declining while knowledge is increasing. Last, we might consider attitudes as a function of knowledge and interest, in the sense that the more we know and the more we are interested in science, the higher are our expectations of science. There is some gradient across the generations, but not a striking one ($0.12 < R < 0.23$). Clearly a more detailed analysis of this hypothesis needs to look at each country separately, and consider the speed with which each context is progressing in this secular transition from an industrial to a postindustrial context of production. Cohort analysis, in combination with longitudinal analysis of each context, will allow us to test this hypothesis beyond speculative cross-sectional comparisons.

CORRELATES OF KNOWLEDGE, INTEREST, AND ATTITUDES

In the last section of this chapter, we explore the correlations of the three indicators of science culture—knowledge, attitude, and interest— and other variables holding each of them constant. We ask the question: to what extent does the overall level of knowledge, interest, and positive attitudes depend on the country, the generational cohort, the period of the survey, the religious outlook of the respondent, his or her level of education, and gender. This is best explored by fitting a binary logistic regression model, where high or low knowledge, attitude and interest are dependent on the levels of each predictor variable, everything else being equal (ceteris paribus). For attitude and interest we use two models, either including or excluding knowledge in the equation. For this purpose, we reduce the knowledge scale from thirteen to five levels to see whether the relation with knowledge is monotonic linear.

Considering the individual characteristics of respondents, gender is a consistent determinant of knowledge, attitudes, and interest. Women are 46 percent less likely to be highly knowledgeable than men, 23 percent less likely to hold high expectations and 34 percent less likely to be highly interested; this holds true even when controlling for knowledge, though the gap is smaller, 15 percent less for attitudes and 26 percent less likely for interest. Everything else being equal, science culture remains more of a male than female domain. Compared to those who declare themselves as religious (see appendix), the not religious are 24 percent more likely to hold positive attitudes, 35 percent more likely to be highly knowledgeable, and 19 percent more likely to be highly interested. Those undeclared on religion are however, 39 percent less likely to hold positive attitudes, 27 percent less likely to be highly knowledgeable, and 26 percent less likely to be highly interested than the religious respondent. When controlling for knowledge, these relations hold, but attenuated.

Education affects knowledge, attitudes, and interest. Compared to those with tertiary education, both secondary and primary education put people at a distance from science and technology on all three dimensions, and this holds for attitudes and interest even when we are controlling for knowledge levels. Education makes an independent contribution to science culture through and beyond scientific literacy.

Table 6.2 The Results of a Binary Logistic Regression of Knowledge, Attitude and Interest (on religion, cohort, country, education, year, and sex; percentage odds changes compared to reference category on high or low knowledge, positive attitude and interest.

	Dep: high-low	attitude	attitude	know13	interest	interest
religion	not religious	24	19	35	19	12
	DK	-39	-34	-27	-26	-19
	religious					
cohort	Born >1977	86	50	250	124	69
	Gen X	34	16	145	44	17
	Baby Boomers	27	13	99	32	11
	Crisis & War	24	15	65	22	8
	Roaring 20s					
country	It	27	7 ns	140	82	47
	NL	32	12 ns	123	91	57
	Lux	19	1 ns	122	91	59
	F	28	6 ns	103	160	120
	UK	63	42	80	87	62
	D-West	72	55	70	8 ns	-8 ns
	Greece	57	42	55	89	71
	B	5 ns	-4 ns	50	18	5 ns
	E	16	9 ns	29	26	18
	DK	19	7 ns	28	53	39
	Eire	16	12	8 ns	11 ns	6 ns
	Portugal					
education	primary	-38	-26	-63	-53	-43
	secondary	-23	-14	-51	-45	-38
	tertiary					
year	1989	-3 ns	-5 ns	14	25	21
	1992	22	20	1 ns	47	46
	2001	-30	-28	-37	11	22
	2005					
sex	female	-23	-15	-46	-34	-26
	male					

(continued)

Table 6.2 (continued)

Dep: high-low	attitude	attitude	know13	interest	interest
k1		-55			-79
k2		-30			-59
k3		-19			-42
k4		-13			-15
k5					
interest low		-31			
interest high					
Nagelk R Sq	0.06	0.08	0.15	0.09	0.13
%classified	73.61	73,2	53.36	66.46	66.46
%model classed	74.2	74.40	62.60	68.10	68.13
model strength	0.6	1.2	9.2	1.6	1.7
	attitude	attitude	know13	interest	interest

Percentage odds changes are defined as % = (Exp(B) -1) x 100. Note that all these figures have a confidence interval of estimation, which is not reported here. N.s. stands for 'not significant' at the level of alpha = 0.05.)

The generational effect on attitudes, knowledge, and interest is consistent. The younger cohorts are closer to science than the older generations. Born after 1977 makes a European 250 percent more likely to be highly knowledgeable of science, 86 percent more likely to hold positive expectations, and 124 percent more likely to be interested in science compared to someone from the roaring 20s generation. If we control for levels of knowledge, the gradient is not as steep.

All three indicators vary considerably on the country of the respondents. All considered Italy, the Netherlands, and Luxembourg are the most knowledgeable contexts. more likely to be knowledgeable of science than a Portuguese of similar standing. A Briton or a West-German are 60 percent or 70 percent more likely to hold positive expectations of science, and a French person is 160 percent more likely to express high interest in science than someone of similar standing in Portugal. When controlling for knowledge, the country differences in attitudes are less significant, only UK, West-Germany, Greece, and Ireland continue to show more positive attitudes than Portugal, all the others are on the same level of expectations. Similarly, for interest the country effect attenuates when controlling for knowledge: West Germans, Belgians, and the Irish show similar levels of interest as the Portuguese, whereas for the others interest remains on a gradient led by France.

IN CONCLUSION

The chapter reported a first exploration of a new micro-integrated database of four rounds of surveys of public understanding of science across twelve European countries. This data allows us to conduct longitudinal (1989 to 2005) as well as quasi-cohort analysis across five generations and compare trends in the European culture of science. The science culture is assessed on three main indicators: knowledge of science, attitudes, and interest in science. The main results of our exploration are as follows:

On a temporal trend between 1989 and 2005, general knowledge of science increases, quality of life expectations remain stable, and interest in science comes and goes again. On a generational trend, general knowledge of science as well as interest increases from older to younger generations, so do expectations of better working conditions, and quality of life expectations are highest among the middle generations. Both temporal and generational trends vary in the different EU countries. Here the emerging story is likely to be complex, reflecting varying historical and economic as well as wider cultural contexts. Clearly, here more work needs to be done.

The gap between men and women persists on all indicators across the generations. Women are more distant from science than men in terms of knowledge, attitudes, and interests. By contrast, the educational gap is closing for knowledge across the generations, less so for attitudes and interests. However, the added value of primary education is evident in the younger generations, where secondary education positions respondents at a larger distance from science. Clearly tertiary education fosters knowledge, positive expectations, and interests across all generations. Secondary education brought people closer to science in the middle generations, less so than among the older and the younger generations. Positive expectations and interest in science is declining among secondary educated Europeans across all generations. The different added value of primary and secondary education on the culture of science in Europe deserves further exploration.

Finally, the postindustrial hypothesis, according to which the interrelations between knowledge, attitudes, and interest depend on context, is evident across generational groups, in particular for interest. Knowledge is a driver of interest among the older generation, but not among the younger ones. An alternative explanation might be that these textbook type knowledge items do not adequately tap into the knowledge that is relevant to stir the younger generation's interest in science. More work is needed to explore these hypotheses in each country context by carefully comparing the temporal and the generational trends the countries' positions in the transition from an industrial to a knowledge society.

APPENDIX: THE EU INTEGRATED DATABASE

Indicators Used in This Report

Knowledge: the knowledge indicator is based on the number of correct responses to 13 knowledge quiz items. The items included are:

1. The centre of the Earth is very hot (true).
2. Radioactive milk can be made safe by boiling it (false).
3. Electrons are smaller than atoms (true).
4. The continents on which we live have been moving for millions of years and will continue to move in the future (true).
5. It is the mother's (father's) genes that decide whether the baby is a boy (false / true).
6. The earliest humans lived at the same time as the dinosaurs (false).
7. Antibiotics kill viruses as well as bacteria (false).
8. Lasers work by focusing sound waves (false).
9. All radioactivity is man-made (false).
10. The Sun goes around the Earth (false).
11. The oxygen we breathe comes from plants (true).
12. Human beings, as we know them, developed from earlier species of humans (true).
13. How long does it take earth to go around the sun? (one year = true).

Interest: interest in new scientific discoveries is coded as single item dummy variable: 1 = very interested; 0 = somewhat or not all interested.

Attitudes: for the present analysis we used mainly the item *'science and technology are making our lives easier, more healthy and more comfortable'* [life quality]; but two other items are also used *'science makes our way of life change too fast'* [pace of life] and *'the application of science and technology will make work more interesting'* [interesting work]. All responses are on a 5-point scale from very much agree (=1) to very much disagree (=5), and were recoded so that a high value signifies a positive attitude to science.

Religiosity: agreement to the item *'we depend too much on science and not enough on faith'* is taken as in proxy for 'religiosity'. Disagreement is taken as an index of 'non-religiosity'. DK or N/N is considered as 'undecided or agnostic'.

Year: the years of the survey studies were 1989, 1992, 2001, and 2005.

Bioage: within each cohort we did a median split and assigned a value of 0 if the age is below the median and 1 if age is above the median. This is a way of controlling for age independently of cohort.

Education: the survey asked each respondent to indicate the age by which they had left full-time education. This is then recoded into primary (1), secondary (2), and tertiary education level (3).

Sex: gender is indicated by a dummy: male (1) and female (0)

Country: EU12 includes Belgium, Denmark, Spain, Germany (West only), Ireland, France, UK (Britain + Northern Ireland), Portugal, Netherlands, Greece, Italy, and Luxembourg.
Some tables report z-value as deviations from the overall mean of all EU12 countries across all four waves (N=50245).

MANOVA model (N=41712)

Design: full-factorial on education, sex, and cohort; Covariates: bioage, year

intercept+year+bioage+cohort+sex+cohort*edu+cohort*sex+edu*sex+ cohort*edu*sex

Knowledge: $F=240$; $df=31$ $p<0.001$; $eta^2=0.151$; power=1.00
Attitude: $F=36$; $df=31$; $p<0.001$; $eta^2=0.026$; power=1.00
Interest: $F=68$; $df=31$; $p<0.001$; $eta^2=0.048$; power=1.00

Impact of predictors and their interactions (criterion: ranking by F)

Knowledge = (intercept) edu, sex, bioage, cohort, year, **cohort*edu**
Attitude = (intercept) bioage, year, sex, edu, **cohort*edu**, cohort
Interest = (intercept) edu, sex, **cohort*edu**, cohort, year, bioage

Not significant interactions: cohort*sex, edu*sex, cohort*sex*edu (p>0.001)

NOTES

1. The items entered Eurobarometer via the United States' indicator series, who took the item from Withey (1958). It is one of the oldest items in this tradition of research. However, the 'quality of life' item is by now considered technically a 'bad question' because it asks, strictly speaking, three different questions. EB 2010 tested this by including a split-half experiment. Half the population were asked the triple question in the original version, and the other half were asked to agree or disagree with the single statement *'science and technology are making our lives healthier'*. The baseline results show

that on the criterion 'health' Europeans (EU27) are slightly more sceptical about the impact of science: only 56 percent tend to agree and 17 percent disagree, compared to 66 percent agree and 12 percent disagree with the original version. It appears that people make a distinction with regard to the impact of science and technology on health (more sceptical) and on comfort and ease of life (less sceptical: see *Europeans, Science & Technology*, 2010). On our longitudinal attitude indicator this distinction is lost.

REFERENCES

Allum N (2010) Science literacy, in: S Hornig Priest (ed) *Handbook of Science and Technology Communication*, Los Angeles, SAGE, 724–27.

Bauer MW, N Allum, and S Miller (2007) What can we learn from 25 years of PUS research? Liberating and widening the agenda, *Public Understanding of Science*, 15, 1, 1–17.

Bauer M, J Durant, and G Evans (1994) European public perceptions of science, *International Journal of Public Opinion Research*, 6, 2, 163–86.

Bauer, MW, R Shukla, and P Kakkar (2008) The Integrated Data on Public Understanding of Science [EB_PUS_1989–2005], London, LSE & NCAER, November.

Europeans, Science & Technology (2010) *Special Eurobarometer* 430 / 73.1, January-February 2010; Brussels, May.

Mason WM and NH Wolfinger (2001) Cohort analysis, in: NJ Smelser and PB Baltes (eds) *International Encyclopedia of Social and Behavioral Sciences*.

Pardo R and F Calvo (2004) The cognitive dimension of public perceptions of science: Methodological issues, *Public Understanding of Science*, 13, 203–27.

PISA (2003) The PISA 2003 Assessment Framework, Paris, OECD, programme for international student assessment.

Rogler LH (2002) Historical generations and psychology: The case of the great depression and WWII, *American Psychologist*, 57, 12, 1013–23.

Rohrbach D (2007) The development of knowledge societies in 19 OECD countries between 1970 and 2002, *Social Science Information*, 46, 4, 655–89.

Withey RC (1958) *The Public Impact of Science in the Mass Media*, Michigan, Survey Research Center.

7 Knowledge of Science and Technology in Japan
IRT Scores for 1991 and 2001
Kinya Shimizu and Takuya Matsuura

SURVEY OF THE PUBLIC UNDERSTANDING OF SCIENCE AND TECHNOLOGY IN JAPAN (J-SCITEK)

A variety of surveys on public understanding of science and technology (PUST) have been conducted in Japan; of these, the three that utilize internationally comparable survey items were conducted by the National Institute of Science and Technology Policy (NISTEP) in 1991, 2001, and 2007.

Planning for the first study (J-SCITEK91) started in 1990. Based on both national need and requests from researchers in the U.S., Canada, and the EU, the NISTEP organized a national research group for the public understanding survey on science and technology to get advice on survey contents and method. Members of the group included the staff of NISTEP, seven university professors, and national cabinet official. Guided by seven meetings of the group and the pilot survey, NISTEP conducted the J-SCITEK91 in November 1991, the target population being adults aged 18 years old or over. The intended sample size was 2000, selected by 2-stage stratified sampling (size of city and geographical regions). The number of respondents was 1457 (response rate: 72.9%).

Some results of the J-SCITEK91 resonated in Japanese society. Although "the young's indifference to science and technology" (Rika-Banare, Is this a phrase? A quote? Who by?) was discussed in Japanese society at that time, science educators, policy-makers, and mass media were struck by a particular result: that the Japanese public has lower interest in and understanding of science and technology than most other industrialized countries.

In fact, although the primary purpose of the J-SCITEK91 is international comparison, some of the key items used in the J-SCITEK91 are not exactly the same as those used elsewhere. For example, questions about level of interest requested that respondents describe their level of interest using a 4-degree scale of "Extremely interested", "Interested a little", "Hardly interested", and "Not at all". The 4-degree scale is useful to distinguish responses for "interested" people ("Extremely" plus "A little") and "not interested" people ("Hardly" plus "Not at all"), but other countries that conduct the same type of public opinion survey have adopted the

3-degree scale: "Very interested", "Moderately interested", and "Not at all". In order to make a comparable index, Miller et al. (1997) converted the 4-degree scales into one of 100 points (Extremely: 100, A little: 67, Hardly: 33, Not at all: 0).

Quiz items to measure civic scientific literacy were modified in the J-SCI-TEK91 from the most frequently used quiz battery in western countries. Most countries have used fifteen quiz items to indicate the level of understanding of science and technology, but only six items were adopted by J-SCITEK91. Although the study conducted by Miller et al. (1997) suggested utilizing the score conversion technique based on the Item Response Theory (IRT), the six knowledge items posed by the Japanese survey were relatively difficult. As a result, the IRT score may underestimate level of understanding of scientific terms and concepts held by Japanese people.

In order to collect data on the public understanding of science and technology comparable to that collected by the international research group, the NISTEP organized the Japanese research group to develop another survey. The new members of the group consisted of only one member of staff from the NISTEP and three university faculty members. In the first part of the questionnaire development, the group prioritized international comparison and carefully translated all survey items used in the US 1999 survey. Second, the wording of questions was carefully modified to suit the Japanese population; the group tried to keep modification to a minimum. Then, some useful items from the other public opinion surveys, such as Eurobarometer in 1992 on "Europeans, Science and Technology—Public Understanding and Attitudes", and the British Survey in 2000 on "Public Attitudes to Science, Engineering and Technology in Britain," were added to the questionnaire. Thus, most items are comparable to the international data set.

The NISTEP conducted their second survey on the public understanding of science and technology (J-SCITEK01) in February to March of 2001. The target population was adults aged from 18 to 69 years old. The intended sample size was 3000, and respondents were selected by 2-stage stratified sampling. The number of respondents was 2146 (response rate: 71.5%). Comparison of the index score among the eighteen surveys of fifteen countries[1] showed the percentage of Japanese public interest in new scientific discovery was the lowest, and the percentage of their interest in new technology was second to lowest. Correct responses to the science and technology quiz were only 51% in Japan, ranked third to lowest among sixteen surveys of fifteen countries.

After six years, the NISTEP conducted a third survey, this time using the Internet. From January 30 to February 2, 2007, requests for survey participation were sent via e-mail to 9245 people registered in an Internet survey company. Valid responses were returned by 2868 monitors (response rate: 31%). The survey collected some comparable data such as issue interests, frequency of museum visits, knowledge quiz items, and so on. However, it lacked the attitudinal questions commonly included in surveys of the other countries. Because of the representativeness of the data, low response rates, and item coverage, it is hard to say the third survey is the comparable to the other international PUST surveys.

REVIEW OF THE SECONDARY ANALYSIS OF J-SCITEKS IN 1991 AND 2001

Although international comparison of the science achievement score of school-aged students shows that Japanese schoolchildren are in relatively good shape in their understanding of science, the adult population did not show the same. Because the Japanese adult public has a lower level of science understanding some researchers explored the cause of the phenomenon.

Okamoto (2007) classified the knowledge items into four categories by using cluster analysis. The first category is a group of items to which more than 70% of the respondents of the J-SCITEK01 could make correct responses. The second category is a group of items to which more than 50%, but less than 70%, of the respondents could make correct responses. The third category is a group of items to which more than 35% of the respondents made an *incorrect* response. The fourth category is a group of items to which more than 45% of the respondents answered "I don't know". Among the knowledge items, the questions that fell into the third category (public misunderstanding of science facts) were those that asked about the coexistence of dinosaurs and humans; the father's gene determination of his offspring's sex; and the effect of antibiotics on viruses. The question asking about the relative size of electrons and atoms and the question about the relationship between lasers and sound waves fell into the fourth category. Okamoto concluded that the Japanese public understood simple knowledge that could be learned both during and after schooling.

Shimizu (2007) conducted simple item analysis of the six common items used in both J-SCIETEK91 and J-SCITEK01. Comparing only the six items revealed an improvement in the science quiz items. Some science facts were already well known at the time of 1991 survey, but knowledge of other facts shifted from an interested population to an uninterested population between 1991 and 2001. On the other hand, knowledge about plate tectonics was widely disseminated among the Japanese population because of the Kobe earthquake in 1995. Studies by Okamoto (2007) and Shimizu (2007) revealed the importance of opportunities for the public to learn science on a daily basis.

In terms of the factors influencing public knowledge of science, schooling is commonly considered as critical. Shimizu (2009) compared the mean correct response percentage of knowledge items across the curriculum cohort. In Japan, we have used changing national syllabuses for elementary and secondary science courses since 1947. Therefore, different generations have received different types of science curriculum. For example, the older generation received pragmatic science courses, the middle-aged generation received post-Sputnik science (more science content taught), and the younger generation received a relatively easier science curriculum. Changes to syllabus allow the comparison between generations to be understood as directly reflecting the difference in type of science curriculum. Shimizu anticipated the younger generation might show lower performance than the older generation because of their less demanding science curriculum. However, the results revealed that the younger generation showed better understanding of science than the older generations. Despite the criticism against the recent simplified science curriculum, the study implies that the effect of school curriculum has only limited impacts.

Another possible explanation of the relatively low level of science understanding among the Japanese public relates to measurement issues and the validity of knowledge items in Japan. Because knowledge items are developed in and for western countries, it is natural to ask if the knowledge items have validity in a Japanese context. Shimizu (2005) examined the predictive and construct validities of the knowledge items for the Japanese public. The study revealed that each knowledge item showed significant and positive correlation to the constructs of "basic facts and concept in science", "scientific method", and "social impact of science", and that each constructed score shows a positive correlation to the respondents' science activities. As long as the constructs measured by knowledge items can be understood as predictive concepts for public involvement in science, the items have some validity.

ITEM ANALYSIS OF J-SCITEK QUIZ BATTERY

Most secondary analyses of the Japanese data are conducted to look for factors to increase the Japanese adult knowledge level of science. However, for the Japanese survey data, item validity has not been checked from the angle of the Item Response Theory (IRT). For example, although Shimizu's (2009) study contained a comparison of 1991 and 2001 data, only the six common items were included in the comparison. In order to increase robustness, it is necessary to calibrate the IRT score and to identify the item characteristics for each commonly used item in the Japanese context.

There are twelve quiz-style items in the J-SCITEK91; twenty in the J-SCITEK01; and six in both surveys. These items are commonly used in the survey on public understanding of science in various countries. The items asking about the meaning of probability are put differently in each survey. In the J-SCITEK91, the probability questions were asked by using the color of flowers, as follows:

[Ask every respondent]

Q13. *A child is interested in the fact that his/her seeds of red flowers and of white flowers will bloom at a ratio of 3:1. He/she asks the father the series of questions below. What is your answer to each question Answer true, not always true, or I don't know.*

   a) *There are no seeds for white flowers if he/she picks up only the three seeds (Item 1).*
   b) *If the first seed produces a white flower, the next seed produces a red flower (Item 2).*
   c) *Each seed has the same probability of producing a white flower (Item 3).*
   d) *If the first three seeds produce red flowers, the fourth will produce a white flower (Item 4).*

To the question of the meaning of 'one in four', the responses of "*not always true*" to each item: *a), b), d)* are coded as correct responses and "true" to item *c)* is coded as a correct response.

The J-SCITEK01 uses the topic of heredity to look for knowledge about probability, and is commonly used in US and European surveys as follows:

Now, think about this situation. A doctor tells a couple that their genetic makeup means that they've got one in four chances of having a child with an inherited illness. Please tell me if each statement is true or false

   1(a). True
   2(b). False
   3. Don't Know

   a) *If their first three children are healthy, the fourth will have the illness (Item 5).*
   b) *If their first child has the illness, the next three will not (Item 6).*
   c) *Each of the couple's children will have the same risk of suffering from the illness (Item 7).*
   d) *If they have only three children, one will have the illness (Item 8).*

As in the J-SCITEK91, the responses of "*not always true*" to each item: *a), b), d)* are coded "correct responses" and "true" to item *c)* was coded as a "correct response".

Both surveys asked the same six items on human evolution, the theory of continental drift, the relationship between laser and sound wave, the effect of antibiotics on viruses, the existence of natural radioactivity, and the relative size of atoms and electrons. However, the item on the method of testing a newly developed medicine was asked differently in J-SCITEK91 and J-SCITEK01. In J-SCITEK91, the respondents are asked to choose the correct answer from four items, as follows:

> Q16. *Suppose a new medicine effective for the treatment of a certain illness is developed. Which is the most frequent method that scientists use to test the medicine?*
>
> 1. *Use the drug for several patients and ask them their opinions.*
> 2. *Use biochemical knowledge about the drug, then decide if it is effective or not.*
> 3. *Separate the group of patients who used the medicine from the group of patients who did not, and compare the results.*
> 4. *I do not know.*

For the purpose of the calibration of the IRT, the choice of "3" is coded as the correct answer, and the other answers, including "I don't know", are coded as incorrect.

On the other hand, J-SCITEK01 used the commonly asked question of testing a high blood pressure drug, as follows:

> Q13. *Now please think about this situation. Two scientists want to know if a certain drug is effective against high blood pressure. The first scientist wants to give the drug to 1000 people with high blood pressure and see how many of them experience lower blood pressure levels. The second scientist wants to give the drug to 500 people with high blood pressure, and not give the drug to another 500 people with high blood pressure, and see how many in both groups experience lower blood pressure levels.*
> *. . . Which is the better way to test this drug?*
>
>> *1(a). All 1000 get the drug*
>> *2(b). 500 get the drug; 500 don't*
>> *3.   DK*

For the purpose of the calibration of the IRT, the choice of "2" is coded as the correct answer, and the other answers, including "DK", are coded as incorrect.

In order to estimate item parameter by BILOG-MG3 (Zimowski et al., 2003), the six common items are used for linking items with adopted 2-parameter logistic model (2PL). The classical item statistics and item parameters of the 2PL model are presented in Table 7.1.

Table 7.1 Classical Item Statistics and Item Parameters of the 2PL Model

No.	Percentile for Correct Response ('91)	Percentile for Correct Response ('01)	Serial Correlation	Item Parameters (2PL)	
				Slope (S.E.)	Threshold (S.E.)
Item01 Probability one in four (Flower) …	.480	--	.688	3.042 (.271)	.274 (.018)
Item02 Probability one in four (Flower) …	.463	--	.656	3.242 (.290)	.303 (.017)
Item03 Probability one in four (Flower) …	.311	--	.527	1.716 (.162)	.602 (.027)
Item04 Probability one in four (Flower) …	.367	--	.576	2.619 (.223)	.463 (.018)
Item05 Probability one in four (Heredity) …..	--	.613	.610	1.850 (.114)	.200 (.022)
Item06 Probability one in four (Heredity) …	--	.697	.681	2.326 (.154)	.022 (.022)
Item07 Probability one in four (Heredity) …	--	.568	.532	.958 (.057)	.237 (.033)
Item08 Probability one in four (Heredity)…	--	.727	.691	2.073 (.131)	-.079 (.025)
Item09 Humans are evolved from earlier species …	.743	.775	.371	.468 (.030)	-1.353 (.108)
Item10 The continents on which we live have been moving their location for millions of years and will continue to move in the future …	.603	.826	.593	.775 (.048)	-.659 (.060)
Item11 Laser works by focusing sounds …	.211	.281	.461	.819 (.052)	1.267 (.051)

(continued)

Table 7.1 (continued)

No.		Percentile for Correct Response ('91)	Percentile for Correct Response ('01)	Serial Correlation	Item Parameters (2PL)	
					Slope (S.E.)	Threshold (S.E.)
Item12	Antibiotics kill viruses ...	.131	.228	.351	.592 (.047)	1.932 (.111)
Item13	Sunlight causes skin cancer ...	.774	--	.527	.659 (.056)	-1.279 (.123)
Item14	All radio-activity is man-made ...	.526	.558	.540	.862 (.048)	.166 (.027)
Item15	Electrons are smaller than atoms ...	.293	.295	.412	.620 (.045)	1.259 (.062)
Item16	Experimental design (Investigation of new medicine) ...	.395	--	.242	.329 (.044)	.847 (.132)
Item17	Experimental design (Investigation of new medicine)...	--	.143	.270	.459 (.056)	2.922 (.272)
Item18	Center of Earth is very hot...	--	.772	.573	.714 (.058)	-.754 (.087)
Item19	Oxygen comes from plants...	--	.666	.267	.302 (.035)	-.973 (.180)
Item20	Father's gene decides baby's sex...	--	.252	.293	.411 (.044)	2.114 (.173)
Item21	The universe started with a huge explosion...	--	.625	.575	.796 (.057)	-.011 (.045)
Item22	Cigarette smoking causes lung cancer...	--	.827	.125	.210 (.033)	-4.053 (.697)
Item23	The earlier humans lived with dino...	--	.402	.462	.693 (.053)	.855 (.046)

(continued)

Table 7.1 (continued)

No.		Percentile for Correct Response ('91)	Percentile for Correct Response ('01)	Serial Correlation	Item Parameters (2PL)	
					Slope (S.E.)	Threshold (S.E.)
Item24	Radioactive milk can be made safe by boiling it...	--	.838	.627	.927 (.068)	-.906 (.081)
Item25	Speed of light and sound...	--	.894	.494	.546 (.050)	-2.148 (.201)
Item26	Period of the earth's revolution...	--	.581	.499	.665 (.048)	.112 (.045)

The percentages of correct answers range from 0.131(Item12) to 0.774 (Item13) in J-SCITEK91 and 0.143 (Item17) to 0.894 (Item25) in J-SCITEK01. The range of item-sub-score serial correlation is 0.125 (Item22) to 0.691 (Item08) in the combined model. The most distinctive quiz item is Item 22 'Cigarette smoking causes lung cancer.' In terms of the parameters of the 2PL model, the threshold (item difficulty) of each item ranged from -4.053 (Item22) to 2.922 (Item17), and the slope (item discrimination) ranged from 0.210 (Item22) to 3.242 (Item02). Translating the threshold into difficulty, the items on human evolution, cigarette smoking, and relative speed of light and sound presents the threshold -1 or less and seem to act as "common knowledge". The item asking about the relationship between sunlight and skin cancer also shows the threshold of less than -1, but it is not asked in the recent survey. Considering the change in media coverage of environmental issues, it is possible that what is being asked about is not "common knowledge" any more. Among the items with the threshold 1 or over, some are related to the scientific principle that supports current technology (i.e., the scientific backbone of the technology in question), such as "laser", "antibiotics", "electrons and atoms" (nanotechonology). In addition, a change in the format of the item that asked about the method of testing a new drug raised the difficulty level.

Based on these parameters, item characteristic curves (ICC) for all items are shown in Figure 7.1. In Figure 7.1, the ICC shown for Item22 (effect of cigarette smoking) is distinct because of the low threshold and slope. It has not only a high percentage of correct answers but also the lowest serial correlation. This means that the Japanese public knows the harm of cigarette smoking regardless of their PUST, perhaps because the item is more related to health rather than science issues. The ICC for the six common items shows the approximate satisfaction trait that covered the wide range

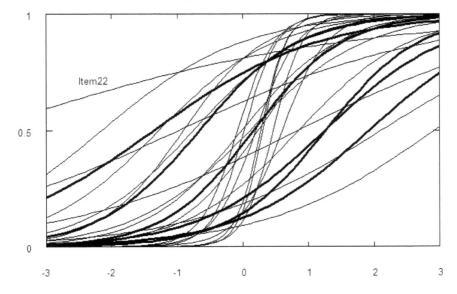

Figure 7.1 The ICC for all items. (Bold lines indicate common six items.)

of threshold. Most items concerning the meaning of a 'one in four probability' show medium difficulty and high discrimination abilities. However, in general, the probability items concerning flower color are slightly easier to answer correctly than the item about heredity.

The total IRT scores of both surveys are estimated and rescaled so that the J-SCITEK91's mean score can be 100, and the S.D. can be 20. Following this rescaling, the IRT score of J-SCITEK01 is 108 (S.D.: 20.3), and there is a significant difference between the two surveys' mean scores (t (3601) =12.24, p <.001). When the samples of both surveys are divided into the six age groups (18–19, 20–29, 30–39, 40–49, 50–59, and 60+), the mean IRT scores in every group are higher in J-SCITEK01 than in J-SCITEK91 (see Figure 7.2). In addition, the younger age groups, the 20–29 in 1991 and 30–31 in 2001, show better understanding in J-SCITEK01 than in J-SCITEK91. On the other hand, older generations, such as 50 through 59, did not show an increase in understanding.

In relation to the other demographic factors, Figure 7.2 presents the mean of each demographic group in J-SCITEK91 and J-SCITEK01. In terms of gender, both men and women scored higher in the recent study. In both studies, the mean score for men is better than for women. In terms of the level of formal science education, there is a significant difference in the knowledge items between two surveys. In 1991, college graduates with a major in any science, medicine, or engineering (SME) failed to show the advantage of science learning in college against non-

120 Martin W. Bauer

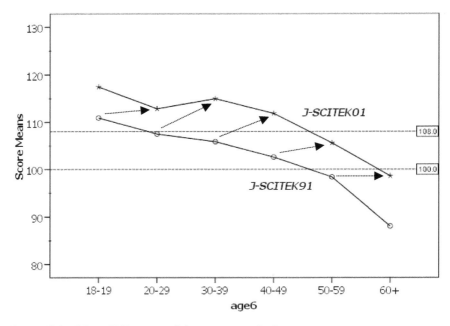

Figure 7.2 Mean IRT scores of six age groups in Japan.

SME college graduates. The authors, at this point, have not reached any explanation. However, SME college graduates show some hesitation to make decisive conclusions. For example, in terms of the item "Laser works by focusing sound", a smaller percentage of SME graduates (27.5%) responded "false" than non-SME graduates (43.8%), but a smaller percentage of SME graduates (22.1%) responded "true" than non-SME graduates (25.0%). The percentage of the response "I do not know" is higher in the SME group (approximately 50%) than among non-SME respondents (approximately 30%). In the J-SCITEK01, college graduates with a SME major improved their score, and this produced significant differences from the non-SME majors.

Because the scale of the issue interest items are different in the two surveys, an index was constructed to identify the relationship between interest in science and technology and the science and technology quiz. In each data set, factor analysis (Extraction Method: Maximum Likelihood) was conducted for variables of the interest in new scientific discovery, new technology invention, medicine, and space explorations. As these four variables can be merged into one factor in each data set, factor scores are calculated. Then calculated scores are combined into four groups (top 25%, upper 25–50%, lower 25–50%, and bottom 25%) in each survey. The reliability analysis for these four variables presents a Cronbach's Alpha of .83 for the 1991 survey and .81 for the 2001 survey. The two-way ANOVA shows a significant increase in the mean score for each group between two surveys; respondents with a higher level of interest demonstrate better understanding of science and technology.

In relation to frequency of visits to social educational facilities, both surveys contained questions asking about the frequency of visits to science museums, natural history museums, and public libraries. Although visiting Science/Technology museums shows a significant positive effect on understanding of science and technology, visits to a Natural History museum failed to show this positive effect.

In terms of the other informal resources for science and technology information, Japanese surveys utilized a unique item. J-SCITEK01 included the following:

> Q25. *Currently, where do you get information about science and technology? Please choose as many as you actually get from the list below*
>
> (a) *TV news*
> (b) *Magazine or weekly magazine articles*
> (c) *TV Documentary*
> (d) *Expo, museum*
> (e) *TV commercial film*
> (f) *Newspaper article*
> (g) *Conversation with family and friends*
> (h) *Newspaper advertisements*
> (i) *Internet*
> (j) *Video, CD-ROM, Tape*
> (k) *Other*
> (l) *None in particular*
> (m) *Don't know*

The choices of the "Internet" and "Video, CD-ROM, tape" were not included in the 1991 survey. By using the other eight items, the two data sets become comparable.

The results of the ANOVA show that all resources and years made significant impact on the public understanding of science and technology. These media have specific characteristics from the viewpoint of the users. Reading science articles in magazines or newspapers requires people to sacrifice their time to read and perhaps shows their specific interest in science and technology. On the other hand, people may make visits to museums, which made a relatively weak impact on science knowledge, for purposes other than education, such as leisure. The characteristics of information resources will determine their impact on science and technology knowledge.

Table 7.2 Mean IRT Score by Year and by Selected Demographic Factors in Japan

		1991	2001	F-value (main effect[1])	F-value (interaction)
Gender					
	Female	96.7	104.9	116.26***	.17 NS
	Male	103.7	112.5	155.93***	
Formal Science Education					
	No College	96.1	103.6	91.60***	11.71***
College W/O SME Major		112.6	115.8	200.27***	
College W/ SME Major		107.9	124.2		
Interest in Science and Technology (Based on Factor Score)					
	0–25%	86.8	97.5		
	25–50%	99.1	105.5	155.14***	3.05*
	50–75%	105.7	111.6	197.90***	
	75–100%	108.2	117.1		
Frequency of Visit to …					
Science/Technology Museum					
	0	99.0	107.5	9.20**	.76 NS
				20.64***	
	1	106.4	115.1		
	2	109.5	116.5		
	3 or more	113.2	113.0		
Natural History Museum					
	0	98.7	107.3	2.27 NS	3.94**
				23.51***	
	1	105.2	114.0		
	2	111.2	113.9		
	3 or more	117.6	107.1		
Public Library					
	0	97.3	104.3	34.53***	.83 NS
				66.23***	
	1	101.4	109.4		
	2	105.6	109.6		
	3 or more	110.3	115.3		
Information Resource					
TV News					
	Not acquired	91.5	101.4	69.42***	1.78 NS
	Acquired	101.9	109.1	76.73***	

(continued)

Table 7.2 (continued)

		1991	2001	F-value (main effect[1])	F-value (interaction)
Information Resource (continued)					
TV Documentary					
	Not acquired	97.6	102.2	73.19***	4.01*
	Acquired	106.3	113.8	203.86***	
TV CF					
	Not acquired	99.1	108.0	51.90***	3.21 NS
	Acquired	105.3	110.7	19.70***	
Newspaper Article					
	Not acquired	93.1	101.6	117.70***	1.83 NS
	Acquired	104.7	111.3	229.93***	
Newspaper Ad					
	Not acquired	98.9	107.9	44.57***	3.31 NS
	Acquired	107.2	112.4	35.96***	
Museum					
	Not acquired	99.2	107.8	16.05***	4.68*
	Acquired	112.6	115.2	56.70***	
Magazine Article					
	Not acquired	97.6	104.3	55.13***	1.06 NS
	Acquired	110.9	116.0	245.59***	
Magazine Ad					
	Not acquired	99.4	107.1	57.44***	.86 NS
	Acquired	106.5	116.2	49.93***	
Family and Friend					
	Not acquired	99.0	107.6	85.43***	.37 NS
	Acquired	104.3	111.8	29.33***	

Notes:
1) Upper: Effect of variables; Lower: Effect of year2)
2) *: .01<p<.05, **: .001<p<.01, ***: p<.001

SUMMARY AND FUTURE TASKS

Unlike western countries, a nationally representative PUST survey has only been conducted twice in Japan. Both studies have affected Japanese society in revealing that Japanese adults do not perform as well in scientific literacy as school-aged children. However, analysis of the integrated data sets of two surveys shows several characteristics that may contribute to the improvement of scientific literacy. First, although adult Japanese performance on the knowledge quiz is not as good as adult performance in other countries, it is improving, especially in the younger generation. By keeping up efforts to promote the public understanding of science and technology, future generations may improve further. Second, the effect of issue interest and the improvement of the quiz item on continental drift imply that relationship between the interests of daily life issues and interests in science may improve our level of the public understanding of sciences. Since the beginning of this century, Japanese scientists have realized the importance of communicating with the public about science and technology. These steady activities give us hope that the Japanese public will have moderate understanding of science and technology.

At this point, these are merely speculations about the future Japanese PUST. In order to have more robust speculation, it is necessary to collect data regularly. The result of analysis of the IRT score called special attention to the younger generation's PUST. Because we have only two data sets, we can only speculate about whether the tendency is temporary, or a consistent characteristic of the younger generation. To discover which, another survey on PUST is necessary.

NOTES

1. The surveys are Eurobarometer 1992, Canada 1989, United Kingdom 2000, United States 1992 and 1999, and Japan 1991.

REFERENCES

Miller, J. D., Pardo, R., and Niwa, F. (1997) *Public Perceptions of Science and Technology: A Comparative Study of the European Union, the United States, Japan and Canada*, Foundation BBV.

Okamoto, S. (2007) Ippan Shimin no Kagakuteki Literacy no Bunseki to Kosatsu [An Analysis and Consideration of Civic Scientific Literacy] *The Journal of Science Policy and Research Management*, Vol. 22, No. 3 / 4.

Okamoto, S., Niwa, F., Shimizu, K., and Sugiman, T. (2001) Kagaku Gijutsu ni Kansuru Ishiki Chosa: 2001nen 2—3 gatsu chosa [The 2001 Survey for Public Attitudes Towards and Understanding of Science and Technology in Japan], NISTEP REPORT No. 72, Tokyo: National Institute of Science and Technology Policy.

Shimizu, K. (2005) Waga kuni ni okeru ippan seijin no Kagakuteki Literacy Sokutei ni Kansuru ichi Kousatsu: Miller no 3 jigen model ni okeru datohsei no kentou [An Examination of Measurement in Civic Scientific Literacy of Japanese Adults: An Examination of the Validity of Miller's 3-dimension Model]. Hiroshima Daigaku Daigakuin Kyoikugaku Kenkyu Kiyo dainibu [Bulletin of the Graduate School of Education, Hiroshima University, Part II], Vol. 54, pp.1–8.

Shimizu, K. (2007) Japanese Survey of the Public Understanding of Science and Technology: Review of results, impact and recent secondary analysis. A paper presented at the Workshop of International Indicators of Science and the Public, London, November 5–6, 2007.

Shimizu, K. (2009) An Empirical Cohort Analysis of the Relationship between National Science Curriculum and Public Understanding of Science and Technology: A Case Study of Japan. *Science, Technology & Society*, Vol. 14, No. 2, pp.365–383.

Zimowski, M., Muraki, E., Mislevy, R., and Bock, D. (2003) *BILOG-MG (version.3)*. Scientific Software International.

8 Adult Scientific Literacy and its Surveys in China Since 1992

Ke Wang, Fujun Ren, Wei He and Chao Zhang

In China the issue of public scientific literacy has been considered an important component of the status of the nation and is valued greatly by the Chinese government. National surveys of public scientific literacy based on internationally adopted methodology have been performed seven times since the 1990s. Trying to find approaches and strategies to enhance Chinese public scientific literacy is the aim of these surveys, which have shown the trends and status quo of citizens' scientific literacy, provided statistical support to government and relevant departments for decision-making, and facilitated the study of the issue. As a consequence, studies on surveys of Chinese public scientific literacy have positively contributed to government policy-making, resulting in the promulgation and implementation of the *Law of People's Republic of China on Popularization of Science and Technology* in 2002[1] and the *Outline of The National Scheme for Scientific Literacy 2006–2010–2020*[2] (hereinafter referred to as *the Outline*) in 2006, which have in turn promoted the development of public scientific literacy. National surveys of scientific literacy have therefore exerted profound influence on Chinese society.

The chapter is divided into two parts. In the first part, it attempts to depict the progress of the measurement of each scientific literacy survey in China; in the second part, it then highlights the major results and gives a brief analysis of those findings.

A BRIEF INTRODUCTION TO NATIONAL SURVEYS OF CHINESE PUBLIC SCIENTIFIC LITERACY

From 1992 to 2007, China conducted seven national surveys mapping public scientific literacy.[3–11] Each survey sampled the population aged between 18 and 69, and respondents were visited in face-to-face interviews. Table 8.1 illustrates the basic information of these surveys. During this period, adaptations were made now and then with the survey method and the questionnaire.

Table 8.1 Technical Parameters of Each Survey

Year	1992	1994	1996	2001	2003	2005	2007
Sampling	5500	5000	6000	8520	8520	8570	10080
Methodology	PPS	PPS	PPS	4-stage stratification PPS (d≤3%)	4-stage stratification PPS (d≤3%)	4-stage stratification PPS (d≤3%)	4-stage stratification PPS (d≤3%)
Weighting	gender	gender	gender	gender, age, education, urban and rural	gender, age, education, urban and rural	gender, age, education, urban and rural	gender, age, education, urban and rural
Items in Questionnaire	70	70	70	70	33	28	30

The Changing of the Questionnaire

The questionnaire for all seven surveys consists of four parts: respondent background, public access to different sources of science and technology information, public understanding of science and technology, and public attitudes toward science and technology.

In the first survey, conducted in August, 1992, the items used in the "public understanding of science and technology" section emulated those developed by J. D. Miller. In the "public access to different sources of science and technology information" and "public attitudes toward science and technology" sections, some items suitable for the Chinese context were added, in order to map a comprehensive picture of public scientific literacy in China and to compare it with that of other countries.[3,4]

In 1994, the second survey used the same questionnaire as in 1992, with a slight cut to the "public understanding science and technology" section.[5]

Some items were added to the 1996 questionnaire; these explored degrees of public knowledge about science and technology, such as radiation, and about basic economic expertise, such as market economy and free trade.[6]

In the "public access to different sources of science and technology information" section, the 2001 questionnaire stressed the public use of electronic media as the media was attracting users rapidly.[7,10]

Big changes were made to the fifth questionnaire, which was used in 2003, based on reflections about the previous ones. First, some items for investigating public engagement in science popularization activities were added in the "public access to different sources of science" section. This was based on an observation of the overall purpose of the survey, national policy needs, in particular the need for science popularization policy, and scholarly interest. Second, the translation of international core items was reworked, and expressions rephrased to sound more like local language, for easier understanding, and some items about "scientific terms" were set as closed questions instead of open ones. Third, option setting of demographic variables was optimized to fully take into account the diversity of Chinese society, facilitating in-depth data analysis. Fourth, items were reduced by one fourth. The revision of the questionnaire in 2003 was a milestone for the survey because it stabilized the framework of the questionnaire and set a template for similar ones in China.[8,11]

In 2005, the sixth survey used the questionnaire of 2003 without any change.[9]

In accordance with the implementation of *the Outline*, demographic information about focus groups was added to the 2007 questionnaire and is being used currently in the eighth national survey. In addition, there were some alterations made to the section on "public attitudes toward science and technology". In order to add attitude indicators for quantitative statistics in further research, we optimized the options setting, with reference to some psychological test methods such as the Likert scale.

The design of the questionnaire for public scientific literacy has been regularly revised in accordance with the Chinese social context, and it has also been a good example of applying theory to practice. Items with local elements were regularly added. Optimization of option setting of demographic variables is useful for data analysis and interpretation based on the current Chinese context. Finally, measuring public scientific literacy has gradually moved away from J. D. Miller's three dimension structure[12,13,14] to "4 sciences and 2 abilities"[1] put forward in *the Outline*.[2] In accordance with the implementation of *the Outline*, studies of the scientific literacy of various social groups have been conducted.

The Improvement of the Sampling Design and Investigative Approach

- The sample distribution becomes more rational by using a four-stage stratification method (see Table 8.1);
- A network of 201 national monitoring points was established during the survey in 2001, helpful for long-term monitoring of scientific literacy in terms of the process control of the investigation;
- To deepen the investigation[15] a qualitative case-study approach was included in 2007;
- A scientific literacy index was used instead of a percentage mark to indicate the quality level of public scientific literacy, which was helpful to the integration and analysis of the data.[19]

Adult Scientific Literacy and its Surveys in China Since 1992   129

Therefore, the focus of the survey or public scientific literacy in China has shifted from theory to practice, from comparing data on China with that of other countries, to contributing to strategy planning and policy-making in order to raise the level of Chinese public scientific literacy.

ANALYSIS OF THE CURVE OF THE CHINESE PUBLIC SCIENTIFIC LITERACY

Chinese Public Scientific Literacy

Similar to J.D.Miller's constructs in the survey, Chinese respondents are considered scientifically literate when they have: (1) basic understanding of science knowledge (scientific terms and concepts); (2) basic understanding of the process of scientific research; and (3) basic understanding of the effect of science upon society. (See appendix 1.)

In the Chinese surveys of public scientific literacy, a scale of four dimensions has been used to measure scientific literacy, namely the public understanding of scientific terms, scientific knowledge, scientific methods, and the relationship between science and society. Specific measurement thresholds are to give correct answers to four items of "scientific terms and concepts", to correctly answer at least ten out of the sixteen items of "scientific views" (knowledge items), to answer correctly all three items of "scientific methods", and to be free of superstitious beliefs (listed in the section on the relation between science and society). If a respondent meets the above four conditions, he or she is considered to have basic scientific literacy.

The construction of scientific literacy can be indicated by mapping the survey results from 2003 to 2007.[8,9,16] Table 8.2 provides the percentage of scientifically literate public (those who meet the threshold for the four dimensions).

According to Table 8.2, although the overall trend of level of public scientific literacy goes upward, growth is rather slow. Respondent scores on scientific terms, scientific knowledge, and the relationship between science and society continue to rise; however scores on scientific methods goes down. Furthermore, as the data shows, public understanding of scientific methods ranks

Table 8.2 Percentage of Scientifically Literate Public and Scientific Literacy Dimensions

	2003	2005	2007
Scientific literacy level (%)	1.98	1.6	2.25
Scientific terms (%)	12.5	13.1	18.4
Scientific concepts (%)	30.0	27.5	33.5
Scienctific methods (%)	8.0	7.4	6.9
Science vs. society (%)	46.7	41.5	59.4

the lowest among the dimensions in all the past surveys; it has become a major factor in the consistently low level of Chinese scientific literacy.

Analysis of the survey data in 2007 demonstrates that there exist considerable differences among those who have basic science literacy.

In those adult social groups specified in *the Outline*, leading cadres and public servants rank high at the top of scientific literacy with a percentage of 10.4%; the urban workforce, which refers to laboring residents in cities, possess a percentage of 3.0%; and the farmers remain at the bottom with a percentage of 1.0%.

A percentage of 2.9% men and 1.6% women tested as scientifically literate.

Citizens at different rungs of the age ladder hold different percentages of scientific literacy. Scientific literacy drops down as the age scale goes up. People aged 18–29 possess the highest percentage of science literacy at 3.5%, with the percentage dropping as age rises (30–39 at 3.0%, 40–49 at 1.9%, 50–59 at 1.5%, and 60–69 at 1.3%).

There is a contrast in scientific literacy between urban and rural citizens. Of urban residents, 3.6% are scientifically literate, compared to 1.0% of their rural counterparts.

It can be drawn from the above description of the data that the Chinese public is making steady progress with regard to their knowledge of scientific terms and scientific concepts, but show declining understanding of scientific methods. Distinctions in scientific literacy are apparent among different social groups: the weight of the scientific literacy balance tilts toward clusters of people who are younger and of higher education backgrounds.

However, the data collected with these methods are not sufficient for in-depth analysis: new forms of data collection in the survey of public scientific literacy are being investigated to fully reflect the Chinese status quo.

COMPREHENSIVE ANALYSIS OF CHINESE PUBLIC SCIENTIFIC LITERACY—2003 AND 2007 SURVEYS AS CASE STUDY

In this section, we will discuss analysis of the level of public scientific literacy, as measured in 2003 and 2007, as well as of its four dimension components, together with demographic variables. Through this approach we assess the impacts of these variables on public scientific literacy and its four dimensions, as shown in Table 8.3.

We considered the effect of gender, age, education, and urban or rural living on scientific literacy. Education and whether one lives in an urban or rural setting are related to public scientific literacy. The level of public scientific literacy is positively related to education, and urban dwellers have higher levels of knowledge of scientific vocabulary and science concepts, whereas age has a negative relationship with scientific literacy level and the levels of each of the four dimensions—younger respondents score more highly. Males hold slightly higher scientific literacy levels in comparison to their female counterparts.

Table 8.3 Correlation of the Scientific Literacy Index and Independent Variables

2003 (No. of valid samples: 8476)	Gender	Age	Education	U&R
Public scientific literacy level	-0.132**	-0.33**	0.627**	0.37**
Level of scientific vocabulary	-0.113**	-0.341**	0.629**	0.373**
Level of science concepts	-0.146**	-0.313**	0.586**	0.352**
Level of understanding of sci. method	-0.1**	-0.224**	0.432**	0.236**
Level of und. of sci. vs. soc.	0.032**	0.101**	-0.154**	-0.079**
2007 (No. of valid samples: 10050)				
Public scientific literacy level	-0.107**	-0.107**	0.449**	0.117**
Level of scientific vocabulary	-0.019**	-0.137**	0.41**	0.132**
Level of science concepts	-0.067**	-0.075**	0.398**	0.121**
Level of understanding of sci. method	-0.009**	-0.053**	0.299**	0.063**
Level of und. of sci. vs. soc.	-0.074**	0.008**	0.094**	-0.028**

**Correlation is significant at the 0.01 level (2-tailed).
Note: Gender: 1.Male, 2.Female;
Age: From 18 to 69;
Education: 1. Illiterate 2. Primary School 3. Middle School 4. High School 5. College 6. University;
U&R: 1 Urban, 2 Rural.

In analysis of the four dimensions of scientific literacy, the effect of demographic variables is strongest on levels of knowledge of scientific terms and scientific concepts, with level of understanding of scientific methods coming next. The effect of demographic variables on the level of engagement is almost negligible. Items measuring engagement are subjective questions, so it is hard to measure this part accurately. From 2003 to 2007, the relationship between the demographic variables, and the level of engagement lacks consistency.

PUBLIC ACCESS TO SCIENCE AND TECHNOLOGY INFORMATION

As indicated in each survey conducted, TV and newspapers are the main channels where the Chinese public acquires science and technology information. Also, people in many regions and places are used to and favor individual conversations to collect information. The 2007 survey data indicate that as many as 34.7% of the Chinese public get their knowledge of science and technology information through conversations with their relatives and colleagues. Compared with other patterns, interpersonal communication was still an important way for the Chinese public to obtain science and technology information. This phenomenon may come from characteristic Chinese economic and cultural contexts. As a result of the economic gap among different districts and social groups in China, people who live in some underdeveloped regions or have lower income cannot afford the cost of televisions or Internet.

Obviously, they face limitations in getting access to science and technology information by mass media, to some extent. Furthermore, historically, Chinese society has been largely agricultural, and in agricultural communities interpersonal relationships are part of a tradition and lifestyle that will likely remain an essential channel for people to obtain all kinds of information.

Given the situation, Chinese people are in fact seeking more channels and diversified approaches for acquiring science and technology information.[17] It is true that television and newspapers are dominant channels for science and technology information, but some other approaches for information acquisition are gaining weight in proportion. The Internet has been experiencing rapid growth as a medium of communication. In 2003, about 5.9% of the population used the Internet to get science information, and in 2007 the figure rose to 10%. Internet is coming into the normal life of Chinese people.[18]

According to Table 8.4, citizens who use the Internet for information-seeking have the highest scientific literacy level, followed by those using newspapers and books. Citizens getting science information through individual conversation have the lowest scientific literacy level, almost none. Comparison between the surveys of 2003 and 2007 shows that more and more citizens have Internet access. Because the Internet is an information channel for those with high scientific literacy level, the enhancement of information technology can be expected to promote public scientific literacy regarding social progress and development.

Table 8.4 Level of Scientific Literacy of Those Reporting the Use of Different Sources of S&T Information

2003 (%)	use %	Literacy	Non-literacy	Chi-Square Value	P Value
Internet	5.9	15.8	84.2	108.41	0.00**
Books	16.2	4.1	95.9	46.77	0.00**
Individual conversation	28.5	1.3	98.7	7.42	0.00**
Newspaper	69.5	3.3	96.7	68.99	0.00**
TV	93.1	2.1	97.9	4.49	0.04*
2007 (%)					
Internet	10.7	4.9	95.1	221.77	0.00**
Books	11.9	4.0	96.0	28.55	0.00**
Individual conversation	34.7	0.8	99.2	4.13	0.00**
Newspaper	60.2	2.9	97.1	18.88	0.00**
TV	90.2	0.8	99.2	2.76	0.09

*Significant at the 0.05 level (2-tailed)
**Significant at the 0.01 level (2-tailed)

PUBLIC ATTITUDES TOWARD SCIENCE AND TECHNOLOGY

The attitude of the Chinese public toward science and technology is consistently measured in the Chinese public scientific literacy survey. As indicated in each survey, the majority of Chinese citizens hold positive attitudes toward the advancement of science and technology. Careers as scientists, doctors, and teachers are most favored and ranked at the first three places throughout the past surveys. Engineers also share that reputation, ranking relatively high. (Shown in Table 8.5.)

According to Table 8.5, citizens who consider engineers and doctors as high reputation occupations hold comparatively high scientific literacy levels; citizens considering teachers as a high reputation occupation hold comparatively low literacy levels.

CONCLUSIONS

The survey questionnaires have become more localized—specific to the indigenous culture—based both on reflections about successive surveys and on the capacity to make international comparisons. Sampling design has included multistage stratification and a large sample size in order to meet requirements for comparison of different regions in China. Meanwhile, the focus of the survey report has moved away from data comparison toward the study of approaches to enhance public scientific literacy.

Table 8.5 Level of Scientific Literacy of Those Who Hold Particular Professions in High Esteem

2003 (%)	Esteem Rank	Percentage	Literacy	Non-literacy	Chi-Square Value	P Value
Teacher	1	57.5	3.4	96.6	0.12	0.76
Scientist	2	46.9	4.4	95.6	26.27	0.00**
Doctor	3	42.0	2.5	97.5	9.76	0.00**
Engineer	5	15.4	5.1	94.9	14.30	0.00**
2007 (%)						
Teacher	1	53.6	2.0	98.0	8.99	0.00**
Scientist	2	51.2	2.9	97.1	50.14	0.00**
Doctor	3	38.4	1.5	98.5	24.06	0.00**
Engineer	5	27.1	3.6	96.4	50.29	0.00**

* Significant at the 0.05 level (2-tailed)
** Significant at the 0.01 level (2-tailed)

As the main findings of surveys of public scientific literacy have indicated, the level of the Chinese public scientific literacy is on the whole increasing steadily but at a relatively slow speed. Differences can be observed in literacy levels, when education, gender, age, and dwelling place are considered. Meanwhile, the attitudes of the majority of the Chinese public toward science and technology, and perceptions of science and technology careers, remain favorable.

Although the main channels through which the Chinese public have access to science and technology information remain the frequently used media (television and newspapers), recent indications are that the public is seeking more channels and diversified approaches. Internet is gaining momentum and creates new ways to improve public scientific literacy.

One of the significant undertakings of the Chinese science popularization enterprise is its civic scientific literacy construction. This requires understanding of the features and peculiarities of scientific literacy on various social ladders and in diverse regions. Past national surveys have provided the Chinese government with empirical data and recommendations for policy-making. By reviewing and analyzing the outcomes of previous surveys, we are seeking measures to improve the conduct of the survey and the sampling and questionnaire design, with a view to enabling a more accurate construction of the level of Chinese public scientific literacy.

NOTES

1. Four sciences and 2 abilities: citizens' basic scientific literacy generally refers to having some necessary **knowledge of science and technology**, mastering basic **science methods**, developing **science thoughts**, advocating **science ethos**, and having **the ability** to apply them to **resolve practical problems** and **participate in public affairs.**

REFERENCES

[1]. 《中华人民共和国科学技术普及法》, 中华人民共和国人大法工委, 2002, 科学普及出版社, p27–34
[2]. 国务院.全民科学素质行动计划纲要（2006–2010–2020）, 2006, 人民出版社 p1–13
[3]. 《中国科学技术指标》（黄皮书）, 1992, 中国科学技术委员会, 科学出版社 p152–167。
[4]. 《中国科学技术指标》（黄皮书）, 1994, 中国科学技术委员会, 中国人事出版社, p113–126。
[5]. 《中国科学技术指标》（黄皮书）, 1996, 中国科学技术委员会, 科学技术文献出版社, p180–187。
[6]. 《中国科学技术指标》（黄皮书）, 1998, 中国科学技术部, 科学技术文献出版社, p155–167。
[7]. 《中国科学技术指标》（黄皮书）, 2002, 中国科学技术部, 科学技术文献出版社, p149–168。

[8]. 《中国科学技术指标》(黄皮书),2004,中国科学技术部,科学技术文献出版社,p121–136。
[9]. 《中国科学技术指标》(2006),中国科学技术部,科学技术文献出版社,p162–174
[10]. 《中国公民科学素养调查报告》,2001,中国科学技术协会 中国公民科学素养调查课题组,科学普及出版社,p71–76。
[11]. 《中国公民科学素养调查报告》,2003,中国科学技术协会 中国公民科学素养调查课题组,科学普及出版社,p30–44
[12]. Miller, J. D. (1983). Scientific literacy: A conceptual and empirical review. Daedalus, 112 (2), 29–48
[13]. Miller, J. D. (1998). The measurement of civic scientific literacy. Public Understanding of Science, 7, 203–223
[14]. Miller, J. D. (2000). Scientific literacy and citizenship in the 21st century, science centers for this century, p369–413
[15]. 任磊 2007中国公民科学素质调查深度访谈报告《科普研究》2008(6)p38–45
[16]. 何薇 张超 高宏斌 中国公民的科学素质及对科学技术的态度《科普研究》2008(6) p8–37
[17]. Ren Fujun, Zhang Chao, He Wei, Channels and Ways for Chinese Public to Obtain Information about Science and Technology. Proceedings of PICMET'08
[18]. 中国互联网络信息中心(CNNIC)第23次中国互联网络发展状况统计报告 2009 http://www.cnnic.net.cn/html/Dir/2009/01/12/5447.htm
[19]. 张超 任磊 何薇 创建中国公民科学素质指数《科普研究》2008(6)p51–58.

Part II
Cross-National Comparisons

9 Comparing the Public Understanding of Science across China and Europe

Xuan Liu, Shukun Tang and Martin W. Bauer

In this chapter we compare the public understanding of science (PUS) across Europe and China on a number of indicators. The results presented are based on an integrated database of public attitude data from the EU and China. This is the first attempt to merge Chinese and European data at microlevel for the benefit of comparison. From a set of structurally or functionally equivalent items, we derived a set of six indicators of PUS: knowledge, interest, attitude and engagement with science, an ecological mindset, and an indicator of superstition. In the following, we present a preliminary analysis of differences in these indicators and their interrelations and construct a five-group typology of the populations combining social standing and PUS.

CHINESE AND EUROPEAN CONTEXTS

There are differences in the status of development of public understanding of science in EU countries and in China and its provinces. These differences not only come from a history in PUS research and science literacy promotion projects, but also from different social and cultural contexts. In Europe, the public understanding of science is an established feature of modern citizenship, as many government decisions are subject to public scrutiny that requires an elevated level of science literacy. The level of science literacy across Europe is the outcome of a sustained historical effort in science education as part of basic literacy, and since the 1980s of concerted and isolated efforts to promote scientific and technological debates. There is a tradition of monitoring public attitudes towards science in many European countries (see chapters in this volume), and the EU central administration has run science-related Eurobarometer surveys since the 1970s, although on an irregular basis (see Bauer, in this volume).

By comparison the situation of science literacy in China can be characterized by three features. As a fast and large developing country,

China increasingly shows a competitive edge in the fields of science, technology, and innovation. Science literacy is considered essential to support this edge. Thus the State Council of the People's Republic of China issued a framework plan 2006–2020 in March 2006 (see *Outline Plan*, 2006). This national framework outlines the activities to develop and the need to build capacity to evaluate the 'quality of the population' (Wu & Mu (2004, 95ff) on the aspect of science literacy. Science literacy is considered an essential human resource in order to acquire and apply knowledge of science and technology, to improve quality of life, and to realize all-round development of the country (ibidem, p2). This outline together with the 'law on the popularization of science and technology' makes it a statutory obligation for all provinces to develop activities as outlined in this framework. The framework in particular diagnoses a gap between urban and rural populations, and a labor force that is insufficiently literate in science. Furthermore, the framework expresses worries about 'ignorance, fatuity and superstition' that still prevail. Science literacy is seen as a bottleneck for sustained economic development. The framework outlines four immediate target populations: young people, farmers, the urban workforce, and public servants. As part of the infrastructure of literacy promotion, the framework guarantees to each provincial capital or every city with more than one million inhabitants the construction of a 'Science Museum'. Equally the technical capacity to establish benchmarks and an indicator system to measure science literacy shall be enhanced. Public science literacy surveys have been conducted biannually since the early 1990s (see Wang Ke, this volume), but the Chinese database is underexploited and has not been integrated nationally or with international databases for the benefits of comparative research.

According to these survey results (see CAST, 2006), 2.25% of the Chinese adult population can said to be scientifically literate (for a definition of literacy, see Wang Ke, this volume). Motivated by this low figure, promoting public understanding of science in China is mostly done in a one-way teaching model. But as most Chinese continue to live with or even rediscover the cultural influence of Confucianism, some of its contents may be contrary to a modern scientific worldview, and this might cause tensions between traditional common sense knowledge and modern scientific ideas. China is a large country with many ethnic minorities living in remote regions where people hold attitudes and values that might clash with modern science and technology, which makes PUS in China a complex phenomenon.

THE INTEGRATED DATABASE

The present study is based on two large-scale surveys conducted in EU countries and China. For Europe we consider Eurobarometer 63.1, conducted in all 27 EU member countries in 2005 (N=26400). For China we consider a series of studies conducted by the authors at various locations, mostly in the Eastern province of Anhui during 2006 and 2007 (N=13000; see appendix for details). The sample covers a cross-section of the Chinese population, representing a kind of 'Middle China' of literacy and attitudes, though with a bias towards the male, rural, and educated population (see appendix). The two surveys were integrated at microlevel on common variables.[1]

The common variables include knowledge (**K**; nine items), interests (**I**; three items), engagement with science and technology (**E**; four items) and basic attitude towards science (**ATT2**; two items), and two further attitude facets. All items are semantically viable translations or made functionally equivalent recodes of the original responses (see appendix, Table 9.5). Socio-demographic information could be matched on **age, gender,** level of **education,** and residential **location**.

KNOWLEDGE OF SCIENCE

The Knowledge (K) indicator consists of the classical items of the knowledge quiz type. Respondents identify a statement to be right or wrong and get a point for correct answer. The index consists of nine items. Response options were 'True', 'False', or 'Don't know' (DK). We counted the ratio of correct answers for each item as shown in Table 9.2.

Table 9.1 Distribution of Responses on Each Knowledge Item

	Question items	EU27	China
		% correct	
K1	Center of the earth's temperature is very high	83.7	53.4
K2	The earth goes around the sun	66.2	78.8
K3	The oxygen we breathe comes from plants	82.7	64.3
K4	Father's genes affect offspring's sex	61.5	36.2
K5	Electrons are smaller than atoms	43.8	33.2
K6	Antibiotics kill bacteria as well as viruses	42.3	39.6
K7	Earth's land slowly drifts and will continue to drift	83.1	48.6
K8	At present, we know that humans are animals evolved from early creatures	64.6	73.3
K9	It takes one day for the earth to go around the sun once	65.5	40.3
	Cronbach's Alpha	0.60	0.63
	Number of observations (N)	26600	12994

To assess the internal consistency of the nine-item questionnaire, we report Cronbach's Alpha, which stands at 0.60 for Europe and 0.63 for China. Below the standard 0.70, the scalar value of these items is not overwhelming, and one should probably explore 'literacy' as a multidimensional scale. However for the present purpose we consider these items as one-dimensional and suitable for a summative scaling and exploratory analysis. The knowledge score is defined as the sum of correct responses for each individual:

$$\text{Knowledge} = K1+K2+K3+K4+K5+K6+K7+K8+K9$$

The indicator K ranges from 0 to 9. In general the ratios of K items are lower for China than for Europe. However, on two items the percentage of correct answer is higher in China: K2 'The earth goes around the sun' (heliocentrism) and K8 'at present, we know, human beings are animals evolved from early creatures' (evolutionary theory). Chinese respondents stand comfortably with both heliocentrism and the theory of evolution. Overall the gap between China and Europe is statistically significant (see Table 9.2).

INTEREST IN SCIENTIFIC TOPICS

The interest indicator consists of three items. The question wording of the interest items is different in Europe and in China, where it read *'Which kind of information about S&T are you interested in?*; in Europe the question read *'for each issue I read out, please tell me, whether you are very interested, moderately interested or not all interested'*. The items and response alternatives can be considered equivalent on environmental issues (**int_environ**), new inventions (**int_invention**), and health related developments (**int_health**). We combined the European 'very interested' and 'moderately interested' with the Chinese 'interested'. Again the interest indicator is the sum of all individual interest scores, counting 'interest' as 1 point and 'no interest' as 0, with a range from 0 to 3:

$$\text{Interest} = \text{int\_environ} + \text{int\_invention} + \text{int\_health}$$

Overall, the European public pays more attention to environment protections (38% versus 19%) and to new inventions (28% versus 15%), whereas on medical developments the difference is smaller (32% versus 28%). Measured on these indicators, the Chinese (i.e., Anhui province) are less interested in developments of science and technology than Europeans.

ENGAGEMENT WITH SCIENCE ON DISPLAY

As for the variable engagement, we use the frequency with which people visited zoos, science museums, science centers, and public libraries over the last 12 months. The sum of these four items defines the engagement indicator. The wording of the engagement items was *"Which kind of following public places have you visited over the last 12 months?"* (multiple response) in the EU and in China 'Which of the following places did you often visit during the past two years?' (multiple response). We constructed equivalences on the common options for Zoo or aquarium (**engage_zoo**), a science and technology museum (**engage_museum**), a science center (**engage_sci_centre**), library and public reading room (**engage_library**). Respondents get 1 point for each relevant item, otherwise they get 0. The index 'engagement' is again the sum of these four items with a range from 0 to 4:

Engagement= engage_zoo+ engage_museum +engage_sci_centre+ engage_library

Europeans generally report more visits to any of the four engagement locations. Comparing European and Chinese the ratios are as follows: for "library and public reading room" (36% versus 7%), followed by "zoo or aquarium" (26% versus 15%), "a science and technology museum" (13.6% versus 6%), "a science center" (8% versus 11%) respectively. Note that for 'science centers' the Chinese report more visits.

The EU public seems more inclined towards active learning in locations where science and technology are on display. But clearly this indicator reflects not only frequency of visits but also the available facilities. We can assume that such facilities are not as easily available in China as in Europe, which is reflected in the ambitions outlines in the Chinese framework plan for 2020. Science and technology activities can be popular in combination with other leisure activities as this is realized in science centers.

ATTITUDE FACETS

The indicator attitude (Att_2) comprises two items: the impact of S&T on the industrial development of the country and on everyday life. Here the two surveys had rather different question formats, but again we made an attempt to create functional equivalences.

In Eurobarometer this facet includes two classical Likert items '*S&T makes our life healthier, easier and more comfortable*' (**atti_life**) and '*S&T do not play a role in industrial development*' (**atti_industry**), to which respondents agree or disagree on a 5-point scale. In China, respondents had to identify the false statement in a set of four (multiple choice). We identify these Chinese choices with European expressions of agreement (1) or disagreement (-1) or DK (0) (see appendix, Table 9.5 for exact equivalences). The value of indicator (Att$_2$) is sum of these two items with a range from -2 to +2:

Att$_2$ = atti_life+ atti_industry

Besides these items, we consider two additional attitude facets: an item regarding belief in a 'lucky number' (Att_lucky) and on the limits of natural resources: 'thanks to scientific and technological advances, the earth's resources will be inexhaustible' (Att_resource). These items are strongly correlated with Att$_2$, but reversed, thus express different facets of people's orientation towards science and technology in China and Europe. Fifty-three percent of Chinese and 20% of Europeans admit to a belief in lucky numbers. Nineteen percent of Europeans utterly reject the idea, and 13% of Chinese do. On the face of it, the Chinese are more 'superstitious' or less rationalistic than Europeans. Indeed, most Chinese seem to think that 6 (liù) means 'everything will go very well', and 8 (ba) means 'making lots of money'. On our index of rationalism, on average Chinese are slightly less rationalist than Europeans (-0.40 versus 0.001 on a scale from -1 to +1: see Table 9.2). We

Table 9.2 General Description of Indices (mean, median, SD; ANOVA, correlations)

	EU: N=26,600			China; N=12294				
	Mean	Med	SD	Mean	Med	SD.	p-value	Eta^2
K	5.93	6	1.96	4.57	5	2.12	<0.001	0.084
I	0.98	1	1.08	0.62	1	0.68	<0.001	0.030
E	0.83	1	0.97	0.38	0	0.63	<0.001	0.056
Att2	0.62	0	0.93	1.34	2	0.93	<0.001	0.171
Ratio	0.00	0	0.61	-0.40	-1	0.70	<0.001	0.076
Ecology	0.16	0	0.59	-0.45	-1	0.77	<0.001	0.147

	EU correlations						China correlations					
	K	I	E	Att2	Ratio	Eco	K	I	E	Att2	Ratio	Eco
K		.20	.32	.20	.13	.12		.18	.12	.38	-.16	-.25
I			.20	.10	.08	.05			.17	.14	-.06	-.10
E				.14	.10	.09				.07	-.07	-.03
Att2					.45	.60					-.58	-.65
Ratio						n.s.						.16

consider a 'belief in lucky numbers' as an index of cultural distance from a purely scientific rationalism.

The third attitude facet considers an ecological mindset: how far people attribute a role to science in extending the limits of growth, an expectation that is less related to science literacy than to other factors (see Bauer, 2009). Here the Chinese express more modernist expectations than Europeans: 63% of Chinese expect science to extend natural resources as compared to only 11% of Europeans. Twenty-six percent of Europeans reject the idea outright compared to 18% of Chinese. We take the rejection of this traditional promise of science as an index of an ecological mindset. On our index the Chinese score more 'negative' than Europeans (-0.45 versus 0.16); a positive value indicates an ecologically minded skepticism towards a science that extends the limits of growth.

Based on these indicators, we end up with six indices of public understanding of science, which are comparable across Europe and China (Anhui province) as shown in Table 9.2.

- Knowledge K (0–9): higher score means higher science literacy
- Interest I (0–3): higher score means more interest in topics of S&T
- Engagement E (0–4): higher score means more visits to locations of S&T display
- Att_2 (-2 to +2): positive score means expectations of positive outcomes of S&T
- Ratio (-1 to +1): positive means rejecting belief in lucky numbers, or being more rationalistic
- Eco (-1 to +1): positive means more ecologically minded, i.e. rejecting the idea that science and technology will extend natural resources forever.

Table 9.2 shows that on all indices the differences between Europe and China are statistically significant, hence beyond what one would expect by chance. Furthermore, considering the parameter Eta^2, we observe that some of the differences are stronger than others. The differences are strongest on utilitarian attitudes Att_2; not too surprisingly Chinese expectations are much higher than European. The other large difference is on ecology-mindedness, more a European than Chinese attitude. The other differences decline in strength from knowledge, to superstition, engagement, and the least differences are on interest in scientific topics.

The intercorrelation between these indices shows remarkable differences between China and Europe. The relationship between knowledge and attitudes is stronger in China than in Europe; whereas in China the relations between knowledge, interest, and engagement is weaker. Most strikingly, both knowledge and positive expectations of science are highly correlated with rationalism, the rejection of superstitious practices, and with ecological thinking in Europe. In China, there is an equally positive relation between knowledge, positive attitudes, and a persistence of superstitious beliefs as well as the rejection of ecological sentiments. In other words, in China the practice of 'lucky numbers' is culturally compatible with science literacy and utilitarian expectations of science and technology, whereas an ecological mind set is not, or maybe not yet. These cultural differences in attitudes between Europe and China deserve further research.

CONSTRUCTING A TYPOLOGY OF PUBLIC UNDERSTANDING OF SCIENCE

For the rest of the paper we will construct and validate a typology of public understanding across Europe and China. We applied a two-step cluster method (with SPSS default) using all six indicators. The analysis results in five clusters as shown in Table 9.3.

Table 9.3 Cluster Analysis of the Resulting Indicators

	Cluster1	Cluster2	Cluster3	Cluster4	Cluster5	
Overall	22%	12%	23%	23%	22%	
EU	26%	15%	21%	15%	24%	
China (Anhui)	8%	4%	24%	46%	17%	
	Educated enthusiasts	*Interested skeptics*	*Rural women*	*Young rural men*	*Young urban mix*	*Eta^2*
K	higher	middle	lower	middle	middle(+)	0.18
I	high	high	Low	low	low	0.05
E	high	middle	low	middle(-)	middle	0.14
ATT2	neutral	negative	negative	positive	neutral	0.05
Ecol	more ecol	more ecol	neutral	less ecol	neutral	0.03
Ratio	less super	less super	neutral	more super	neutral	0.03
Edu	higher(3rd)	higher mixed	low	middle	middle	0.09
Age	middle(-)	oldest	Middle(+)	younger	younger	0.05
Sex	mixed	mixed	more female	more male	mixed	0.01
Location	mixed more urban	mixed more urban	more rural	totally rural	urban	0.05

For the cluster analysis solution, we included Knowledge (K), Att2, Interests, Engagement, Atti-lucky, and Atti-resource as continuous variables and edu, age, sex ,and location as categorical variables.

In terms of sociodemographics markers the five clusters profile as follows:

- C1: middle age, higher education, urban areas, mix sex:
    educated enthusiast;
- C2: older age, higher education, urban areas, mix sex:
    interested skeptics;
- C3: less education, rural area, more female, middle age:
    rural women;
- C4: younger, more male, mainly rural, middle education:
    young rural men;
- C5: younger, middle education, urban, mix sex:
    young urban mix.

The five clusters can be profiled on the six indicators and four sociodemographic variables as shown in Table 9.3 to create a typology. Considering the Eta^2, we observe that education, age, and location are the factors that mostly distinguish the groups; the other variables explain less of the variance: Cluster1 (22%), who we might call **'educated enthusiasts'** are high in knowledge, highly interested, highly engaged, and neutral in utilitarian attitudes, but ecologically minded, and less superstitious. In this group we find many more Europeans than Chinese. Cluster2 (12%), let us call them the **'interested skeptics'**, is characterized by average knowledge, high interest, average engagement, but less expectations from science, an ecological mindset, and less superstition. Cluster3 (23%) are mainly **rural women** of middle age with low knowledge, low interest, low engagement, and negative attitude; they are more ecologically minded, and not particularly superstitious, though with an interesting difference between China and Europe (see below). Cluster4 (23%), who we might call the **young rural men**, have average knowledge, low interest, average engagement, and low positive attitude, are more superstitious than others and less ecologically minded. Finally, Cluster5 (22%), called the **young urban mix**, has medium knowledge, low interest, average engagement, and neutral attitudes, and is also neutral with regard to superstition and ecology.

Table 9.3 shows that European countries distribute more widely among five types, but with some concentration in Cluster1 (26%), the educated urbanites with high knowledge and interest in science, low superstition, and an ecological mind set, and Cluster5 (24%), the young urban mix, with low interest in science and neutral attitudes. By contrast, the Chinese samples cluster mainly in Cluster4 (46%), young rural males who have great expectations of science, but little ecological orientation, and are also superstitious, and Cluster3 (24%), the rural uneducated, mainly

Table 9.4 The Distributions of PUS Clusters across EU27 and China (Anhui province)

Country	Educated enthusiast	Interested skeptics	Rural women	Young rural male	Young urban mix
Belgium	38%	14%	13%	17%	16%
Denmark	**61%**	16%	5%	5%	11%
Germany WEST	26%	14%	2%	12%	23%
Germany EAST	21%	18%	14%	13%	**33%**
Greece	23%	10%	33%	9%	23%
Spain	22%	7%	42%	8%	19%
Finland	42%	18%	13%	8%	16%
France	30%	17%	18%	17%	17%
Ireland	17%	11%	17%	17%	36%
Italy	23%	7%	30%	7%	31%
Luxembourg	28%	20%	20%	13%	17%
Netherlands	40%	**21%**	10%	12%	16%
Austria	15%	13%	26%	15%	28%
Portugal	12%	4%	**59%**	6%	18%
Sweden	48%	16%	8%	14%	12%
United Kingdom	17%	18%	23%	10%	34%
Cyprus	16%	14%	33%	14%	22%
Czech Republic	16%	19%	6%	18%	39%
Estonia	28%	21%	10%	13%	26%
Hungary	11%	18%	33%	12%	24%
Latvia	25%	20%	9%	18%	26%
Lithuania	26%	16%	16%	9%	31%
Malta	9%	13%	36%	27%	14%
Poland	22%	13%	14%	19%	30%
Slovakia	14%	14%	9%	**32%**	28%
Slovenia	23%	16%	19%	21%	19%
Bulgaria	22%	16%	19%	11%	30%
Romania	19%	11%	25%	18%	25%
China (Anhui)	8%	4%	24%	46%	17%
Total	**21%**	**12%**	**23%**	**23%**	**22%**

women, with low scientific literacy and low expectations from science, though less superstitious than other types.

To achieve an even stronger sense of comparison, Table 9.4 lists the distribution of the five PUS types across all countries, EU27 and China. On all clusters we can find considerable variation across Europe, and no EU country matches China in its mix. This shows that China is clearly a world of its own, in particular Cluster4 dominates, the young rural men probably on the move in search of work in urban centers. We also note that in all clusters the differences between China and Europe persist: within each cluster, Europeans are more knowledgeable, more interested and engaged, but hold less expectations towards science, are more ecologically minded, and more rationalist than Chinese people of similar standing.

In present day China, there are still significant gaps in education and mentality between the urban and rural populations; they differ in income level and access to public resources. This huge gap between urban and rural China has a long history, which, due to a variety of reasons, continues in such matters as household registration, public infrastructure construction, compulsory education, social welfare, and access to public services. As for PUS, most new initiatives go to large cities and urban areas. It is easy to understand that, under the present conditions of poor life quality and even survival issues in rural China, there is less attention to science and technology matters, but people hold high expectations.

Another phenomenon is very notable: the 46% of Chinese in Cluster4, younger men of secondary education, are currently the main blue-collar working class in Chinese society. According to recent surveys, this group comprises the middle to lower level of the social structure and is mostly composed of service staff, industrial workers, and agricultural labor. The Chinese sample in Cluster4 represents the working class of China. This group scores far lower in both the interest and engagement with science than similar groups in the EU. To understand this situation, we must consider its sociological explanation. China's labor force has experienced a rapid process of transformation since the 1980s. Consequently, a large number of these people are migrants from rural to urban cities and are a floating population, who do not benefit from the welfare and public services of urban residents. Working in jobs that urban residents generally find inferior, this floating population is often denied the privileges enjoyed by permanent urban residents who have city residency permits giving access to subsidized housing, medical care, and schooling for their children. We must imagine a typical case as a migrant worker in urban area. At present, migrant workers usually live in shabby districts of the city, where the rent is low. Recent research shows that the migrant workers living in shantytowns make few demands on the environment and society despite the fact that their living conditions are extremely poor. This huge floating population in China's urban areas provides the needed

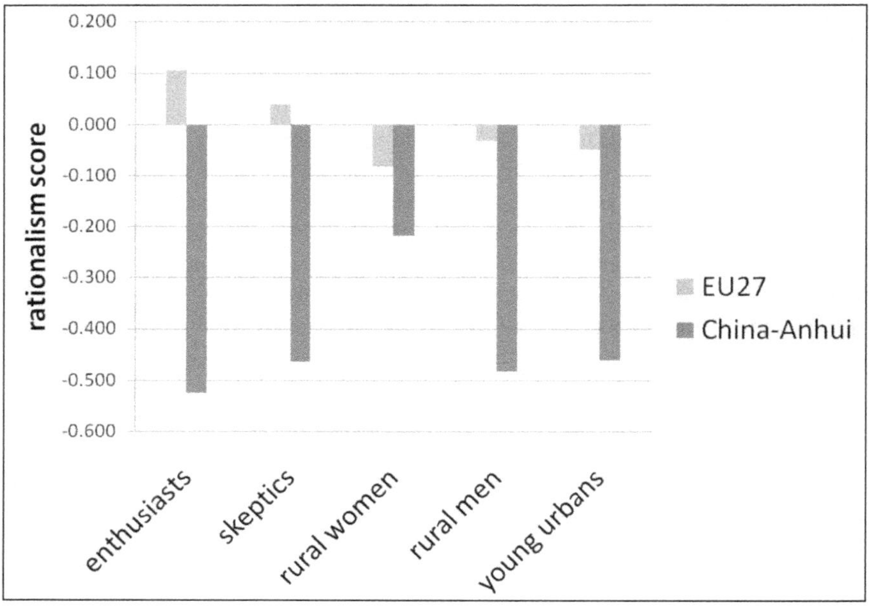

Figure 9.1 Marginal means for rationalism (i.e. rejecting belief in lucky numbers) for PUS clusters and EU versus China from a MANOVA model that includes all six indicators (K, I, E, att_2, Ecol, and Ratio). The interaction effect for rationalism across cluster and country is significant (F=111, P<.001, df=4, R^2 = 0.10).

labor for the explosion of urban construction and performs many service functions, including sales and domestic work. They add vitality to the current economic development.

We can observe a notable result with re gard to Cluster3, the rural women, where we observe a difference between Europe and China with regard to superstition and rationalism in everyday life. Figure 9.1 shows the marginal means for 'rationalism' from a model, including all indicators as dependent variables and clusters and region (EU versus China) as predictors. The gap between Europe and China remains consistent across all PUS types with the curious exception of C3, the rural, less educated, mainly women of middle age. In the Chinese sample, they display a rather sober outlook, the least superstitious of all, while in Europe this group is the most superstitious of all types. We consider this a surprising result for two reasons: if 'lucky number' is indeed a good indicator of 'rationalism', we must then conclude that rural life is not the center of Chinese irrationalism. To the contrary, irrationalism is most accentuated in urban life, among the most educated (C1 and C2) as well as among the medium educated (C4 and C5), among the old as well as among the young. It might well be that this kind of 'irrationalism' (lucky numbers) is correlated with the uncertainties of modern urban life and its rapid transitions. And as C1 and C2 show, for those highly knowledgeable and interested in science, these superstitions are not in contradiction with scientific rationalism. The urban Chinese seem to live very comfortably with a kind of 'cognitive polyphasia' (see also Allum this volume), i.e., being able to handle both scientific literacy and traditional irrationalities in everyday life, and expressing this compatibility seems easier in urban than in rural life. This is interesting because it contrasts with the European scene, where the traditional belief in a 'lucky number' is mostly found in rural areas covered by C3 and clearly negatively correlated with science literacy.

CONCLUSION ON COMPARISON AND DISCUSSION

We have merged a European and a Chinese survey database on the microlevel and conducted a first comparative exploration of PUS in these very different contexts. From this comparison analysis, we can make a number of observations as follows:

> **First,** Chinese levels of science literacy are lower than Europeans, so is interest, engagement with science on display, and so is the ecological mindset. However, in China basic attitudes to science are more positive, indicating higher expectations of science and technology with regard to economic development and improvements to personal life. Also more prevalent is everyday superstition, as indicated by belief in lucky numbers.

China and EU follow a different model of public understanding of science. The more people know, the more positive they are in China whereas in Europe this is reversed: the more people know, the more skeptical they are and the less they expect from science. This observation is consistent with the two-culture model of public understanding of science, which predicts different knowledge-attitude-interest relations in different development contexts (Bauer, Durant & Evans, 1994).

> **Second,** in European countries the most important sociodemographic correlates of science literacy, interest, engagement, and attitudes are education and age. In China the most relevant variables are education and the urban–rural divide. This is a reflection of differences between urban and rural life of China, in particular the imbalance in the allocation of resources between urban and rural population.

Third, we constructed a typology of public understanding of science in relation to different socioeconomic groupings, and we could identify and profile five clusters and their distribution across these diverse populations. The majority of Europeans fall into Cluster1 and Cluster5. Cluster1 comprises the highly educated urban population with good scientific knowledge, high interest and engagement with science, rationalist but with skeptical attitudes and an ecological mindset. Cluster5 comprises the moderately educated, younger urbanites, with medium science literacy, but little interest in science, neutral attitudes, and no a particular sense for ecology. By comparison the large majority of Chinese belong in Cluster4: young males with low to medium science literacy, low interest and little engagement with science on display, but high expectations of science, no ecological mindset, and fairly superstitious. Cluster3 is also quantitatively important in China, comprising the less educated rural population, typically middle-aged women, with little expectations of science, but some sense of ecology and displaying a surprisingly sober rationalism. The latter is particularly surprising, as in the European context, rural life is the harbor of superstition, wheras in China this seems to be more a feature of urbanity.

As the fight against 'ignorance, fatuity and superstition' is a major concern of the Chinese framework for public understanding of science (*Outline*, 2006), our surprising observations regarding urban irrationalism, and its compatibility with other aspects of science literacy, such as knowledge and attitudes, and its relation to urban or rural life, deserve further investigations.

APPENDIX I: EUROPEAN AND CHINESE SURVEY SAMPLES

The European Sample (32 Countries)

Between 3 January and 15 February 2005, the TNS Opinion & Social, a consortium created between Taylor Nelson Sofres and EOS Gallup Europe, carried out wave 63.1 of Eurobarometer, on behalf of the European Commission, DG Research.

The sample design uses a multistage, random probability procedure. In each country, a number of sampling points were drawn with probability proportional to population size (for a total coverage of the country) and to population density.

In order to do so, the sampling points were drawn systematically from each of the "administrative regional units", after stratification by individual unit and type of area. They thus represent the whole territory of the countries surveyed according to the Eurostatnuts II (or equivalent) and according to the distribution of the resident population of the respective nationalities in terms of metropolitan, urban, and rural areas. In each of the selected sampling points, a starting address was drawn, at random. Further addresses were selected as every Nth address by standard "random route" procedures, from the initial address. In each household, the respondent was drawn, at random (following the "closest birthday rule"). All interviews have been conducted face-to-face in people's homes and in the appropriate national language. As far as the data capture is concerned, CAPI (Computer Assisted Personal Interview) was used in those countries where this technique was available.

The Chinese Sample

The Chinese science literacy research project 2007 was planned and executed by a team from Hefei, led by Professor Shukun Tang of Hefei Technical University. It was a three-stage empirical study comprising a preliminary sample in Hefei City, a sample from Anhui Province and a test in eastern, central, and western regions of China.

The Hefei Pilot Study (December 2006)

The Hefei city was a pilot study testing the questionnaire, identifying problems and making amendments. The sample was **purposive** and not intended to be representative of the Hefei City population, but to cover different types of people across urban or rural areas, gender, age, and levels of education. Two hundred interviews were conducted in neighborhood committee; others included university students, higher education staff, civil servants, and migrant workers. The Hefei sample consisted of n= 247 respondents.

Province of Anhui (January–March 2007)

This survey was conducted in a four-stage stratified sampling method: Probability Proportionate to Size Sampling (PPS).

The first stage: Cities, districts, and counties were selected from the provincial regions. A stratified sort equidistant sample method was adopted, considering average education and the total population. First, we allocated the sample size for each city according to the city's population; then we selected counties or districts at random in each city and decided the sample size for each district relative to population. Seventeen cities and 57 counties were selected.

The second stage: In each unit, a stratified sample was extracted in streets, towns, and villages on equidistance. The sample size was pro rata of the population and resulted in 256 streets/townships.

The third stage: committees were selected. Each committee provided the 'local authority' for the interviews. Thus there were very few nonresponses. During this stage simple random sampling (SRS) or equidistant sampling (systematic sampling) of the committees was adopted. 101 neighborhoods (urban) and 364 village committees (rural) were selected.

The fourth stage: households were selected in each committee area. The interviewer followed the right hand rule selecting the households.

Interviewers used a two-dimensional table of random numbers to select one person in each household **older than 15 years and younger than 69 years** for the interview. The final sample comprised 12994 respondents from 17 cities and 57 counties, comprising 101 urban neighborhoods (3.8% of 2627 in total) and 364 rural villages (3.5% of 19367 in total).

Samples from Eastern, Central, and Western Regions (July-August 2007)

For further research, the project team designed the eastern, central, and western comparison samples by purposive sampling in Pudong (in Shanghai, the east), Hefei (in the center), and Tianshui City (in the west). Around 350 interviews were undertaken in each city.

Comparing the Anhui Sample with All China

In the Anhui sample, male respondents accounted for 57.2% (51.5%, China, 2005). Compared with the 2005 census, the proportion of men in the sample is higher than expected for All China. As for location, 28% (37%, China 2005) respondents in the sample reside in urban areas. The urban population of the sample was lower than for All China. As for education, 25.6% are illiterate or primary educated (47.7%, China 2005), 64.3% have secondary schooling (51.5%, China 2005), and 10.1% are college and university educated (3.8%, China 2005). The Anhui sample is clearly biased towards the more educated population. On age, we cannot directly compare the Anhui survey and the Census of 2005, because the age brackets are not consistent. The Anhui sampled 15–26 (24.8%), 27–36 (26.2%), 37–46 (23.2%), and 47–69 years (25.8%); the Census of 2005 reports 15–24 (20.5%), 25–34 (31.3%), 35–44 (24.3%), and 45–69 years (19%). There seems to be little difference between the Anhui sample and the All China Census, but on average the Anhui group should be older than the all China group. The sample has a bias towards the younger populations.

In conclusion, we must conclude that compared to the All China Census of 2005, the Anhui database of 2007 is **more educated, more rural, younger, and more male**. For any direct comparisons between EU27 and China on results depending on the levels of variables, the Chinese data will be biased towards the more educated, young, rural, and male respondent. For structural comparisons between EU and China we expect these biases to be less of a problem.

APPENDIX II: CONSTRUCTION OF EQUIVALENCE FOR ATTITUDES AND INTEREST

There are differences in the Eurobarometer and Chinese questionnaires in the wording of the question for attitude and interest. We recoded the variable as shown in Table 9.5.

Table 9.5 Equivalence Recoding for Attitude (Att2) and Interest (I) Items

	Options	Codes	EU: Science & technology are making our lives healthier, easier and more comfortable		China: Which of the following do you think can not be true	
			Options	Codes	Options	Codes
Atti_life	agree (positive)	1	strongly agree	1	development of science & technology cannot improve the quality of life	3
			tend to agree	2		
	neutral	0	neither agree nor disagree	3	do not know	5
	disagree (negative)	-1	tend to disagree	4	development of science & technology to improve traffic	1
					development of science & technology to make communication easier	2
			strongly disagree	5	contributing to scientific research ca make life healthier	4

			EU: Science & technology do not play an important role in industrial development		China: Which of the following do you think cannot be true	
	Options	Codes	Original options	Codes	Original options	Codes
atti_industry	agree	1	strongly disagree	5	some industrial development has nothing to do with science and technology	4
			tend to disagree	4		
	neutral/dk	0	neither agree nor disagree	3	Don't know	5
	Disagree/DK	-1	tend to agree	2	scientific and technological progress to improve the skills of workers	1

(continued)

Table 9.5 (continued)

atti_industry (continued)			EU: Science & technology do not play an important role in industrial development		China: Which of the following do you think cannot be true	
	Options	Codes	Options	Codes	Options	Codes
		-1	strongly agree	1	scientific and technological progress brought about by the improvement of means of production	2
		-1			progress of S&T can promote industry development	3

Interest(I)			EU: For each issue I read out, please tell me if you are very interested, moderately interested or not at all interested in it?		China: What kind of information about S&T are you interested in?	
	environment protection (int_environment)		environmental pollution(vqa1_4)		environmental protection(a3)	
	new invention (int_invention)		new inventions and technologies(vqa1_5)		new inventions(a3)	
	health care (int_health)		new medical discoveries(vqa1_3)		health care(a3)	
	Options	Codes	Original options	Codes	Original options	Codes
	interested	1	very interested	1	interested	1
			moderately interested	2		
	uninterested/dk	0	not at all interested	3	uninterested/dk	0

NOTES

1. The data sets were integrated during a year of study in 2009, during which the first author, Xuan Liu, spent 12 months at LSE STePS following training courses and working closely with Martin W Bauer in examining the two surveys, assessing the potentials, and constructing comparisons. The data microintegration is documented in Liu & Bauer (2009).

REFERENCES

Bauer MW (2009) The environmental competence of science and the public understanding of science, in: Shimizu K (ed) The Public Understanding of Science and Technology for Sustainable Development, Symposium Proceedings, Hiroshima University, 7 September 2009, p28–45.

Bauer M, J Durant, and G Evans (1994) European public perceptions of science, International Journal of Public Opinion Research, 6, 2, 163–186.

CAST (2006) Study on Measurement Indicators of Scientific Literacy of Chinese Citizens and its Demonstration Survey (2006DCYJ11-A).

EC (2005) Europeans, Science and Technology, Eurobarometer Special 63.1, technical report, no 224, Brussels, April 2005.

Liu Xuan and MW Bauer (2009) Science Literacy, Public Understanding of Science in China 2007 and EU 2005, Integrated Codebook, LSE-STePS-ISP, July 2009.

Outline Plan (2006) Outline of the National Scheme for Scientific Literacy—2006–2010–2020, Beijing, Popular Science Press (English edition, 2008).

Wu Cangping and Mu Guangzong (2004) China's Population Situation and Policies, Beijing, Foreign Language Press—Focus on China Series.

10 Information and Attitudes towards Science and Technology in Iberoamerica

Carmelo Polino and Yurij Castelfranchi

The studies generally branded as part of the Public Understanding of Science (PUS) 'movement' encompass diverse traditions and analytical perspectives.[1] Demoscopic surveys of attitudes towards science and technology (S&T) have been a widely applied tool in the PUS tradition. These surveys were first developed in the US, some central countries (UK, France), and Europe (Eurobarometer), although later they were adopted in other countries and regions of the world (Canada, Japan, New Zealand, India, Italy, China, Spain, Russia, South Korea, Argentina, Brazil, Mexico, Colombia, etc.). Along with this expansion—and as a consequence of it—there has been a proliferation of works on the topic, including debates about indicators (Allum et al., 2008; Raza et al., 2002), statistical consistency (Pardo, Calvo, 2002), analytical dimensions (Osborne, 2003; Wynne, 1995; Michael, 1998; Miller, 1998), methodologies (Pardo, Calvo, 2004), models (Wagner, 2007; Pardo, Calvo, 2006; Sturgis, Allum, 2004; Godin, Gringas, 2000) and interpretations derived from field work carried out at the local, regional, and international levels (Allum et al., 2008; Miller, 2004; Bensaude-Vincent, 2001; Miller, Pardo, Niwa, 1997; Bauer, Shoon, 1993; Bauer et al., 1993). Currently, many analysts are convinced that it is time to find integrating perspectives that might allow a greater convergence between the traditions. An ambitious agenda is presently being canvassed along these lines, aiming to tackle problems concerning PUS using a unified and international perspective (Bauer, Allum, Miller, 2007), integrating national and international surveys into a global database, and developing global indicators and alternative data streams, such as media monitoring and longitudinal qualitative research (Bauer, 2008:125).

The general aim of this 'new international agenda' coincides with the way in which some institutions in Iberoamerica have come to think about this topic, through a work itself based on collaborative strategies in networks with a view to cross-regional comparisons. Mechanisms of methodological integration and pilot experiences of measurement have been proposed (Vogt, Polino 2003) since the first studies presented by the Iberoamerican Network of Science and Technology Indicators (RICYT), and the Organization of Iberoamerican States (OEI)—with the cooperation of the Fundação de Amparo à Pesquisa do Estado de São Paulo (FAPESP), and, later, the Spanish Foundation for Science and Technology (FECYT), Colciencias (Colombia), as well as other institutions.[2] Actually, in some countries the consolidation of national surveys was possible in many ways thanks to the configuration of this regional network.[3] Likewise, the need emerged to critically revise the tradition of studies of PUS, and to reflect on theoretical models and the concepts behind these models (Vogt, Polino, 2003; Vaccarezza, et al., 2003a, 2003c). Also a search was begun for indicators and indexes that better fit local realities and political expectations, but which also allow comparability and currency in the international context (Polino et al., 2006; Vogt, Polino, 2003; Vaccarezza et al., 2003b). Finally, we can highlight the development of a program that went beyond demoscopic empirical studies to incorporate qualitative studies that consider the analysis of scientific culture as a structural feature of society, irreducible to individuals (Polino et al., 2006; Vaccarezza, 2007).

The attractiveness of the Iberoamerican region for comparative studies and new perspectives is due to two main aspects. On one hand, Iberoamerica is a large region of the world that encompasses many nations, cultures, and languages, but in which two languages dominate: Spanish and Portuguese. Shared languages and similar historical circumstances are precious for the researcher, as they simplify the task of standardizing protocols for surveys and interviews, and allow for fruitful comparing and benchmarking. On the other hand, the region is characterized by huge ecological, cultural, and linguistic diversity;[4] by extreme social stratification and differentiation; and by very different macroeconomic and political situations. Such diversity might allow researchers to formulate complex and heterogeneous data sets, which in turn might yield useful insights and test certain hypotheses on the factors that influence perceptions and attitudes.

Table 10.1 displays selected indicators from international and regional institutions (UNESCO, PNUD, World Bank, RICYT), which allows us to illustrate some comparable data between countries. The information also includes specific indicators on S&T systems. The history of S&T in Iberoamerica is peculiar. In the so-called developing countries, the institutionalization and professionalization of science, as well as the formation of national S&T systems and policies, followed quite unique and specific trajectories, very different from those observed in Western Europe and the US (RICYT, 2008; Velho, 2005; Saldaña, 1996; Albornoz, 2001).

Certain factors that had wide-ranging repercussions in many 'advanced capitalist' countries (the rise of financial capitalism, globalization, neoliberal policies, etc.) also had crucial effects in most of the Iberoamerican region, influencing the shape of socio-technical systems and S&T policies; yet these effects were manifested in different ways, and varied across the countries that compose the region. In Iberoamerica phenomena such as the

Table 10.1 Comparative Indicators by Country

Comparative indicators by country	Argentina	Brazil	Chile	Colombia	Spain	Panama	Portugal	United States	Ibero America
Population (million people). Source: RICYT (2007)	39,36	189,30	16,60	43,93	45,20	3,30	10,60	301,62	600,90
Life expectancy at birth, 2005–2010 (years) Source: Unesco (2008)	75.3	65.6	78.6	73.3	80.1	73	77.9	77.9	
GNP per capita–PPP U$S (2005). Source: World Bank (2007)	13.800	8.140	10.920	6.970	17.830	7.050	15.370	31.600	
Human Development Index (HDI). * Source: PNUD (2007)	.866	.813	.878	.807	.955	.840	.909	.956	
Expenditure on R&D related to GDP. Source: RICYT (2007)	0.51%	1.11%	0.67%	0.16%	1.27%	0.20%	1.21%	2.66%	0.87%
Expenditure on R&D by funding source. Source: RICYT (2007)									
Government	67.6%	52.9%	44.4%*	37.7%	47,0%	47.9%	44.6%	27.6%	51,0%
Enterprises	29.3%	44.7%	45.8%*	27.2%	45.5%	0.3%	47.6%	66.6%	40.9%
Higher Education	1.4%	2.4%	0.8%*	25.6%		0.5%	0.7%	2.7%	4,0%
Non-profit organizations	1.1%		0.3%*	5.4%	0.5%	1.3%	2.3%	3.2%	0.5%
Foreign	0.6%		8.7%*	4.1%	7.0%	50,0%	5.4%		3.5%
Total	100.0%	100.0%	100.0%	100.0%	100.0%	100.0%	100.0%	100.0%	100.0%

(continued)

Table 10.1 (continued)

Comparative indicators by country	Argentina	Brazil	Chile	Colombia	Spain	Panama	Portugal	United States	Ibero America
Expenditure on S&T by type of activity. Source: RICYT (2007)									
Basic Research	29.3%		35.7%	24.0%**	20.2%	29.0%	18.5%	17.5%	
Applied Research	42.7%		49.0%	47.0%**	43.6%	38.0%	35.7%	22.1%	
Exp. Development	28.0%		15.3%	29.0%**	36.3%	33.0%	45.8%	60.4%	
Total	100%		100%	100.0%	100.0%	100.0%	100.0%	100.0%	
Granted patents. Source: RICYT (2007)									
Residents	445	3 468***	67	12	2 603	15		93 665	6 070
Not residents	2 324	3 628***	515	210	19 220	243		89 236	36 998
Total	2 769	7 096***	582	222	21 823	258		182 901	43 068
Papers in Science Citation Index (SCI) SEARC. Source: RICYT (2007)	6.479	23.109	3.559	1.239	40.594	369	7.466	388.160	90.388

* Last available year, 2004
** Last available year, 2001
*** Last available year, 2006

increasing importance of privatization and appropriation of research; the growth of international and/or interdisciplinary networks; the stronger links between university and business enterprises; the social accountability of scientific institutions; and sensible public debate and conflicts linked to technoscience are observed in the context of different societal needs and social and political conflicts. They thus end up generating diverse local configurations.

The conjunction of these arguments makes it interesting to investigate the meanings attributed in Iberoamerica to keywords and slogans such as sustainability, interactivity, engagement, social participation, e-democracy, bottom-up governance, participatory decision-making, precautionary principle, etc. Measurement studies must be accompanied, therefore, by analyses of the socio-cultural and political itineraries of S&T in such a way that the indicators and indexes are put in perspective and explained from a wider sociological outlook.

This paper examines some results of the first large-scale Iberoamerican comparative survey on public perceptions and attitudes toward science and technology. The survey was undertaken in the context of a regional pilot project coordinated by RICYT, OEI, and FECYT and was supported by local institutions.[5] The main objective of the project was to make a methodological contribution to the survey field of PUS in the Iberoamerican region, proposing the development of a core set of common questions and indicators to be used in national surveys.[6] We present a summary of three selected topics: public interest and information, attitudes to S&T in terms of risks and benefits, and attitudes to civic participation in decision-making processes related to S&T.[7] We also compare these findings with other Iberoamerican databases and with data from European surveys.

IBEROAMERICAN SURVEY (2007)

The Iberoamerican survey was carried out at the end of 2007 in seven large cities: Bogotá (Colombia), Buenos Aires (Argentina), Caracas (Venezuela), Madrid (Spain), Sao Paulo (Brazil), Santiago (Chile), and Panama City (Panama). A representative sample of the population of people over 16 years of age was collated, distributed according to gender, age, and education (margin of error: 3 per cent; level of confidence of 95 per cent). The size of the sample was about 1.100 per city (7.800 in total).

Information and Interest

The axis of the analysis of interest and information was drawn using direct and indirect strategies. The former was implemented at the beginning of the questionnaire when interviewees were asked to say with what frequency they watch television or read a newspaper and then what type of TV programs, newspaper sections, or news they preferred. Here, S&T figured as one item among a wider set of preferences. The second tactic consisted of deploying a battery of specific indicators to measure the declared information seeking patterns: television; radio; newspapers; Internet; books and magazines on popular science; visits to museums, centres, and exhibitions; casual conversations with friends on these topics; and finally participation in events such as protests, signed declarations, forums of debate, letters to newspapers, etc., that might have been motivated by a topic concerning science, technology, or the environment.

The questionnaires yielded coherent results across both strategies. In general terms, in all the cities (with the exception perhaps of Bogotá) there is no predominant or salient interest in topics of science and technology; this is found in combination with auto-evaluations that indicate absence of information about these topics (in this case Bogotá is not different from the rest). This applied to the reading of science news in newspapers and to the televised contents of science, technology, and nature. And this low information profile is accentuated, as it is to be expected, having evaluated other habits: seven out of ten interviewees did not look for scientific information on the Internet; also seven out of ten never read popular science magazines; the proportion of those who never read books is nine out of ten; seven out of ten never visit museums, centres, or exhibitions of science and technology. It is important to point out that this distribution generally coincides with that obtained by national surveys—Argentina (SECYT, 2007), Brazil (MCT, 2007), Spain (FECYT, 2006), Colombia (Colciencias, 2005), or Venezuela (MICYT, 2007). Concerning conversations with friends on topics of science, technology, and the environment, the wide band that chose "yes, occasionally" probably did so because they were not sure what to include (or not) in this category. In any case, the outcome tallies with a general tendency to show low interest in S&T topics in the interviewees' lives. The same pattern comes up in the following question, regarding participation in events related to science and technology. The answer "no, never" was predominant in all the cities, with the exception of Bogotá and Madrid.

Some comparisons can now be established with the information coming from Europe. Globally speaking, a parallel can be observed between the information from both regions, which indicates that, like most European adults, Iberoamericans share a general tendency in most of the comparable indicators and a relative homogeneity in the frequency distribution for each indicator.

More specifically, it is possible to assert that, in the case of TV, proportions remain stable in the average consumption range, although the Iberoamerican group that declared to be "very interested" is proportionally

Table 10.2 S&T Information Indicators for Different Media

	Watch TV programs bout scientific research*		Read general science articles in newspapers and magazines**		Listen to radio programs about scientific research		Search Internet for information about scientific research		Buy specialized press about scientific research***	
	IB (2007)	EU (2007)	IB (2007)	EU (2007)	IB (2007)	EU (2007)	IB (2007)	EU (2007)	IB (2007)	EU (2007)
Regularly	23.4 %	16 %	14.1 %	12 %	5.4 %	4 %	11.4 %	7 %	5.9 %	5 %
Occasionally	52.6 %	45 %	37.9 %	37 %	18 %	22 %	22.1 %	21 %	21.1 %	17 %
Hardly Ever****	-	21 %	-	18 %	-	26 %	-	14 %	-	19 %
Never	23.5 %	17 %	47.1 %	32 %	75.7 %	47 %	65.4 %	57 %	72 %	58 %
DK	0.45	1 %	0.9 %	1 %	0.8 %	1 %	1.1 %	1 %	1 %	1 %
Total	100 %	100 %	100 %	100 %	100 %	100 %	100 %	100 %	100 %	100 %

* In the Iberoamerican survey, nature and technology are mentioned, as well as science.
** In contrast to Eurobarometer (2007), the Iberoamerican study separates newspapers and magazines in two different questions.
*** In the Iberoamerican case, the people are directly asked if they "read" this type of product.
**** This category in the scale was not contemplated in the Iberoamerican survey. That is why certain differences found in specific items must not be exaggerated. In these cases, the results should be understood in terms of general tendencies.

greater than their European counterparts. In the case of newspapers, declared consumption is lower than that of TV in both regions. The same happens with radio, Internet, and specialist magazines.

S&T Information Index

To obtain a condensed vision of Iberoamerican interviewees' consumption of scientific information, an "index of scientific information consumption" was constructed (ICIC, according to the Spanish acronym) on the basis of an unweighted average of eight variables that measure declared information-gathering behaviour concerning S&T topics.[8]

All the items in this set of questions are positively correlated, with correlation coefficients of moderate magnitude, typically in a range from 0.31 to 0.49—with the exception of the pair "popular science magazines–popular science books", which show a coefficient of 0.65. Factorial analysis reveals the one-dimensionality of these items.[9]

The ICIC estimate allows us to observe an asymmetric distribution that indicates the existence of a small segment of the population that is highly informed about topics of science and technology. Certainly, this type of distribution is common in surveys measuring the level of political information of the population; that is to say, a consistent segment of a slightly informed majority is present, and smaller segments show higher information levels.

After crossing the index with socio-demographic variables, it is possible to appreciate that women and men do not have substantially different information-gathering behaviours. The age variable shows some moderate differences, to the extent that we can say that the information is less present among the elder adults (55 years and up), without going so far as to produce marked contrasts. On the other hand, the educational level correlates in a positive manner with the index (with moderate statistical correlation). This means that, in general terms, more education corresponds with a greater tendency to be informed. For example, among university students the proportion of people at the top end of the band is twice the average, and more than three times the level of the segment with an average schooling level.

Looking at ICIC index scores according to the cities included in the study yields notable similarities and differences. In all cities, the group of people more accustomed to seeking information about science and technology ("High" and "Very High") does not exceed ten per cent of the population. On this matter, the cities that are better positioned are Bogotá (8.9 per cent) and Panama (7.9 per cent), whereas the lowest proportion is registered in Santiago (3.4 per cent) and then São Paulo (2.2 per cent). The latter city shows a perceptibly higher than average concentration of interviewees in the "Null" and "Very Low" categories.

Table 10.3 ICIC (S&T Information Index) by City

	Bogota	Buenos Aires	Caracas	Madrid	Panamá	Santiago	São Paulo	Total
Null level of S&T information	116 10.9%	112 10.4%	231 22.5%	110 10.2%	143 13%	145 13.7%	237 22.1%	1094 14.6%
Very low level of S&T information	280 26.3%	227 21.1%	239 23.3%	229 21.2%	252 22.9%	382 36.1%	393 36.7%	2002 26.8%
Low level of S&T information	404 38%	505 47%	371 36.2%	457 42.3%	421 38.3%	345 32.6%	331 30.9%	2834 37.9%
Medium level of S&T information	170 16%	193 18%	117 11.4%	222 20.6%	197 17.9%	149 14.1%	86 8%	1134 15.2%
High level of S&T information	73 6.9%	33 3.1%	48 4.7%	55 5.1%	76 6.9%	30 2.8%	18 1.7%	333 4.5%
Very high level of S&T information	21 2%	5 .5%	19 1.9%	7 .6%	11 1%	6 .6%	5 .5%	74 1%
Total	1064 100%	1075 100%	1025 100%	1080 100%	1100 100%	1057 100%	1070 100%	7471 100%

Risks and Benefits from Science and Technology

Over the last twenty years, different definitions of the concept of risk, and its dissimilarities with 'danger', have been formulated. The idea of "risk societies" was proposed by authors such as Beck (1998, 2008), Giddens (1990), and Luhmann (2005), and it helps us understand why decisions made in the modern world imply solutions as well as threats. Risk and risk perception have become important themes in the contemporary social sciences, and are especially relevant in the field of Social Studies of Science and PUS.

The survey included two sets of indicators to evaluate benefits and future risks of S&T. In all the cities—with the exception of Caracas—most of the people point out that in the next twenty years it will be necessary to manage risks derived from S&T. The most pessimistic vision can be found in Bogotá. At the same time, the ample majority of interviewees (76 per cent) point out that science and technology will produce benefits. Again, Bogotá stands out, and shares the most optimistic vision with Buenos Aires.

The correlation between risks and benefits allows for the substantive identification of three segments that are statistically significant (Table 10.4). The first and major one shows that 46 per cent of interviewees highlight

Table 10.4 Balance between Future Risks and Benefits from S&T

		Attitude to future benefits of S&T				Total
		Many benefits	Quite a few benefits	Few benefits	No benefits	
Attitude to future risks of S&T	Many risks	919	509	473	111	2012
		13.2%	7.3%	6.8%	1.6%	28.9%
	Quite a few risks	671	1115	379	55	2220
		9.6%	16.0%	5.4%	.8%	31.8%
	Little risks	1028	802	226	18	2074
		14.7%	11.5%	3.2%	.3%	29.7%
	No risks	445	167	36	19	667
		6.4 %	2.4 %	.5%	.3%	9.6%
	Total	3063	2593	1114	203	6973
	% of Total	43.9%	37.2%	16.0%	2.9%	100.0%

risks as much as benefits. The second group tends to emphasize the benefits of science and technology and to minimize the risks, and encompasses more than one third of the interviewees (35 per cent). The third group, representing 15 per cent of the population, sees, on the contrary, few benefits and a lot of risks. The segment of people who consider science and technology of little relevance—both few benefits and few risks—or who are not in a position to give an opinion, is quantitatively insignificant. The cross-tabulation with socio-demographic variables (sex, age, education) or constructed variables (such as the ICIC index) shows that there are no substantial differences between the identified segments.

Table 10.5 allows us to appreciate an attitudinal typology, distributed by city, with regard to the future balance of risks and benefits. The definition of the six "target groups" was performed by computing a new target variable from the two questions presented above.[10] Four out of ten interviewees on average think that there will be more benefits than risks. The proportion is higher in Panama and Caracas (in both it represents half of the population) and smaller in Santiago and São Paulo. In the second segment, two out of ten interviewees point out that there will be more risks than benefits; here, the citizens of São Paulo and Santiago share a more cautious vision, in contrast to Caracas or Madrid. The third segment, comprising three out of ten interviewees on average, values the future existence of risks and benefits in the same measure. In this case the perception is even across cities.

As shown above, the perception of risks and benefits does not differ substantially when considering a bivariate correlation with variables such as age, gender, or education. However, this picture changes when data are contemplated in the light of the attitudinal typology introduced in Table 10.5. In this case, attitudes towards risks and benefits turn out to be affected by education levels. The idea of benefits is more present in those with higher levels of education, and the consideration of risks diminishes gradually as the instruction level increases. On the other hand, and leaving aside persons without schooling, the equivalence between risks and benefits is distributed in a relatively uniform manner according to educational level (basic, average, tertiary, and university education). These tendencies can also be appreciated after applying the ICIC indicator as a cross-tabulation variable. In the more informed segments, a higher proportion values the benefits of S&T, and the consideration of risks diminishes. Again, the position that balances out benefits and risks is distributed in a quite uniform way in segments with different levels of information.

Social Participation and Public Policies

In recent years, the political rhetoric of contemporary democratic societies has incorporated debate on the generation of mechanisms of greater transparency and social participation in the management of public matters, by

Information and Attitudes towards Science and Technology   169

Table 10.5 "Attitudinal Typology" Considering Risks and Benefits from S&T by City

	Bogota	Buenos Aires	Caracas	Madrid	Panamá	Santiago	São Paulo	Total
More benefits than risks	396 35.9%	426 39.3%	519 49.1%	435 39.5%	520 47.9%	349 32%	337 32.2%	2982 39.4%
More risks than benefits	214 19.4%	237 21.8%	129 12.2%	172 15.6%	192 17.7%	287 26.3%	314 30%	1545 20.4%
Risks and benefits are the same	399 36.1%	320 29.5%	234 22.2%	320 29.1%	342 31.5%	339 31.1%	325 31.1%	2279 30.1%
Values the benefits, but no answer on risks	52 4.7%	49 4.5%	51 4.8%	65 5.9%	18 1.7%	35 3.2%	22 2.1%	292 3.9%
Values the risks, but no answer on benefits	131 .2%	131 .2%	212 .0%	181 .6%	4 .4%	22 2%	11 1.1%	102 1.3%
No answer on either benefits or risks	30 2.7%	40 3.7%	102 9.7%	91 8.3%	10 .9%	58 5.3%	37 3.5%	368 4.9%
Total	1104 100%	1085 100%	1056 100%	1101 100%	1086 100%	1090 100%	1046 100%	7568 100%

virtue of the evolution of political systems towards a 'bottom-up' democracy. In this frame, the role of expert systems and the legitimacy of technocratic decisions have been discussed. These discourses have also seeped into other, different ambits, and S&T has been no exception (Bucchi and Neresini, 2008; Irwin, 2008; Fiorino, 1990; Funtowicz and Ravetz, 2000, 1997; López Cerezo and Luján, 2000).

As the research agendas of the PUS movement were extended to encompass scenarios of consensus and dialogue between science and society (M. Bauer et al., 2007), surveys have introduced questions relating to social participation and the making of decisions pertaining to S&T. The European public, for example, is demanding greater opportunities for involvement (Eurobarometer, 2005; 2005.224). The same tendency is evident in the cities included in the Iberoamerican survey: there is wide agreement on the fact that citizens must be listened to and their opinions considered (seven out of ten interviewees demand greater access to the decision-making sphere), whereas variations across cities are not significant.

Movements in favour of participative democracies in the developed world do not imply the extinction of classical technocratic approaches either. This is especially clear in the case of S&T: sciences and technologies are simultaneously perceived as important and difficult, and most of society is not accustomed to reflect on questions that affect its development and impact (social, political, economic, cultural, ecological, etc.). One consequence of this is that, although there is a demand for a space for participation in public policy, it is often preferred that the judgements of experts prevail. The Eurobarometer survey (2005.224) is a good example of this: it asks people whether decisions about S&T should be based primarily on the advice of experts about risks and benefits involved or on the general public's views of these risks and benefits. The report concludes that, at the European level, two in three citizens (66 per cent) believe that S&T decisions should be based mostly on the advice of experts (see also the discussion in Luján and Todt, 2007, 2008).

The Iberoamerican citizens interviewed are on the whole also inclined to make the same evaluation, and in the same proportion. Six out of ten think that the experts should make decisions on social problems relating to S&T.[11] The evaluation remains stable when associated with various sociodemographic indicators: no differences are observed according to age, gender, and educational level or information habits on S&T. It does not depend, either, on subjects' positions on the future consequences of S&T in terms of risks and benefits. In this regard, we could have expected that the most critical people (that is, those who believe that the risks will outweigh the benefits) would be more clearly in favour of social intervention and decreasing the role of the technocratic expert, but this is not a clear tendency.[12] Some differences between the cities are observed in comparison to the overall average (65 per cent), but they are only slightly significant or very close to the margin of sampling error; we would have to say, then, that in general the tendency is quite uniform, considering the different political and cultural systems.

The concordance between attitudes towards expert decision-making and public participation requires an in-depth investigation within a framework of qualitative analysis and stronger quantitative indexes and indicators. This concordance might be a sign of the co-presence of two strong narratives in contemporary scientific culture. On one hand, S&T are seen by most people (and depicted by most of the media) as absolutely crucial for competitiveness, innovation, and development: techno-scientific progress, social progress, and wealth are seen as strongly linked; in this sense, S&T are seen as the territory where "the future is at stake". This discourse is strongly linked to the 'deficit' narrative, which goes as follows: people do not know enough; therefore they are not able to make rational, informed decisions. Thus, S&T should be allowed to run (and accelerate) almost automatically, and governance should be delegated to the experts. On the other hand, recent reconfigurations in the capitalist system (linked to financial globalization, the "digital revolution", and the so-called "knowledge-based economy") has led to new practices and discourses that function side-by-side with classical ones (understandably, with some frictions) both in the market and in politics. One of these reconfigurations has introduced the notion that products and policies should be always manufactured after listening to people: cars are custom-built; policies and reforms are made, or dams and roads built, after a process of consultation in which 'local' is taken into account, and so on. In this sense, it is logical that many people expressed the need for both expert (or minimal) governance of S&T and the desire for social participation in decision-making.

Index ICIC Regression Analysis

In this last part, we present results from a regression analysis that models the ICIC index as a function of demographic (gender, age, education, city of residence) and attitudinal variables (opinion on "citizen participation" and "risks and benefits") commented on in this paper. The ICIC index, with sixteen categories, is close to an interval variable, which is what is assumed in a linear least squares regression.

Table 10.6 shows that women have an ICIC below men (even after controlling for education and other variables in the model). Besides, it shows that age has no significant impact, and that education, as expected, has a strong impact on the ICIC: for each increase of one category in the educational ordinal variable, the dependent variable increases more than half a point. São Paulo and Caracas appear somewhat below Bogota, while Madrid is above it. Santiago is also above Bogota, although by a smaller magnitude and only marginally significant. Buenos Aires does not differ significantly from Bogota.

Table 10.6 Model: ICIC Determinants' Coefficients[a,b]

	Non-standardised coefficients		Standardised coefficients	t	Sig.	Confidence interval: 95% for B	
	B	Tip. error	Beta			Lower limit	Upper limit
(Constant)	,156	,169		,919	,358	-,177	,488
Experts should make the decisions on social problems relating to S&T	,018	,019	,011	,961	,336	-,019	,056
Citizens should play a more important role in decisions relating to S&T	-,070	,021	-,039	-3,435	,001	-,111	-,030
More benefits than risks	,771	,109	,219	7,110	,000	,559	,984
More risks than benefits	,540	,112	,127	4,809	,000	,320	,761
Risks and benefits are the same	,653	,110	,175	5,938	,000	,438	,869
Values the benefits, but no answer on risks	,279	,143	,030	1,951	,051	-,001	,560
Values the risks, but no answer on benefits	,291	,201	,019	1,448	,148	-,103	,686
Gender (male=1)	-,227	,038	-,066	-5,897	,000	-,302	-,151
Age grouped by 6 segments	-,008	,012	-,008	-,656	,512	-,033	,016
Education grouped by 5 segments	,537	,019	,333	27,636	,000	,499	,575
Buenos Aires	-,111	,073	-,022	-1,516	,130	-,255	,033
Caracas	-,257	,074	-,050	-3,463	,001	-,403	-,112
Madrid	,258	,072	,053	3,565	,000	,116	,399
Panamá	,072	,070	,015	1,025	,305	-,065	,209
Santiago	-,122	,073	-,024	-1,674	,094	-,264	,021
São Paulo	-,388	,073	-,079	-5,307	,000	-,532	-,245

a. Dependent variable: ICIC (eight indicators).

With respect to attitudinal variables, only the item asking if citizens should play a more important role in decisions about S&T has a statistically significant effect on the ICIC. This is a negative effect, and indicates that the higher the agreement with the participation of citizens the greater level of information.

Furthermore, citizens who have opinions about risks and benefits have more information than people within the category "no answer on either benefits or risks". However, the group "values the benefits, but no answer on risks" does not reach the conventional level of confidence (.05) for a minimum margin. In turn, in the case of one who "values the risks, but no answer on benefits," the coefficient is clearly not significant.

The coefficients of the three significant categories at the level of .05 are sorted as expected: the bigger one is for those people who believe that there are "more benefits than risks", followed by those who consider that "risks and benefits are the same", then for those who declare that there are "more risks than benefits". Nevertheless, one should not overestimate the significance of these differences, which appear to be small and not statistically relevant.

CONCLUSIONS

We analyzed part of the data from the thus far biggest international Iberoamerican survey on public perception of S&T. We tested ICIC, an indicator of the consumption of scientific information (discussed in FECYT-OEI-RICYT, 2009; SECYT, 2007) and showed that it seems a consistent and valid variable, related to attitudes, interests, and values. On one hand, our data indicate the existence of a globalized substrate of scientific culture: some perceptions of S&T seem to be shared by both rich and developing countries. On the other hand, the data revealed some interesting features that differentiate Iberoamerica from other regions, and also some differences between the countries (such as Brazil and Colombia). Such peculiarities deserve deeper investigation and, in some cases, pose interesting questions about possible complex, non-linear relations between S&T infrastructure and public perceptions and attitude in a country. Besides this, our data analysis also supported criticisms of some simplified assumptions linked to "deficit models": our interviews showed that relations between knowledge, educational level, interests, and attitudes towards S&T are actually complex and often non-linear. For example, although the idea that benefits of S&T outweigh risks is more diffused in those with high levels of education; the claim that science today poses as much benefit as risk is distributed in a quite uniform manner in population. Two narratives of techno-scientific progress seem to coexist, with some friction and contradictions, in public representations: the laissez-faire, technocratic one (experts decide, we should let

S&T run semi-automatically), and the "engagement" one (people should participate more in decision processes and policy-making).

At the same time, we think that recent works and the emergence of networks in Iberoamerica support the view that a new international agenda is needed, which should include: database integration; the formulation of new indicators; fresh interpretations of existing data; new conceptual developments, and cross-fertilization between public perception, media, and cultural indicators. We hope that the Iberoamerican survey will contribute to the debates on indicators and their construction and validation. We also expect it to be helpful in further research on different aspects of science, technology, culture, and civic participation. Stronger interactions are needed in order to produce integrated data sets and to discuss conceptual problems.

NOTES

1. PUS expressions, such as "public perception", "scientific literacy", "scientific culture", "public awareness", "public engagement", "public participation", "public attitudes", etc., became more and more common in areas as varied as science policy, public opinion research, journalism, education, science studies, and cultural studies. Yet, these terms have their origins in different theoretical backgrounds, and the fact that many of them are used indistinctively reveals a lack of conceptual clarity within the PUS field.
2. The first comparative pilot survey in the region was implemented in November 2002 by RICYT and OEI, in collaboration with FAPESP, in cities in Argentina, Brazil, Uruguay, and Spain, based on non-representative samples of population (see Vogt, Polino, 2003).
3. The Iberoamerican region comprises twenty-one countries. Eleven of these countries' national S&T councils or analogous institutions have conducted at least one national survey on public perception of science from 1987 to 2009: Argentina (2004, 2006), Brazil (1987, 2006), Chile (2007), Colombia (1994, 2004), Ecuador (2006), Mexico (1997, 2001, 2003, 2004, 2005), Panama (2001, 2006), Portugal (2002), Spain (2002, 2004, 2006, 2007, 2008), Uruguay (2007), and Venezuela (2004, 2006).
4. In Brazil alone, about 170 different native languages are known. Several regions of South America are considered, from the ecological point of view, areas of "mega diversity". In a typical European forest, about a dozen species of trees can be found. But it is possible that *more than 300 different species of trees* live in just half of the Atlantic rainforest.
5. Colciencias and Observatorio de Ciencia y Tecnología (Colombia); Comisión Nacional de Investigaciones Científicas y Tecnológicas (Conicyt, Chile); Fundação de Amparo à Pesquisa do Estado de São Paulo (Fapesp, Brazil); Ministerio de Ciencia y Tecnología (Mct, Venezuela); Programa Acerca, Agencia Española de Cooperación Internacional (Aecid); and Secretaría de Ciencia y Tecnología (Senayct, Panamá). The questionnaire was developed by a team of researchers from more than fifteen institutions and nine countries: J.A. López Cerezo (Spain), C. Polino (Argentina), C. Cabello (Spain), C. Vogt (Brazil), T. Arboleda (Columbia), M. Verdugo (Chile), M. Cámara Hurtado (Spain), Y. Castelfranchi (Brazil), S. Daza (Columbia), ME Erazo (Ecuador), ME Fazio (Argentina), D. Chiappe (Argentina), A. Firmino (Portugal), J.L. Luján (Spain), L. Massarani (Brazil), C. Moreno (Spain), L. Palma (Panama),

G. Romero (Venezuela), C. Palma (Portugal), R. Sequera (Spain) and M. Solé (Spain).

6. The development of surveys in the region has followed, in many ways, methodological imperatives arising from experiences of the European Union with the Eurobarometer and the NSF reports. At the same time, the studies have incorporated local approaches and measurements. The current situation shows a significant dissemination of surveys and some methodological convergences. Nonetheless, there are still many important methodological divergences at different levels that, in some cases, limit the comparability of the indicators collected (conceptual dimensions, questionnaires and sample design, fieldwork organization, etc.).

7. The book compiling general survey results was published at the end of 2009 (FECYT-RICYT-OEI, 2009).

8. The indicators that compose the index are: 1) "science and technology programs on TV"; 2) "reading news about science and technology in newspapers"; 3) "radio programs on science and technology"; 4) "reading popular science magazines"; 5) "reading popular science books"; 6) "use of Internet as source of information on scientific topics"; 7) "visits to museums, centres and scientific exhibitions"; 8) "conversations with friends about topics to do with science and technology". Each indicator can take three possible values: "Regularly", "Occasionally", or "Never". The answer "Regularly" gets one point. The answer "Occasionally" equals 0.5 points; the answer "Never" gets no points; and finally "doesn't know / doesn't answer" is considered as a "Missing value". This approach makes the index show values that fluctuate between "0" and "8.0".

9. The first factor accounts for 47% of total variance, and the second only reaches 10%.

10. The segments were defined by using criteria of intensity on the response given: the "more benefits than risks" group comprises respondents who answered positively about *many* or *quite a few benefits*, or *few benefits* and *little risks* from S&T. This group tends to reject or minimize risks; the balance is clearly favourable to the benefits. The contrary occurs with the segment "more risks than benefits". This group thinks about *many* or *quite a few risks* and *little benefits* from S&T. They are inclined to reject or minimize benefits. Therefore, the balance is oriented to the accentuation of future risks. The third group, "risks and benefits are the same", indicates those people who evaluated risks and benefits with the same intensity: for instance, one could have said *many benefits* and *many risks*, or *little benefits* and *little risks*. Finally, groups responding "values the benefits, but no answer on risks", "values the risks, but no answer on benefits", "no answer on either risks or benefits" have labels from which it is easy to deduce their characteristics.

11. If we observe the composition of the group of individuals that opposes the taking of decisions by experts (somewhat below 20% of the total), we can see that gender representativeness is even; that there is a higher representation of people with a middle education level (four out of ten), and a third that has university education; and that overall they tend to profess a measured position on the impact of science and technology, identifying with the idea that it poses as much risk as benefits.

12. It is interesting to note that this kind of discussion and oscillation between positions tending towards 'bottom-up' democracy and plenary social participation, on one hand, and those where technocratic values prevail, on the other, are open questions in the context of science studies. An example is the debate opened up by the proposal of Collins & Evans (2002).

BIBLIOGRAPHY

Albornoz, M. (2001), "Política científica y tecnológica: una visión desde América Latina", *CTS + I, Revista Iberoamericana de Ciencia, Tecnología, Sociedad e Innovación*, OEI, N°1, septiembre-diciembre.
Allum, N., Sturgis, P., Tabourazi, D., and Brunton-Smith, I. (2008), "Science knowledge and attitudes across cultures: A meta-analysis", *Public Understanding of Science*, 17, 1, 35–54.
Bauer, M. (2008), "Survey research on public understanding of science", in M. Bucchi, B. Trench (eds.), *Handbook of Public Communication of Science and Technology*.
Bauer, M., Allum, N., and Miller, S. (2007), "What can we learn from 25 years of PUS survey research? Liberating and expanding the agenda", *Public Understanding of Science*, 16, 1, 79–95.
Bauer, M., Durant, J., and Evans G. (1993), "European public perceptions of science", *International Journal of Public Opinion Research* 6, 2, 164–186.
Bauer, M. and Shoon, I. (1993), "Mapping variety in public understanding of science", *Public Understanding of Science*, 2, 2, 141–155.
Beck, U. (2008), *La sociedad del riesgo global mundial. En busca de la seguridad perdida*, Barcelona, Paidós.
Beck, U. (1998), *La sociedad del riesgo: hacia una nueva modernidad*, Barcelona, Paidós.
Bensaude-Vincent, B. (2001), "A genealogy of the increasing gap between science and the public", *Public Understanding of Science*, 10, 1, 99–113.
Bucchi, M. and Neresini, F. (2008), "Science and public participation", in E. Hackett, O. Amsterdamska, M. Lynch, and J. Wajcman (eds.), *The Handbook of Science and Technology Policies (third edition)*, Massachusetts Institute of Technology, 449–472.
CNPQ (1987), *O que of brazileiro pensa da ciencia e da tecnologia?* Brazilia/Rio de Janeiro: CNPq.
Colciencias (2005) *La percepción que tienen los colombianos sobre la ciencia y la tecnología*, Bogotá, COLCIENCIAS.
Collins, H.M. and Evans, R. (2002), "The third wave of science studies", *Social Studies of Science*, 32, 2, 235–296.
Etzkowitz, H. and Leydesdorff, L. (2000), "The dynamics of innovation: From national systems and "Mode 2" to a triple helix of university-industry-government relations, *Research Policy*, 29, 109–123.
Etzkowitz, H. and Leydesdorff, L. (1996), "Emergence of a triple helix of university-industry-government relations", *Science and Public Policy*, 23, 279–286.
Eurobarometer (2007), "Scientific research in the media", Special Eurobarometer 282/Wave 67.2, TNS Opinion & Social.
Eurobarometer (2005) *Europeans, Science & Technology*, 224 / Wave 63.1, TNS Opinion and Social, European Commission.
FECYT (2006), *Percepción social de la ciencia y la tecnología en España*, Madrid.
FECYT-OEI-RICYT (2009), *Cultura científica en Iberoamérica. Encuesta en grandes núcleos urbanos*, Fecyt, Madrid.
Fiorino, D. (1990), "Citizen participation and environmental risk: A survey of institutional mechanisms", *Risk Analysis*, 9, 293–299.
Funtowicz, S. and Ravetz, J. (2000), *La ciencia posnormal: ciencia con la gente*. Barcelona, Icaria.
Funtowicz, S. and Ravetz, J. (1997), "Ciência pós-normal e comunidades ampliadas de pares face aos desafios ambientais", *História, Ciências, Saúde—Manguinhos*, IV 2, 219–230.

Gibbons, M., Limoges, C., Nowotny, H., Schwartzman, S., Scott, P., and Trow, M. (1994), *The New Production of Knowledge: The Dynamics of Science and Research in Contemporary Societies*, London, Sage.
Giddens, A. (1990), *The Consequences of Modernity*, Cambridge, Polity.
Godin, B. and Gringas, Y. (2000), "What is scientific and technological culture and how is it measured?" *Public Understanding of Science*, 9, 1, 43–58.
Irwin, A. (2008), "STS perspectives on scientific governance", in E. Hackett, O. Amsterdamska, M. Lynch, and J. Wajcman (eds.), *The Handbook of Science and Technology Policies (third edition)*, Massachusetts Institute of Technology, 583–607.
López Cerezo, J.A. and Luján, J.L. (2000), *Ciencia y política del riesgo*, Madrid, Alianza.
Luhmann, N. (2005), *Risk: A Sociological Theory*, New Brunswick, NJ: AldineTransaction.
Luján, J.L. and Todt, O. (2008), "A new social contract for technology? On the policy dynamics of uncertainty", *Journal of Risk Research*, 11, 4, June, 509–523.
Luján, J.L. and Todt, O. (2007), "Precaution in public: The social perception of the role of science and values in policy making", *Public Understanding of Science*, 16, 1, 97–109.
Michael, M. (1998), "Between citizen and consumer: Multiplying the meanings of the 'public understanding of science'", *Public Understanding of Science*, 7, 4, 313–327.
Miller, J. (2004), "Public understanding of, and attitudes toward, scientific research: What we know and what we need to know", *Public Understanding of Science*, 13, 3, 273.
Miller, J. (1998), "The measurement of civic scientific literacy", *Public Understanding of Science* 7, 3, 203–223.
Miller, J., Pardo, R., and Niwa, F. (1997), *Public Perceptions of Science and Technology: A Comparative Study of the European Union, the United States, Japan, and Canada*, Chicago, Chicago Academy of Sciences.
MCT (2007), "Percepção Pública da Ciência e Tecnologia", http://www.mct.gov.br
MICYT (2007), *Venezolanos participan y opinan. Segunda encuesta nacional de percepción pública de la ciencia, cultura científica y participación ciudadana*, I. La Rosa, J.M. Cruces, Caracas (eds.).
Nowotny, H., Scott, P., and Gibbons, M. (2001), *Rethinking Science: Knowledge in an Age of Uncertainty*, Cambridge, Polity Press.
Osborne, J. (2003), "Attitudes toward science: A review of literature and its implications", *International Journal of Science Education*, 25, 9, 1049–1079.
Pardo, R. and Calvo, F. (2006), "Mapping perceptions of science in end-of-century Europe", *Science Communication*, 28, 1, 3–46.
Pardo, R. and Calvo, F. (2004), "The cognitive dimension of public perceptions of science: Methodological issues", *Public Understanding of Science*, 13, 3, 203–227.
Pardo, R. and Calvo, F. (2002), "Attitudes toward science among the European public: A methodological analysis", *Public Understanding of Science*, 17, 1, 155–195.
Polino, C., López Cerezo, J.A., Castelfranchi, Y., and Fazio, M.E. (2006), "New tools and directions toward a better understanding of social perception of science in Ibero-American countries", The 9th International Conference on Public Communication of Science and Technology, South Korea.
Raza, G., Singh, S., and Dutt, B. (2002), "Public, science and cultural distance", *Science Communication* 23, 293–308.
RICYT (2008), *El Estado de la Ciencia—Principales Indicadores de Ciencia y Tecnología Iberoamericanos / Interamericanos*, RICYT-OEI, Buenos Aires.

Saldaña, J.J. [Editor] (1996), *Historia social de las ciencias en América Latina*, México, Porrúa.
SECYT (2007), *La percepción de los argentinos sobre la investigación científica en el país. Segunda encuesta nacional*, Buenos Aires, SECYT/ Ministerio de Educación, Ciencia y Tecnología.
Sturgis, P. and Allum, N. (2004), "Science in society: Re-evaluating the deficit model of public attitudes", *Public Understanding of Science*, 55–74.
Vaccarezza, L. (2007), "The public perception of science and technology in a peripherical society: A critical analysis from a quantitative perspective", *Science, Technology & Society*, 12, 1, 141–163.
Vaccarezza, L., Polino, C., and Fazio, M.E. (2003a), "Hacia una medición de la percepción pública de la ciencia en los países iberoamericanos", en *El estado de la ciencia. Principales Indicadores de Ciencia y Tecnología Iberoamericanos / Interamericanos*, Buenos Aires, RICYT/CYTED.
Vaccarezza, L., Polino, C., and Fazio, M.E. (2003b), "Medir la percepción pública de la ciencia en los países iberoamericanos. Aproximación a problemas conceptuales", en *Revista Iberoamericana de Ciencia, Tecnología e Innovación*, número 5 enero/ abril, ISSN: 1681-5645, OEI, España, http://www.campus-oei.org/revistactsi/numero5/articulo1.htm
Vaccarezza, L., López Cerezo, J.A., Luján, J.L., Polino, C., and Fazio, M.E. (2003c), "Indicadores iberoamericanos de percepción pública, cultura científica y participación ciudadana (2001–2002). Documento de base", Documento de Trabajo N°7, Buenos Aires, Centro REDES, http://www.centroredes.org.ar/template/template.asp?nivel=documentos&cod=00
Velho, L. (2005), "S&T institutions in Latin American and the Caribbean: An overview", *Science and Public Policy*, 32, 2, 98–108.
Vogt, C. and Polino, C. [Editors] (2003), *Percepción pública de la ciencia. Resultados de la encuesta en Argentina, Brasil, España y Uruguay*, FAPESP, LABJOR/ UNICAMP, OEI, RICYT/CYTED, San Pablo.
Wagner, W. (2007), "Vernacular science knowledge: Its role in everyday life communication", *Public Understanding of Science*, 16, 1, 7–22.
Wynne, B. (1995), "Public understanding of science", in S. Jasanof et al., *Handbook of Science and Technology Studies*, Thousand Oaks, Sage.
Ziman, J. (2000), *Real Science. What It Is, and What It Means*, Cambridge, Cambridge University Press.

11 The Science Culture Index (SCI)
Construction and Validation
Rajesh Shukla and Martin W. Bauer

CULTURAL INDICATORS

Science performance indicators have come a long way since early discussions of a 'science of science' (Price, 1963) and the launch of journals like *Scientometrics* in the late 1950s. Figures on R&D expenditure, manpower, patents, and high-tech balance of payments have become routine national statistics in many countries; numbers of publications and citations are widely used to evaluate research institutions and teams of researchers.

Science indicators became the remit of a few national and international actors seeking to standardise data and data collection; a history that is excellently charted by Godin (2005). Following precursors in government agencies of the US, UK and Canada that reach back to the First World War, UNESCO (since 1960) initiated indicators on 'science activities' (STA) and the NSF set the trend with annual 'science indicators' reports (first report in 1973). The OECD pursued a narrow agenda, first (in Frascati in Italy, 1963) focussing on R&D and scientific manpower, later adding high-tech balance of payments (1990), innovations and patents, human resources (in Canberra 1995), and opening its scope to general literacy attainment with PISA. The EC made an important contribution with its innovation surveys in the 1990s.

Most of these global efforts focussed on **'objective'** input and output measures. OECD standardises statistics from its more than thirty industrial member countries, its economist-bias favours data with a monetary metric, although the notion of 'intangible assets' (e.g. creativity, marketing) is in evidence. UNESCO's attempts are wider in coverage and scope (see for example the World Science Reports, e.g. UN 2003). On the other hand it battles with a lack of basic information in many countries. Africa and the Arab world are found lacking in basic statistical information (Butler, 2006).

'Subjective' culture such as the **climates of opinion, belief, attitudes, interests, engagements, values, images, and the 'semiophere' of science** were never an integrated part of these efforts at science indicators. UNESCO operated with the concepts of 'related science activities' (RSA) that included communication, education, exposition, product testing, and improvement.

Since 1971, NSF reports on "public attitudes" along with funding streams, education, PhDs, publications, patents, citations, and impacts. Similar attempts to report on 'public attitudes' in a regular or irregular manner have since proliferated (see Bauer, 2008). Recently, we have seen new momentum in national and international efforts to consider the intangible assets of a subjective culture of science. But only the US NSF (since 1979), FAPESP (2004) in Brazil, and NCAER in India (see Shukla, 2005) report objective and subjective indicators simultaneously.

Inonu (2003), among others, showed that **scientific production is poorly explained by economic facts** (e.g. GDP or GDP per capita in purchasing power parities). There are poor countries with rich science, and rich countries with poor science. Similar arguments can be found in research on democracy and development. The process of democratization across the world is not a direct function of economic power, i.e. the richer the more democratic. Research shows that civic culture, such as the 'emancipatory desire' for autonomy and self-expression, and the care for a public sphere, mediates the relation between economic power and democratisation (e.g. Welzel, 2006). Similarly, if scientific productivity is not explained by economic prowess, it is likely that the dispositions of citizens play a role.

Non-economic conditions need to be considered to understand a country's science base. The culture is both condition and catalyst of human activities. The 'subjective' side of human action, the dispositions, attitudes, imagination, and moods are not mere epiphenomena of objective structures. They explain how people act in contexts not of their own making. Cultural comparisons feature prominently in explaining the long-term path of developments both within the Western world and between Western and other world regions (see Berg and Bruland, 1998). The 'objective world' is not a good proxy for the 'subjective world', and vice-versa.

If scientific production is not only a matter of economic strength, we need to consider other factors that determine scientific culture, and for the purpose of comparative measurement we want to construct indicators of such a culture. The term 'cultural indicator' has several meanings. *First,* the term refers to the **creative sectors of** the economy, including design, architecture, advertising, cinema, arts, music, museums, and the production and consumption of its products and performances. Indicators report added value to GNP, share of employment, and export value (for example, the Work Foundation in the UK, 2007). Its growth is closely tied to the economic cycle (Chang and Chan-Olmstead, 2005). *Second,* UNESCO uses the term to report **cultural diversity,** which includes languages, religions, festivals, nature and heritage sites, museums, communication and translation efforts, and the consumption of cultural goods such as cinema, museum, and concert going (see http://www.unesco.org/culture/worldreport). *Third,* it is used to mean the records of local knowledge and traditional agricultural practices as for instance the 'Cultural Indicators for SARD' (e.g. sustainable agricultural development; FAO, 2003). *Fourth,* the term has a history in mass media effects research where it refers to the 'cultivation' research programme focussed on the mid-range power of the mass media to foster ideological worldviews using TV exposure (see Gerbner, 1969). Here 'culture' denotes the 'unrealistic world of television' as a driving force of everyday beliefs. *Fifth*, the term refers to the **'longitudinal content' analysis of print media** (for example Klingemann, Mohler and Weber, 1982) and distinguishes 'cultural' indicators from 'social' indicators. 'Social' indicators are **indicators of actions**, metrics that evaluate the success or failure of a social agenda. By contrast, 'cultural' indicators are **indicators for action**. They map contexts relevant for certain agendas, but not of one's own making (see Melischek, Rosengren and Stoppers, 1984; also Bauer, 2000). *Sixth,* the international assessment of mathematical literacy (TIMSS) characterises **cultures of mathematics:** the US focus on procedures, the French emphasise concepts, Swedes solve practical problems, and Germans can handle graphics and tables (for example Klieme and Baumert, 2001). *Finally,* there is the **subjective indicators movement** that gathered pace in the 1970s as it set up monitors of the 'subjective state of the nation'. The term also covers **'world value surveys'** with orientations of survival, life style, well-being, and happiness (see Inglehart, 1990 and http://www.worldvaluessurvey.org). Here, the inferences on change in perceptions must be based on large differences in order not to be misleading (Turner and Krauss, 1978).

CONSTRUCTION OF A COMPARATIVE PUS DATABASE

The efforts of measuring the subjective culture of science have come some way. The US NSF has published a comprehensive annual Science Indicators Report since the 1970s. However, hitherto there is little attempt to integrate public sentiment with the more objective indicators of science, either nationally or internationally. Nor is there an academic research stream to discuss this idea. Only recent late-comer activities point in this direction, one where both objective and subjective indicators are constructed jointly, for India (Shukla, 2005) and for Latin America (Polino et al., 2005; FAPESP, 2004). It seems that the two exercises, science indicators and indicators of public understanding, develop in parallel universes, the former widely institutionalised, the latter depending on fragile and shifting cooperations between civil servants and interested academic researchers.

We will explore the hypothesis that the 'culture of science' is an independent driver of scientific society. There might be two versions of this hypothesis, a weaker and a stronger formulation. The weaker one take the following mathematical form for the sake of brevity:

Model: SCI = STS + PUS ⇔ $a^*(STS) + b^*(PUS)$ + error

The model stipulates an additive function, a linear combination of the objective science base (STS) and perceptions of public understanding (PUS).

Empirically their contribution is weighted and retains a residual of unexplained variance. The model suggests that, in extreme cases when PUS = 0, science culture will be fully explained by the structural variance (STS), or in the absence of an objective science base, culture is explained entirely by perceptions. Or in a dynamic model, if either PUS or STS remains unchanged, this could still result in a change of science culture. Neither the science base nor PUS is necessary for science culture; they are logically substitutable.

Two survey efforts in India and Europe, by Eurobarometer in 2005 and NCAER 2004 (Appendix 11.1A), were partially coordinated and integrated at micro level. The common core of variables includes **engagement** (visits to science expositions and fairs), **interest** and **informedness** of science, nine items measuring **science literacy**, and seven **attitude** items, together with socio-economic information about the respondent such as **gender, age,** and level of **education**. These items had either exactly or functionally equivalent formulations. Scalar analysis of all these items allows us to construct the following scales with fairly reliable characteristics across the two populations: (i) Knowledge or literacy, (ii) Attitudes (AttA and AttBc), (iii) Interest, (iv) Confidence (admitted informedness), and (v) Engagement with expository science.

Attitude is compared on two different scales. This arises from the fact that EB63.1 includes a split-half design. One set of items (AttA) is comparable only to half of the EU sample; a second set (AttBc) is comparable to the other half. AttA includes the two items '*Science & technology are making our lives healthier, easier and more comfortable*' and '*Scientists should be allowed to do research on animals*'. AttBc includes the items '*New technology makes work interesting*' and '*Modern science and technology will create better opportunities for the next generation*'.

Plotting knowledge against each of these indicators allows us to examine the linearity and non-linearity of aggregate measures across the fifty-five units of analysis in Europe and India. For any statistical analysis these (linear and non-linear) relationships among the PUS indicators need to be addressed appropriately. In the context of construction of the Cultural Index, non-linearity suggests that we introduce a conditional transformation beyond a certain threshold level of knowledge. In order to apply principal component analysis, which is used in the index construction, a basic assumption is the linear relationship among component variables. Examining the linear or non-linear relations that 'knowledge' holds with 'attitude' (attA and AttBc) and 'engagement' the two latter variables were conditionally transformed to fulfil the linearity assumption before undertaking the suggested statistical framework (i.e. considering both main effects of indicators and their interactions).

THE SCIENCE CULTURE INDEX (SCI): METHODOLOGY AND BENCHMARKING

In the light of the above discussion, we have conceptualised a composite index named "Science Culture Index (SCI)" developed by combining

"STS" performance indicators as well as "PUS" indicators to determine the level of "Science culture" of thirty-two European countries and twenty-three Indian states. The index reflects the level of science culture of a country/state considering indicators related to input, output, and impact (public understanding) of STS progress and focusses on how well the country/state as a whole is participating in creating and sustaining a scientifically cultured society.

Quantified objective and subjective indicators were integrated in the construction of SCI to capture the **multidimensional nature of science culture** (Godin and Gingras, 2000). The relationships among these determinants, which themselves are composed of a number of subindices, are complex, mutually interacting, and multidirectional, so that each of the components is both a cause of change in others and an outcome of the influences of the latter. Figure 11.1 presents the conceptual framework of the SCI. Details of the indicators are given in Appendix 11.1.

What factors go into the complex interplay of SCI? This question was posed while selecting the indicators, as the objective is to construct an index that focusses more on cultural outcomes and achievements rather than on effort or inputs such as numbers of scientists, R&D expenditures, or policy environments. It was not easy to capture the interactions among the constituent parts of SCI in a single numerical figure. The choice of indicators and methodology assumes special significance in this regard. Therefore, the methodological challenge, which is the focus of this section, is how to put these complex concepts into operation.

The validity and reliability of all available indicators were assessed during the early stages of quantitative analysis, and a 'short-list' of

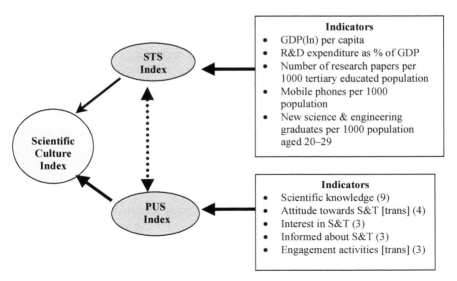

Figure 11.1 Conceptual framework of SCI.

indicators was produced. As for the "STS Index", an extensive literature survey was conducted to select possible indicators for inclusion in the framework. Regression analyses were carried out using a generalised linear model to find coefficients of these candidate indicators, capturing the strength of their relationship with a SCI made up of a Human Development Index (HDI), which served as a screening device. The selection process yielded the following five indicators: GDP per capita, R&D expenditure, scientific productivity, mobile phone use, and science education graduates per cohort.

We used five key variables in the calculation of the 'PUS Index', derived from the integrated data sets of Eurobarometer (EB63.1) and National Science Survey (India): science knowledge, attitude towards science, interest in S&T, level of informedness about S&T, and engagement with expository science.

Science knowledge comprised nine standard items—each judged as correct or incorrect—taking standard science wisdom as a baseline. Summing each individual's correct answers (score 1 for a correct answer, 0 for a wrong answer) gives his total score for science knowledge. We constructed a serviceable one-dimensional scale of science knowledge: the scores for both the EU and India samples are normally distributed, and results show an acceptable level of internal consistency.

Attitude towards S&T was quantified using four questions used to measure science attitude and were compared on two different scales. This arises from the fact that EB 63.1 includes two different versions of the questionnaire (split-half designs). One set of attitude items (AttA) is therefore comparable only to one half of the EU population, whereas a second set (AttBc) is comparable to the other half. Attitude A includes the following two items: 'science & technology are making our lives healthier, easier and more comfortable' and 'scientists should be allowed to do research on animals'. AttBc includes the following two items: 'new technology makes work interesting' and 'modern science and technology will create better opportunities for the next generation'. Scaling was done in same way as in the case of science knowledge thereby obtaining higher internal consistency between answers on different items within the scale for India than for the EU, particularly for AttA.

Considering the interaction effects and non-linearity between knowledge and attitudes, we incorporated the conditional transformation of attitudes into the PUS index as discussed in the previous section. All the results here reported are based on conditional transformed indicators for AttA and AttBc.

Interest in and informedness about S&T were measured by respondents' responses to questions about their interest in and informedness about issues including new science inventions and discoveries. The average score for both indicators were significantly higher for the EU than India.

Engagement in S&T activities was measured by asking three questions on a two-point scale: on visits to zoos/aquariums, museums, and exhibitions/science fairs. The scaled measure for the EU is 1.26, which is significantly higher than that of India (0.61). Like attitudes, engagement enters the index in a conditional transformation mode.

The average level of knowledge about science concepts was found to be very high in both groups. However, values for the three indicators: interest, information, and engagement were quite low for India. All indicators were highly correlated showing that they were appropriately chosen to measure PUS index.

VALIDATION OF THE INDEX

All three indices, namely, 'STS Index', 'PUS Index', and 'Science Culture Index' (SCI) were calculated for thirty-two EU countries and twenty-three Indian states. These indices are conceptualised as having a positive relationship with science's development in society. In other words, a higher value of SCI reflects a more scientifically cultured society. So, ranking of any country/state based on its index value gives an assessment of its relative performance to the whole sample.

The results show a huge disparities and diversities within as well as between the European countries and Indian states. For instance, the average value of SCI for the EU is 0.720 (ranging from 0.378 for Turkey to 1.000 for Sweden), which is significantly higher than India at 0.196 (ranging from 0.000 for Bihar to 0.459 for Chandigarh). A similar trend is observed in the case of the STS and PUS Indices. Variability among Indian states is much higher than that of EU countries for all three indices (Figure 11.2).

What are the relative contributions of different indicators to the sub-indices 'STS Index' and 'PUS Index'? To respond to this question, coefficients of the different dimensions of the 'STS Index' and 'PUS Index' were obtained. These coefficients make it possible determine the relative importance of corresponding dimensions in determining the ultimate index score. A straightforward rearrangement of the weighted components of two indices helps to express it as a weighted sum of the actual value of their constituent indicators. Thus,

$$STS_{Est.} = 0.065*\ln(GDP) + 0.069*R\&D + 0.035*Paper + 0.0002*Mobile + 0.012*SE\ Graduate$$

$$PUS_{Est.} = 0.0576*Knowledge + 0.1240*Attitude\ (A) + 0.1249*Attitude\ (B) + 0.0973*Interest + 0.0941*Informedness + 0.1186*Engagement$$

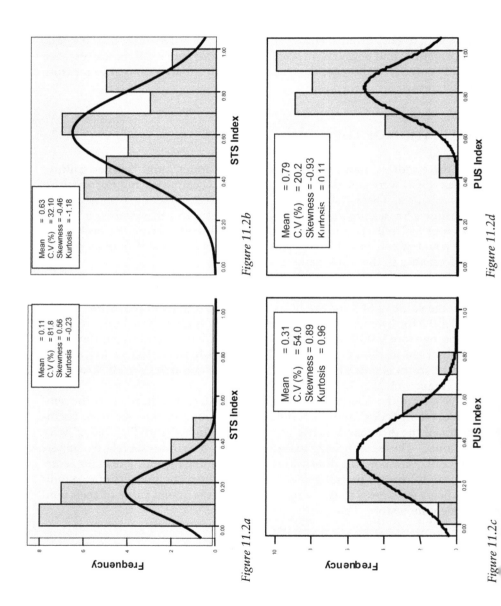

Figure 11.2a

Figure 11.2b

Figure 11.2c

Figure 11.2d

The Science Culture Index (SCI) 187

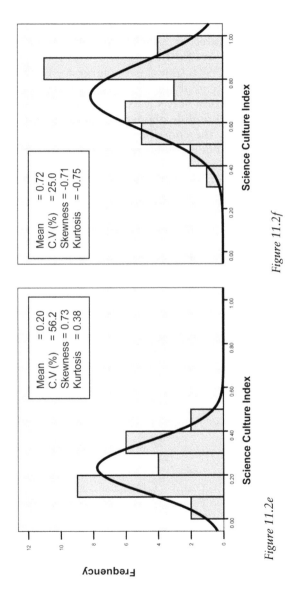

Figure 11.2e

Figure 11.2f

Figure 11.2 Estimates of PUS index, STS index, SC index for India-23 on the left and EU-32 on the right.

These coefficients, however, should not be interpreted as partial regression coefficients as the left-hand side variables are not observable. For instance, the coefficient of knowledge does measure the increase in PUS value per unit increase in knowledge assuming no change in other variables. The model can be used to compute the share of each dimension in the indices for each state/country and for the average value for the whole sample.

Table 11.1 presents shares of each component in the average 'STS Index' and 'PUS Index' scores for EU countries and India. The contribution of GDP per capita to 'STS Index' is the largest and explains almost 61 per cent and 81 per cent of the 'STS' score for the EU and India respectively. The contribution of R&D expenditure is the second highest followed by penetration of mobile phones for Indian states. However, in the case of EU countries, the contribution of R&D expenditure is the lowest. The second highest contributor in the EU is penetration of mobile phones followed by population of science education (SE) graduates.

The pattern of contribution of indicators to the 'PUS Index' score is similar for EU countries and India. However, it shows that the importance of knowledge, attA and attBc. is greater for India than EU countries. For instance, their contribution to 'PUS Index' is around 80 per cent for India as a group, but it falls to about 60 per cent for EU countries. In other words, these indicators played a much larger role in explaining the 'PUS Index' score for India than the EU.

Table 11.1 Percentage Contribution of Indicators to Subindices

Index	EU	India
STS Index		
GDP per capita (ln)	61.1	80.6
R&D expenditure	7.0	10.8
Publication (paper)	8.6	1.9
Mobile telephone	13.7	4.1
SE graduate	9.6	2.5
PUS Index		
Scientific knowledge	28.5	39.8
AttA	17.7	22.1
AttBc	14.8	18.8
Interest	14.2	11.6
Informedness	11.5	7.7
Engagement	13.1	12.4

All three indices showed significant positive rank correlation with the Human Development Index (HDI) and the Technology Advancement Index (TAI). However, the correlation of the HDI with SCI was slightly higher (0.90) compared to its correlation with STS (0.87) and PUS (0.78). This implies that a polity with a good science culture is bound to excel in health development too—Kerala, in India, being the well-known example. Not surprisingly, the HDI was found to be highly correlated with In(GDP) as well.

The entire sample of EU countries (32) and Indian states (23) has been grouped into four clusters[1] on the basis of SCI values to facilitate a proper understanding and meaningful interpretation. Twelve EU countries belong to the first cluster with SCI values greater than 0.80. The second cluster comprises fifteen EU countries, which is a step below with SCI values ranging between 0.56 and 0.80. The third cluster comprises those with SCI values between 0.22 and 0.56, which include five EU countries and nine Indian states. Finally, there is the cluster comprising the remaining fourteen Indian states with SCI values less than 0.22.

It is observed that the average per capita GDP of the 'leaders' (first cluster) and 'competent' (second cluster) are much higher than those of the two lower categories. They also spent a significantly higher percentage of their total GDP on R&D. The elite clusters also show higher scores for all PUS indicators than the two bottom clusters.

Regional disparity is a major concern for policy makers and such disparity is evident among the four categories of states in this research as well. Does it follow that states that perform better in terms of the STS indicators are also the ones to have a higher level of public understanding of science and, finally, a higher level of science culture? Or are there interstate differences, within each of the four categories, with respect to S&T and economic growth? To understand the complexities, index values for each constituent of the entire sample are standardised, taking the EU at 100, aiming to know how they differ in terms of their overall scientific development. Table 11.2 (see Appendix 11.2) gives the scores and corresponding ranking of the total sample grouped into four categories, which provides a relative assessment of a state/country's performance and could be considered as an indicator of changes over time.

Summary/average scores for the three indices in terms of the weighted scores[2] for all four clusters of state/country were computed. It was evident that 'leaders' and 'competents' accounted for much higher scores compared to 'potentials' and 'aspirers'—the top two clusters scored 114 and 99 on SCI, compared to scores of 53 and 22 for the two bottom clusters. A similar trend was observed in the cases of STS and PUS indices as well.

The Index also reveals that it does not hold that if a country/state scores high on the 'STS Index', its 'PUS Index' will be of equal ranking. Take the example of the United Kingdom, which ranks third in the 'STS Index', but in terms of 'PUS Index' and 'SCI' its rank is 13th and 6th. Though Luxembourg is ranked 13th on the 'STS Index', it is ranked 3rd on the 'PUS Index'.

Most Indian states at the bottom group of the table—in terms of all three indices—have more or less similar rankings and scores for all indices. However, some of the developed states such as Delhi, Kerala, Chandigarh, and Himachal Pradesh perform fairly well with respect to all three indices and rank closely to EU countries such as Poland, Hungry, Bulgaria, and Romania.

The highest SCI scoring countries tend to score uniformly high in indicator variables. Consequently, these countries display a relatively low variability among contributions of individual indicators. Variability is defined by the coefficient of variation. Variability increases as one moves down the list in decreasing order of SCI scores. The highest variability is found among the 'Aspirers' scores. This pattern is evident from Figure 11.3.

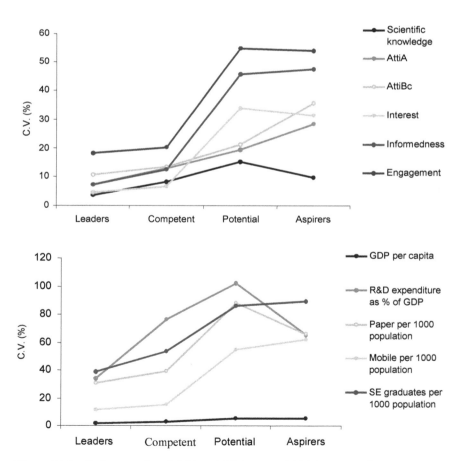

Figure 11.3 Estimates of variability—PUS (top) and STS inciators (bottom).

DISCUSSION

In this chapter we outlined the rationale and methodology of the construction of a composite index of science culture, the SCI. This is the first attempt to construct such an index with the **ambition of global validity**. Such an index should be able to depict the full range of variability across the globe. Several considerations went into the construction.

Science indicators have come a long way in definition and global standardisation since the 1960s, mainly under the auspices of international organisations like the OECD. However, these efforts focus entirely on objective indicators of input (R&D, number of personnel) and output (patents, innovation, balance of hi-tech trade) with an economist bias. They neglect the more subjective and symbolic features of a scientific culture, the attitudes, morale, imagination, and public sentiment vis-à-vis science and technology. Culture is both a precondition and a catalyst of science and technological development. **Objective structures and subjective meaning** are two sides of the same coin, to take one as a proxy of the other is cutting short a potentially complex reality of compensation, substitution, and potential functional equivalences.

The definition of subjective indicators such as literacy, attitudes, and interest has its own history, one that progresses institutionally isolated from mainstream science indicator developments. By 'subjective', we mean that data is collected at the level of individual interviews with respondents who are selected from a nationally representative sample. With few exceptions, subjective indicators are not part of the science indicator routine. Such data, if collected at all, is reported in different chapters without any cross-references to other science indicators. Our SCI seeks to re-open this discussion by combining both objective (STS input and output) and subjective indicators (PUS) into a meaningful **composite index: SCI = f [STS, PUS]**. Two versions of this model are possible: a 'strong', multiplicative or a 'weak', additive function for this index, with implications as to the relative importance of the objective and subjective element of culture. The SCI discussed in this paper is currently based on the 'weak' additive model of PUS and STS.

There is a global corpus of data emerging that addresses issues of knowledge and literacy, attitudes, interests, and engagement. For the moment, we are bracketing the partially valid critique addressed to these measures and offer an **innovative usage of these existing indicators**. We define an index of PUS using five subjective indicators: PUS = f [knowledge, attitudes-transformed, interest, informedness, engagement]. The higher PUS, the stronger the subjective science culture of a particular context.

The discussion of subjective indicators of science under the headings 'adult science literacy' and 'public understanding of science' has had its fair share of debate and polemic. In particular the relationship between knowledge and attitude has been controversial. The standard model suggests that knowledge is a driver of positive attitudes, but there is little evidence for this in general. More valid on a global scale is the two-culture dynamic model, which suggests that **knowledge and attitudes relate in a non-linear, inverted U-shape**. On lower levels of development, the standard model applies; on higher levels the correlation between knowledge and attitude is likely to be negative: the more we know, the more sceptical we tend to become. Enthusiastic attitudes are an asset in lower than average levels of literacy, but with higher than average levels of literacy, critical attitudes are the asset; the science ideal, of a 'community of sceptics' is generalised. Our SCI index thus makes productive use of PUS research over the last twenty-five years: attitudes enter the model after a **conditional transformation**. Above a certain level of literacy, negative attitudes score positively. A similar conditional transformation applies for 'engagement'. High engagement below a certain average level of literacy scores negatively.

For the construction of the index we built a database that **combines twenty-three Indian and thirty-two European states**. STS data (GDP, R&D, phones, education, science education) and PUS data (knowledge, attitude, interest, informedness, engagement) is aggregated and analysed for each of these fifty-five units of analysis. PUS data derives from the integration of the most recent PUS surveys, EB 63.1 of 2005 (EU, N=32,000) and the National Science Survey of 2004 (India, N=30,000) (Shukla, 2005). These fifty-five units cover a wide range of contexts, from very underdeveloped states in India to very developed countries in Europe. This allows us to simulate the global range of contexts and to validate the index within that range.

The SCI is constructed based on principle component analysis and standardised on a scale from 0 to 1. For the moment the index is a **relative measure:** the lowest unit is defined as 0, the highest unit as 1. In this definition, the index is not suitable for measuring changes in absolute level of science culture over time. It is, however, possible to assess changes in the rank order between the units over time.

The **reliability of the index** is tested through the internal consistency of each component (Cronbach's Alpha) and principle component loading. The intercorrelation between the single indicators is sufficiently high for both STS and PUS to justify their combination into an index. Indeed, it was also shown that the conditional transformation of attitudes and engagement scores improves the internal consistency of the PUS index. STS and PUS are correlated ($r = 0.86$), but make an independent contribution to SCI. Both PUS and STS are highly correlated with the combined SCI (rank correlations 0.95 and 0.98 respectively). This shows that ranking of the countries for science culture will be more or less the same whether it is based on the composite index or sub-indices. But in view of global applicability, the composite index is to be preferred.

For each subindex, STS and PUS, the **relative contribution** of each indicator is calculated. This gives a clearer sense of what drives the index, and that the drivers have a slightly different weight in India and in Europe. STS is dominated by GDP per capita, the other components making a much smaller contribution. For PUS the weights are more evenly distributed, knowledge being the most important input in both contexts. These different columns show that some countries' overall standing on the SCI is raised or lowered by their standing on PUS relative to STS. For example the UK, 3rd on STS drops to 6th on SCI, because of being 13th on PUS. Scores for Italy and Austria behave similarly. On the other hand, Luxembourg, Norway, and the Netherlands, for example, improve their standing on SCI with a good PUS score relative to their STS.

We validated all indices on existing measures of development, such as the HDI and the TAI. The positive correlation among them suggests that they measure some common aspect of 'development'.

We also grouped the fifty-five units into four **subgroups of regions** or countries and provided a differential profile for each group. The four groups differ in the variability of the indicators. Generally we observe the higher the SCI, the less variability on the component indicators. These profiles are **diagnostic** and suggest that, both to increase the level and to reduce the variability of the SCI and its components, they should be targets of policy interventions; depending on the level of SCI in a region, a different intervention might be indicated.

A final consideration might be the **presentation of the SCI index**. There are different bases for comparison. Results show versions of the SCI relative to different bases, taking as a baseline for comparison the EU or India, or the overall mean or median. The ranking among units is not affected by this variation, but the presentation of the index might be. What changes is the range between the lowest and the highest of index scores. Here the **rhetoric of numbers** might come into play, and suggest a particular choice among these alternatives for particular purposes.

With this research we hope to re-open and stimulate the discussion on the place of public understanding of science, the subjective side of the culture of science, within the science indicator enterprise. And, instead of brushing aside existing measures of PUS as 'snow of old days' we suggest a new and innovative way of using existing data, which might also suggest future uses.

What is needed in the near and mid-term future is a more global collaboration, bringing existing data into comparisons of this kind to make it speak as a diagnostic tool. A comparison of India and Europe, with a vast divergence of contexts, is clearly a step in the right direction. In a globalised world, science is not just a matter of and for developed countries.

Too much of the existing data on PUS is underexploited and remains hidden in often unknown places once it has been presented as the 'latest news' in the mass media and created a buzz for researchers at the time. The existing PUS data clearly has more value than just hitting the news headline of the day, although that might be important for the individual researcher and the PUS enterprise as a whole. What is needed is a considered and concerted approach that collects, integrates, and analyses data on a global scale, and develops their diagnostic power through systematic comparison. SCI is our opening gambit for a more considered and technical discussion, so that in the near future we might be in the fortunate position to evaluate and compare several alternative indices with a similar ambition: to assess science culture, its objective and subjective features, with a globally reliable and validated instrument. There is still some way to go.

APPENDIX 11.1: DATA SOURCES

A. PUS Indicators

India (National Science Survey, NCAER, 2004)

"National Science Survey 2004" was an all-India field survey undertaken by the National Council of Applied Economic Research, New Delhi. Sample respondents, individuals over 10 years of age, were selected by adopting a multistage stratified random sampling design from a wide cross-section of people (age, education, and sex) in the country. In view of India's diversity in terms of languages and locations, the sample size and selection procedure were designed to provide state level estimates.

Respondents were selected from the entire country by covering both rural and urban areas, with the objective of enhancing the precision of the estimates. The rural sample was selected from a representative number of districts from across the country, and the urban sample was selected from big metropolitan cities to small towns with populations below 5,000. A total of about 347,000 individuals (115,000 rural and 232,000 urban) were listed, covering 553 villages in 152 districts as rural and 1128 urban blocks in 213 towns as urban. Over 30,000 individuals were selected from the listed individuals to collect detailed information through a questionnaire approach involving face-to-face interviews. For detailed survey methodology, please refer to Appendix III of the **India Science Report**, available at *www.insaindia.org/India%20Science%20report-Main.pdf*

Europe (Eurobarometer 63.1, 2005)

The special Eurobarometer N⁰224 is part of wave 63.1 and covers the population of the European Member States (EU-25), aged 15 years and over, EU candidate countries (Bulgaria, Romania, Croatia, and Turkey), and EFTA countries (Iceland, Norway, and Switzerland). The sample design in all states is multistage probability based. In each country, a number of sampling points was drawn with probability proportional to population size and density.

In order to do so, the sampling points were drawn systematically from each of the "administrative regional units", after stratification by individual unit and type of area. In each of the selected sampling points, a starting address was drawn, at random. Further addresses were selected as every nth address by standard "random route" procedures from the initial address. In each household, the respondent was drawn at random (following the "closest birthday rule"). All interviews were conducted face-to-face in people's homes and in the appropriate national language. *Computer Assisted Personal Interview (CAPI)* was also used in a few countries where this technique was available.

B. S&T Performance Indicators

- GDP per capita ($PPP)
  EU: OECD key figures, 2004
  India: Economic Survey, 2004/05

- R&D expenditures as % of GDP
  EU: Public and Business expenditure as reported in EIS 2005 for the year 2003; it is the simple addition of these components.
  India: Share R&D expenditure for covered states were estimated using total R&D expenditure available in the State Sector R&D expenditure, DST. It is matched with national share (0.78%).

- Mobile phones (penetration per 1,000 population)
  EU: CIA statistics
  India: National Survey of Household Income and Expenditure (NCAER, 2004/05)

- Scientific publications by 1000 tertiary educated population
  EU: Total numbers of SCI articles for 1999: E Inonu (2003), The influence of cultural factors on scientific production, Scientometrics, 56 (1), 137–146. Tertiary educated population per 1,000 population aged 25–64, EIS 2005.
  India: Publication for 1998: Basu A and R Aggarwal (2006) India Science Literature in Science Citation Index: A report. http://itt.nissat.tripod.com/itt0104/scirep.htm. Tertiary educated population per 1,000 population aged 25–64: National Science Survey, NCAER, 2004.

- New science and engineering graduates 20–29 per 1,000 population
  EU: European Innovation Scoreboard (EIS) 2005
  India: National Science Survey 2004 (NCAER)

APPENDIX 11.2

Table 11.2 The Index Scores for STS, PUS and SC to the Basis of EU=100

Grouping	States	STS Index		PUS Index		SC Index	
		Scores	Rank	Scores	Ranks	Scores	Ranks
Leaders	Sweden	149	1	114	6	132	1
	Finland	137	2	111	8	124	2
	Switzerland	125	4	120	1	123	3
	France	122	6	120	2	121	4
	Denmark	123	5	112	7	118	5
	United Kingdom	125	3	104	13	114	6
	Luxembourg	103	13	120	3	112	7
	Netherlands	102	15	116	4	109	8
	Austria	114	9	103	14	108	9
	Italy	121	7	94	23	107	10
	Norway	103	14	110	9	107	11
	Belgium	103	12	109	10	106	12
Competent	Ireland	117	8	95	20	106	13
	Germany	104	11	107	12	105	14
	Iceland	110	10	101	15	105	15
	Slovenia	91	17	115	5	103	16
	Czech	88	19	108	11	98	17
	Greece	94	16	100	16	97	18
	Spain	89	18	91	26	90	19
	Slovakia	69	23	99	17	85	20
	Estonia	72	22	93	24	82	21
	Croatia	65	24	99	18	82	22
	Portugal	80	20	82	29	81	23
	Cyprus	64	25	94	22	79	24
	Lithuania	79	21	78	31	78	25
	Latvia	56	29	94	21	75	26
	Malta	57	28	92	25	75	27
Potential	Poland	60	27	88	28	74	28
	Hungary	49	32	95	19	72	29
	Bulgaria	51	31	74	32	62	30
	Romania	46	35	79	30	62	31

(continued)

Table 11.2 (continued)

Grouping	States	STS Index Scores	Rank	PUS Index Scores	Ranks	SC Index Scores	Ranks
Potential (continued)	Chandigarh	32	40	89	27	60	32
	Delhi	47	33	65	35	56	33
	Kerala	35	38	70	33	52	34
	Himachal Pradesh	54	30	50	38	51	35
	Turkey	46	34	54	36	50	36
	Karnataka	26	43	68	34	47	37
	Pondicherry	36	37	50	37	43	38
	Punjab	36	36	47	40	41	39
	Uttaranchal	62	26	21	54	40	40
	Haryana	30	42	39	42	34	41
Aspirers	Madhya Pradesh	12	48	47	39	29	42
	Maharashtra	23	45	35	43	28	43
	Tamil Nadu	24	44	33	44	28	44
	Assam	32	39	22	53	26	45
	Gujarat	30	41	23	51	26	46
	Uttar Pradesh	4	53	46	41	25	47
	Andhra Pradesh	23	46	23	52	22	48
	West Bengal	15	47	29	48	21	49
	Jharkhand	9	49	32	45	20	50
	Chattisgarh	8	50	30	46	19	51
	Orissa	8	51	30	47	18	52
	Rajasthan	6	52	24	49	14	53
	Meghalaya	1	55	23	50	11	54
	Bihar	2	54	19	55	8	55
EU (All 32 countries)		100	-	100	-	100	-
India (All States)		17	-	35	-	25	-

NOTES

1. Leaders (SCI >0.80), Competent (0.56 < SCI 0.80), Potential (0.22 < SCI 0.56), and Aspirers (SCI 0.22)
2. Since EU countries and Indian states have different populations, weighted scores are calculated by taking population into consideration. For example,

to calculate the weighted SCI score for 'Leaders', the SCI score of each constituent of the group is multiplied by its corresponding population and the weighted average is taken.

REFERENCES

Allum, N, D Boy, and MW Bauer (2002) European regions and the knowledge deficit model, in Bauer MW & G Gaskell (eds) *Biotechnology—The Making of a Global Controversy*, Cambridge, CUP.

Allum, N, P Sturgis, D Tabourazi, and I Brunton-Smith (2008) Science knowledge and attitudes across cultures: A meta-analysis, *Public Understanding of Science*, 1.

Bauer M (2008) Survey research on public understanding of science, in Bucchi M and B Trench (eds) *Handbook of Public Communication of Science and Technology*, Oxon, Routledge, 111–130.

Bauer M (2000) Science in the media as a cultural indicator: Contextualising surveys with media analysis, in Dierkes M and C vonGrote (eds) *Between Understanding and Trust: The Public, Science and Technology*, Amsterdam, Harwood Academic Publishers, 157–178.

Basu A and R Aggarwal (2006) India Science Literature in Science Citation Index: A report http://itt.nissat.tripod.com/itt0104/scirep.htm

Berg M and K Bruland (eds) (1998) *Technological Revolutions in Europe, Historical Perspectives*, Cheltenham, Edward Elgar.

Butler D (2006) The data gap: Statistics on science investment and performance are lacking across the Muslim world, *Nature*, 444 (2 Nov), 26–27.

Chang BH and SM Chan-Olmstead (2005) Relative constancy of advertising spending, *Gazette*, 67 (4), 339–357.

EIS (2005) European Innovation Scoreboard, Brussels.

FAO (2003) Final report on an indigenous people's initiative to establish cultural indicators for SARD, Rome, FAO. [ftp://ftp.fao.org/SD/SDA/SDAR/sard/Cult_ Indic_FAO_REPORT_FINAL.pdf]

FAPESP (2004) Science Indicators, São Paulo, [see chapter 12 on public perceptions]. [http://www.fapesp.br/english/materia.php?data[id_materia]=463

Gerbner, G. (1969) Towards 'cultural indicators': The analysis of mass media mediated message systems, *AV Communication Review*, 17, 137–148.

Godin B (2005) *Measurement and Statistics on Science and Technology—1920 to the Present*, London, Routledge (vol 22 of series).

Godin B and Y Gingras (2000) What is science and technological culture and how it is measured? A multi-dimensional model, *Public Understanding of Science*, 9, 43–58.

Inglehart A (1990) Culture Shift in Advanced Industrial Society, Princeton, NJ, PUP.

Inonu E (2003) The influence of cultural factors on science production, *Scientometrics*, 56 (1), 137–146.

Klein LR and S Ozmucur (2002/2003) The estimation of China's economic growth, *Journal of Economic and Social Measurement*, 62 (8), 187–202.

Klieme E and J Baumert (2001) Identifying national cultures of mathematics education: Analysis of cognitive demands and differential item functioning in TIMSS, *European Journal of Psychology of Education*, 16 (4), 385–402.

Klingemann, HD, PP Mohler, and RP Weber (1982) Cultural indicators based on content analysis: A secondary analysis of Sorokin's data on fluctuations of systems of truth, *Quality and Quantity*, 16, 1–18.

Melischek G, KE Rosengren, and J Stoppers (eds) (1984) *Cultural indicators: An international symposium*, Vienna, Austrian Academy of Sciences.

NSF (2006) Science Indicators Report, Washington, NSF.
OECD (2004) *S&T Statistical Compendium 2004*, Paris, OECD.
Polino C, ME Fazio, and J Castelfranchi (2005) Surveys on the public perception of science in Ibero-American countries, the RICYT's experience, paper presented at AAAS meeting, Washington, 17–21 Feb 2005.
Price DJ de Solla (1963) *Little Science, Big Science*, New York, Columbia University Press.
Shukla, R. (2005) *India Science Report: Science education, human resources and public attitudes*, Delhi, NCAER.
Shukla R and MW Bauer (2007) The science culture index (SCI): construction and validation. A comparative analysis of engagement, knowledge and attitudes to science across India and Europe—a concept paper, Delhi and London, NCAER & LSE, September 2007.
Turner CF and E Krauss (1978) Fallible indicators of the subjective state of the nation, *American Psychologist* (May), 456–470.
UN (2003) Report on statistics of science and technology, UN Economic and Social Council, E/CN.3/2004/15.
Welzel C (2006) Democratisation as an emancipative process: The neglected role of mass motivations, *European Journal of Political Research*, 45, 871–896.
Work Foundation (2007) *Staying Ahead: The Economic Performance of the UK's Creative Industries*, London, Work Foundation [www.work.foundation/].

12 A Comparative View on Adolescents' Attitudes towards Science[1]

Svein Sjøberg and Camilla Schreiner

The position of science and technology (S&T) in a society changes through time and from one society to another. In developing countries, many young people would like to opt for a career in S&T, whereas many rich, highly developed countries notice declining recruitment of students to science and technology studies. "Europe needs more scientists!" is the title of the final report from a large EU project addressing the condition of science and technology in the EU, which gives special attention to the number of people entering S&T educations and careers (EU, 2004). The title of the report reveals the point: falling recruitment to most S&T educations is seen as a big problem in most European countries. The same tendencies are noted in the US (NSB, 2008) and in most other OECD-countries. There is a great political concern about the decline in S&T recruitment in nearly all OECD countries (OECD 2006).

The lack of relevance of the S&T school curriculum is seen as one of the greatest barriers to good learning and as the reason for young people's low interest in the school subject and lack of motivation for pursuing the subject in their higher education. ROSE, The Relevance of Science Education, is an international comparative project meant to shed light on affective factors of importance to the learning of science and technology. The target population is students towards the end of secondary school (age 15). The research instrument is a questionnaire mostly consisting of closed questions with 4-point Likert scales. The rationale behind the project, including the questionnaire development, theoretical background, procedures for data collection etc., is described in Schreiner and Sjøberg (2004) and is available in print from the authors or from the project website.[2] In this article, we will present the ROSE project and include a few general results from analysis of the data material.

ROSE IN BRIEF

The key feature of ROSE is to gather and analyse information from the learners about several factors that have a bearing on their attitudes to S&T

and their motivation to learn S&T. Examples are: A variety of S&T-related out-of-school experiences, interests in learning different S&T topics in different contexts, prior experiences with and views on school science, views on and attitudes towards science and scientists in society, future hopes, priorities and aspirations, as well as young people's feeling of empowerment with regard to environmental challenges, etc.

ROSE has, through international deliberations, workshops and piloting among many research partners, developed an instrument that aims to map out attitudinal or affective perspectives on S&T in education and in society as seen by 15-year-old learners. The ROSE advisory group comprises key international science educators from all continents.[3] We have tried to make an instrument that can be used in widely different cultures. The aim is to stimulate research cooperation and networking across cultural barriers and to promote an informed discussion on how to make science education more relevant and meaningful for learners in ways that respect gender differences and cultural diversity. We also hope to shed light on how we can stimulate the students' interest in choosing S&T-related studies and careers—and to stimulate their life-long interest in and respect for S&T as part of our common culture.

About forty countries are taking part in ROSE, and many more have shown an interest in the project. The ROSE instrument is used for many different educational purposes in these countries. ROSE research partners (persons and institutions) were "recruited" through international networks and organizations for science education research and have met at conferences such as ESERA (European Science Education Research Association) and IOSTE (International Organization for Science and Technology Education), and special ROSE workshops have been hosted in several European countries and in Malaysia. The data from the following thirty-four countries were found to meet the criteria for data quality, and are included in the comparative analysis: Austria, Bangladesh, Botswana, Czech Republic, Denmark, England, Estonia, Finland, Germany, Ghana, Greece, India (Gujarat), Iceland, India (Mumbai), Ireland, Japan, Latvia, Lesotho, Malaysia, N Ireland, Norway, Philippines, Poland, Portugal, Russia, Scotland, Slovenia, Spain, Swaziland, Sweden, Trinidad, Turkey, Uganda, Zimbabwe. In most countries the target population is the whole national cohort, but in some countries the ROSE target population is defined as the students in one *region* of the country (e.g. Karelia in Russia, Gujarat in India and the central region in Ghana). In addition, many countries (e.g. Brazil, Taiwan, Italy, France and Israel) have published national reports, although their data have not been incorporated in the international data file.

The participating researchers in different countries were requested to apply random sampling methods as described in a handbook that was developed by the Norwegian organisers in cooperation with the international advisory team. For various reasons, e.g. due to limited financial resources, some countries have not been able to comply with the request. This means that not all of the forty participating countries have samples that without reservation can be regarded as representative for their 15-year-old students.[4]

ROSE has been supported by *The Research Council of Norway, The Ministry of Education in Norway, The University of Oslo*, and the newly established *National Centre for Science Education*. Industrialised countries have covered their own expenses, and some funding for data collection was provided for developing countries and countries with less available resources. Participation in the project has in many countries led to the release of local funding for the participants.

The ROSE material may illuminate a range of important and topical discussions in the science education community, for example issues such as curricular content vs. students' interests, cultural diversity, students' disenchantment with their science classes and students' perceptions of science in society and gender differences. Discussions on such issues have taken place in many papers and conference presentations based on the ROSE material (see e.g. Jenkins, 2005; Jidesjö & Oscarsson, 2004; Lavonen, Juuti, Uitto, Meisalo & Byman, 2005; Ogawa & Shimode, 2004; Trumper, 2004). About ten PhD students and several Master students are basing their theses on ROSE data. The first PhD thesis based on ROSE was presented in Norway (Schreiner, 2006), the second in Ghana (Anderson, 2007).

In the following, we will report some results from analysis of the ROSE material. All diagrams show mean scores for 14–16-year-old girls and boys from a number of countries in the ROSE sample. The countries are sorted partly geographically, with neighbouring countries together, and partly by level of development, using the Human Development Index as a proxy.

The Likert scales have four response categories, and the response categories vary from one question to another. Question groups A, C and E have the heading: *What I want to learn about*. These questions are inventories of possible topics to learn about, each with a 4-point scale. The extreme categories in the scale are labeled *Not interested* (coded 1) and *Very interested* (coded 4). It is a rather lengthy question with a total of 108 items. In order to avoid fatiguing the students, the items were put into three groups: question A, C and E; from now on referred to as *question ACE*.

Other questions have included lists of statements, and the students were asked to indicate on a 4-point scale whether they *Disagree* (coded 1) or *Agree* (coded 4). In the following graphs, this scale is collapsed by presenting the code 3 and 4 as "agree" and the responses are given in percent of the total.

SIMILARITIES BETWEEN COUNTRIES

In question ACE we ask the students to indicate how interested they are in learning about a variety of topics. One underlying hypothesis for this question is that in spite of few students choosing S&T educations and careers, and in spite of research findings that many students do not like school science, many young people find aspects of S&T interesting. The

A Comparative View on Adolescents' Attitudes towards Science 203

ACE question provides empirical data on what topics groups of students are interested in learning about. This insight can inform our discussions on how S&T curricula can be constructed in order to meet the interests of different groups of learners. Asking the students how interested they are in various topics is one approach for getting in touch with science lessons' potential for engagement.

We do of course not argue that science curricula should be determined from student opinion polls on what they find interesting. But, on the other hand, we believe that the teaching of school science has the potential to enliven, motivate, enrich, engage and inspire the students. To achieve this, we need to be aware of the interests, hopes and priorities of the learners.

For exploring similarities between countries in the ACE items, hierarchal cluster analysis is a useful explorative statistical tool. Results from the hierarchical cluster analysis can be presented in *dendrograms*. The dendrogram in Figure 12.1 below shows how similar or close the countries and country clusters are to each other: The branches illustrate how clusters are formed at different stages in the analysis and the distances between the clusters.

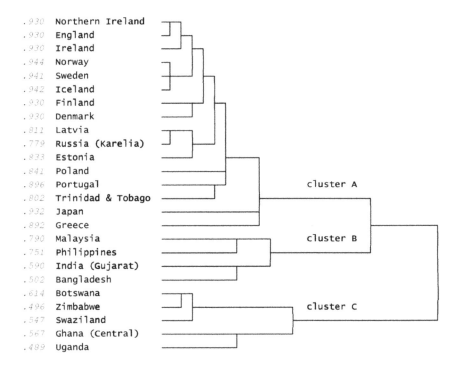

Figure 12.1 Hierarchical cluster analysis of residual ACE mean scores for all countries. Proximity measure: squared Euclidean distance; clustering method: between-groups linkage. To the left, we have inserted a column showing the national HDI (Human Development Index) values (UNDP, 2004). (Source: Schreiner, 2006)

The distance along the horizontal axis from the point at which the clusters come into existence to the point at which they aggregate into a larger cluster represents the distinctness of the clusters. The distinctness tells us how different one cluster is from its closest neighbour. The more compact a cluster is, i.e. the further to the left the branches merge, the more similar to each other the countries are.

Annually, the United Nations Development Programme (UNDP) publishes a Human Development Report (HDR). In each HDR, the countries are ranked according to a Human Development Index (HDI). The index monitors average national achievement in three dimensions of human development: income, education and health.[5] In this article, the HDI-value (based on data from 2004) will be used as an indicator for the level of development in a country. To the left in Figure 12.1, we have inserted a column showing the national HDI values.

By reading the dendrogram from the left towards the right, we see that the meta-cluster contains three main clusters: (A) High HDI countries including all the European countries plus Japan and Trinidad and Tobago, (B) Medium HDI Oriental countries and (C) Low HDI African countries. As the length of the branches for all these three clusters are relatively long, they can be perceived as three distinctive clusters of countries. Cluster B is more similar to cluster A than cluster C is.

One noticeable result from the analysis above is that similarities between countries in this part of the questionnaire seem to be determined by two properties: geographical closeness and level of development. The general pattern is that first the countries merge with geographically neighbouring countries, and next, groups of neighbouring countries merge with groups of countries having comparable levels of development.[6] But the unifying effect of geographical closeness only works within a certain limit of diversity in development. For example, Japan is geographically closer to the Philippines and Malaysia than to Europe, but the Japanese students seem to have more interests in common with European students. This may possibly be explained by the relatively high level of development and industrialisation in Japan. The response profiles of students in the Oriental countries (like Malaysia, Philippines, India and Bangladesh) appear as relatively similar to each other. We should note that the Russian students' orientation towards science and science education appear comparable to the profiles of the students in the Baltic countries (Latvia and Estonia). Keep in mind that the Russian students in ROSE come from Karelia, a region quite close to the Baltic countries and Finland.

MOST YOUTH APPRECIATE S&T IN SOCIETY

A possible explanation for young people's lack of interest in studying S&T could be that they hold a negative view of the role that S&T play in society,

A Comparative View on Adolescents' Attitudes towards Science

and that they blame S&T for the unintended catastrophes and risks (e.g. the Chernobyl disaster in 1986, Bovine Spongiform Encephalopathy or "mad cow disease", the depletion of the ozone layer, global warming and overpopulation) following in the wake of the technological development (Beck, 1998; Sjøberg, 2009).

Contrary to such expectations, the ROSE results indicate that youth express a positive view of S&T. Average scores for girls and boys in nearly all countries show strong agreement with statements like these:

- *Science and technology will find cures to diseases such as HIV/AIDS, cancer, etc.,*
- *Science and technology are important for society,*
- *Thanks to science and technology, there will be greater opportunities for future generations,*
- *New technologies will make work more interesting,*
- *The benefits of science are greater than the harmful effects it could have,* and
- *Science and technology make our lives healthier, easier and more comfortable.*

Figure 12.2 illustrates this with one example. The diagram shows responses to *Science and technology are important for society*. In average, girls and boys in all countries agree that S&T are important for society, and the gender differences are negligible.

In general, youth in developing countries are very positive, whereas young people in some of the wealthy countries are more hesitant. Gender differences are rather small in most countries on this question.

The response is rather different for the statement *The benefits of science are greater than the harmful effects it could have*, shown in Figure 12.3. Here we note that the responses are much less positive in wealthier countries, and that the gender difference is considerable, girls are much more skeptical than the boys. The most remarkable results are, however, the responses from Japanese youth. Also, on other questions of the same nature, they indicate a considerably higher degree of skepticism towards the role of S&T in society than youth from other countries.

The rather negative or reluctant attitudes of Japanese youth towards S&T has gained considerable attention, for example in the Japanese newspaper *Ashanti* in December 2004. ROSE data with a focus on Japan were also presented at the Science Agora in Tokyo in 2008.[7]

A FUTURE JOB IN SCIENCE OR TECHNOLOGY?

As mentioned, there is a widespread concern about recruitment to the S&T sector. Many questions address this issue, and here we only present

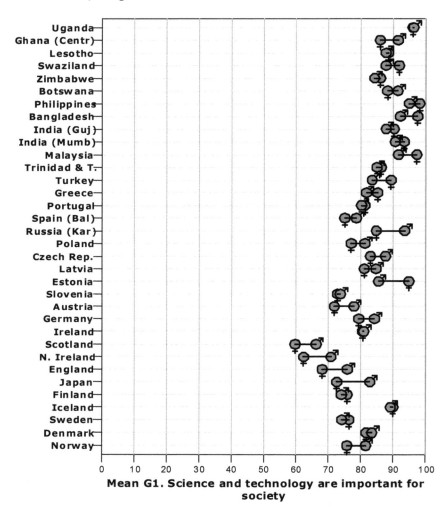

Figure 12.2 Science and technology are important for society. Percentage who "strongly agree" and "agree" for boys (male symbols) and girls (female symbols). Countries are sorted partly by level of development (HDI), partly by geographical proximity.

results from single items on this issue. We see from Figure 12.4 that there are large cross-national differences when it comes to students' agreement with the statement *I would like to become a scientist*. The mean scores in the developed countries are extremely low, and the girls are even more negative than the boys. Japan has particularly large gender differences.

Responses to the item *I would like to get a job in technology* are illustrated in Figure 12.5. Also in this diagram, we note pronounced differences

A Comparative View on Adolescents' Attitudes towards Science   207

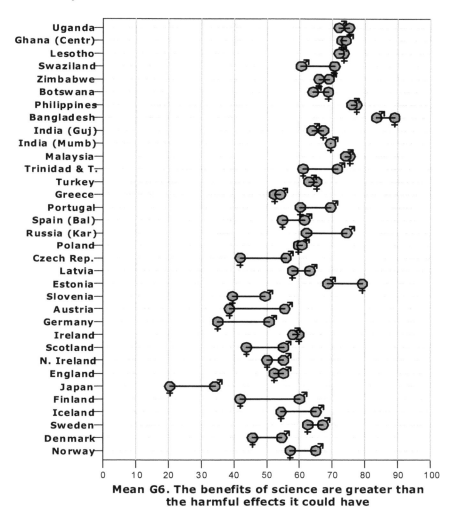

Figure 12.3 The benefits of science are greater than the harmful effects it could have. Percentage who "strongly agree" and "agree" for boys (male symbols) and girls (female symbols). Countries are sorted partly by level of development (HDI), partly by geographical proximity.

between countries and between girls and boys in each country. Boys in more developed countries give average scores close to the neutral value, but most girls in these countries do *not* want to work with technology. In developing countries, both girls and boys agree with the statement. Also in these countries, there are some gender differences, but they are by no means as large as in the developed countries.

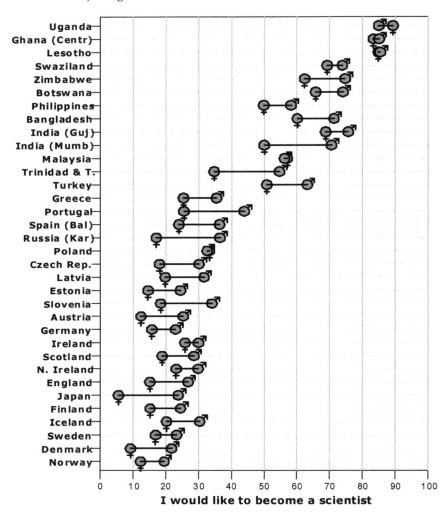

Figure 12.4 I would like to become a scientist. Percentage who "strongly agree" and "agree" for boys and girls. Percentage who "strongly agree" and "agree" for boys (male symbols) and girls (female symbols). Countries are sorted partly by level of development (HDI), partly by geographical proximity.

ORIENTATIONS TOWARDS S&T: LINKED TO THE LEVEL OF DEVELOPMENT IN A COUNTRY

Young people's values, views and ways of understanding themselves, their surroundings and the world are products of the culture in which they are

A Comparative View on Adolescents' Attitudes towards Science

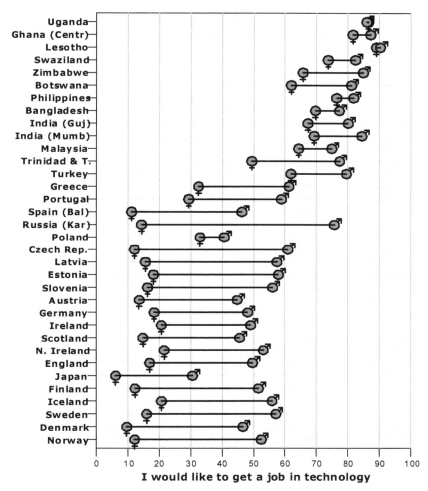

Figure 12.5 I would like to get a job in technology. Percentage who "strongly agree" and "agree" for boys and girls. Percentage who "strongly agree" and "agree" for boys (male symbols) and girls (female symbols). Countries are sorted partly by level of development (HDI), partly by geographical proximity.

growing up. Our data show a strong relationship between the HDI for a country and the responses in the ROSE questionnaire. For example, the national average score across all items in the ACE question (*What I want to learn about*), the Pearson product moment correlation coefficient with HDI is -.85 (p < .01), which indicates a very strong inverse relationship: The higher the level of development in a country, the lower the interest the students express in learning about S&T-related topics, although there are some interesting outliers.

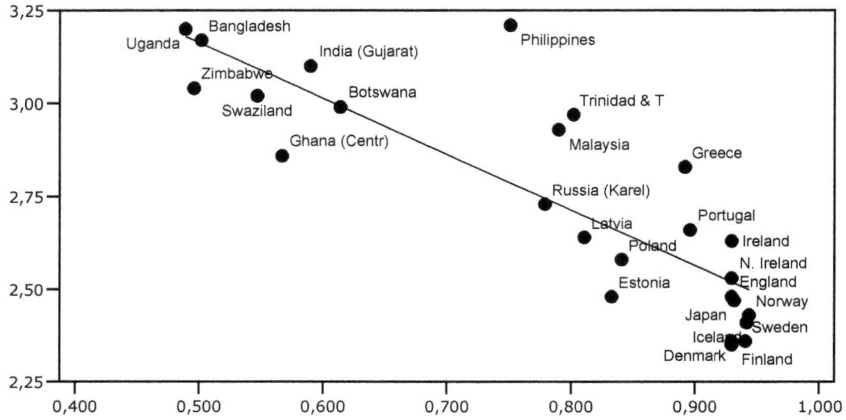

Figure 12.6 Country means of interest in learning science vs. level of development measured by Human Development Index (HDI). Scatter-plot with regression line: HDI values (horizontal axis) and the national average score across all items in the ACE question (vertical axis) for all countries. Pearson product moment correlation coefficient is -.85.

Figure 12.6 shows that the ACE mean for some of the countries in the extreme low end of the HDI scale lies one unit (in a scale with a span of three units) above many countries in the extreme high end of the HDI scale. In most of the ACE items, students in countries like Uganda and Bangladesh express much more interest in learning about the topics than students in more developed countries like Norway, Iceland, Finland and Japan. In the Philippines, youth also express a general high level of interest.

The correlation coefficients for the four questions reported in Figures 12.2 to Figure 12.5 are given in Table 12.1. The table indicates the general pattern appearing from most analysis of the ROSE material: The more developed a country, the less positive young people are towards the role of S&T in society.

Table 12.1 Pearson Product Moment Correlation Coefficient of Mean Agreement with Statements with HDI ($p < .01$)

I would like to become a scientist	-0.94
I would like to get a job in technology	-0.91
Science and technology are important for society	-0.78
The benefits of science are greater than the harmful effects it could have	-0.73

We see from our data that whether young people wish to opt for a career in S&T is closely related to the country's level of development. An important challenge in poor countries is of course related to the betterment of material conditions and economic growth and to the improvement of health and the welfare system. Further material development of the society is naturally a main political and public issue, and in this respect S&T are seen as fundamental driving forces. One may assume that in such societies, a job in S&T is perceived as important for society and thereby as meaningful for the individual.

When today's modern societies were in the era of early industrialisation, the focus was directed towards *progress, growth* and *building the country*. Consequently, exactly this—to build the country—was perceived as important for the society and meaningful for the individual. It may be that we now have passed the era in which the work of physicists, technicians and engineers are seen as crucial for people's life and well-being in the more developed countries. Other studies also indicate that in poorer countries, young people have a rather heroic image of scientists as persons, whereas this is not the case in highly developed Western societies (Sjøberg, 2002). In modern societies, neither scientists nor engineers are heroes or attractive role models for the young generation.

Obviously, the level of development influences people's expectations of the benefits of developments in S&T (Sicinski, 1976). The Eurobarometer (EU, 2005) also shows that belief in the benefits of S&T is much stronger in the less developed EU countries than in the wealthier and more developed.[8] According to Inglehart (1990), late modern societies can be characterised as post-materialistic societies emphasising values like environment, democracy, care for others, self-actualisation etc. The recruitment of Western students to medicine, biology and environment studies are *not* falling, and in these subjects the girls often outnumber the boys. This may indicate that youth in more developed countries believe that the most important challenges facing our society are related to health and environmental issues, and, consequently, that these fields can offer meaningful jobs.

NOTES

1. This article draws on material from Schreiner, 2006; Schreiner & Sjøberg, 2004, 2005 and 2007.
2. http://www.roseproject.no/
3. The group had, in addition to the Norwegian team, the following members: Dir. Vivien M. Talisayon (the Philippines), Dr. Jane Mulemwa (Uganda), Dr. Debbie Corrigan (Australia), Dir. Jayshree Mehta (India), Prof. Edgar Jenkins (England), Dir. Vasilis Koulaidis (Greece), Dr. Ved Goel (The Commonwealth, now India), Prof. Glen Aikenhead (Canada) and Prof. Masakata Ogawa (Japan).
4. National reports on how the survey was organised in each country are available from the ROSE website. http://www.roseproject.no/

5. The HDI is a summary measure of human development based on the weighted average of three indices: (1) a long and healthy life, as measured by life expectancy at birth, (2) education, as measured by the adult literacy rate (two-thirds weight) and the combined primary, secondary and tertiary education gross enrolment ratio (one-third weight) and (3) a decent standard of living, as measured by GDP per capita (PPP US$). (For details on how the index is calculated, see e.g. Technical note 1 in UNDP, 2008).
6. In spite of non-random sampling procedures, countries that are commonly considered as similar to each other (for example African, Baltic or Asian countries) do in most instances show similar or related response patterns. This can be seen as some validation of the data.
7. Available on http://www.ils.uio.no/english/rose/network/countries/norway/eng/nor-sjoberg-japan2008.pdf
8. The Eurobarometer (EU, 2005) has collected data in 32 countries: the 15 "old" EU countries; the 10 new member states (previously Eastern Europe); the four "candidate countries" (Turkey, Croatia, Bulgaria and Romania); and the three EFTA countries (Iceland, Norway and Switzerland).

REFERENCES

Anderson, I.K. (2007). *The relevance of science education—As seen by pupils in Ghanaian junior secondary schools*. PhD thesis Cape Town: University of Western Cape Town.
Beck, Ulrich (1998). *World Risk Society*. Cambridge: Polity Press.
EU (2004). *Europe needs more scientists!* Brussels: European Commission, Directorate-General for Research, High Level Group on Human Resources for Science and Technology in Europe. Available from: http://europa.eu.int/comm/research/conferences/2004/sciprof/pdf/final_en.pdf (accessed 2005-10-24).
EU (2005). Europeans, science and technology. *Special Eurobarometer224*, Brussels: European Commission. Available from: http://europa.eu.int/comm/public_opinion/index_en.htm
Inglehart, Ronald (1990). *Culture Shift in Advanced Industrial Society*. Princeton, NJ: Princeton University Press.
Jenkins, Edgar W. (2005). Important but not for me: Students' attitudes towards secondary school science in England. *Research in Science & Technological Education, 23*(1), 41–57.
Jidesjö, Anders & Oscarsson, Magnus (2004, 25–30 July). *Students´ attitudes to science and technology. First results from The ROSE project in Sweden*. Paper presented at the XI IOSTE (International Organization for Science and Technology Education) symposium: Science and Technology Education for a Diverse World, Lublin, Poland.
Lavonen, Jari, Juuti, Kalle, Uitto, Anna, Meisalo, Veijo, & Byman, Reijo (2005). Attractiveness of science education in the Finnish comprehensive school. In A. Manninen, K. Miettinen, & K. Kiviniemi (Eds.), *Research Findings on Young People's Perceptions of Technology and Science Education. Mirror results and good practice*. Helsinki: Technology Industries of Finland.
NSB (2008). *Science and Engineering Indicators 2008*. Arlington, VA: National Science Board, National Science Foundation.
OECD (2006). *Evolution of Student Interest in Science and Technology Studies*, Policy Report from OECD's global science forum. Available from: http://www.oecd.org/dataoecd/16/30/36645825.pdf

Ogawa, Masakata & Shimode, Shoko (2004). Three distinctive groups among Japanese students in terms of their school science. Preference: from preliminary analysis of Japanese data of an international survey The Relevance of Science Education (ROSE). *Journal of Science Education in Japan, 28*(4).

Schreiner, Camilla (2006). *Exploring a ROSE-garden. Norwegian youth's orientations towards science—seen as signs of late modern identities. Based on ROSE (The Relevance of Science Education), a comparative study of 15-year-old students' perceptions of science and science education.* Doctoral thesis, University of Oslo, Faculty of Education, Department of Teacher Education and School Development, Oslo. Available from: http://www.ils.uio.no/english/rose/network/countries/norway/eng/nor-schreiner-thesis.pdf

Schreiner, Camilla & Sjøberg, Svein (2004). Sowing the seeds of ROSE. Background, rationale, questionnaire development and data collection for ROSE (The Relevance of Science Education)—a comparative study of students' views of science and science education. *Acta Didactica 4.* Oslo: Dept. of Teacher Education and School Development, University of Oslo. Available from: www.ils.uio.no/forskning/publikasjoner/actadidactica/index.html (accessed 2005-08-06).

Schreiner, Camilla & Sjøberg, Svein (2007). Science education and youth's identity construction—two incompatible projects? In D. Corrigan, J. Dillon & R. Gunstone, (Eds.), *The Re-emergence of Values in the Science Curriculum*: Sense Publications.

Sicinski, Andrzej (1976). The future: A dimension being discovered. In H. Ornauer, H. Wiberg, A. Sicinski & J. Galtung (Eds.), *Images of the World in the Year 2000* (p. 121–159). Atlantic Highlands, NJ: Humanities Press.

Sjøberg, Svein (2002). Science for the children? Report from the science and scientists project—*Acta Didactica 1.* Dept. of Teacher Education and School Development, University of Oslo.

Sjøberg, Svein (2009). *Naturfag som allmenndannelse: en kritisk fagdidaktikk [Science as General Education. A critical approach].* (third ed.). Oslo: Gyldendal Akademisk.

Sjøberg, Svein & Schreiner, Camilla (2010). The ROSE project. An overview and key findings. Available from: http://www.roseproject.no/network/countries/norway/eng/nor-Sjoberg-Schreiner-overview-2010.pdf

Sjøberg, Svein & Schreiner, Camilla. (2005). How do learners in different cultures relate to science technology? Results and perspectives from the project ROSE (The Relevance of Science Education). *Asia-Pacific Forum on Science Learning and Teaching, 6*(2), Foreword.

Trumper, Ricardo (2004, 25–30 July). *Israeli students' interests in physics and its relation to their attitude towards science and technology and to their own science classes.* Paper presented at the XI IOSTE (International Organization for Science and Technology Education) symposium: Science and Technology Education for a Diverse World, Lublin, Poland.

UNDP (1990–2009). *Human Development Report.* New York: United Nations Development Programme. Available from: http://hdr.undp.org/

Part III
Measurement Issues

13 The Sources and Impact of Civic Scientific Literacy

Jon D. Miller

In the 21st century, the health of democratic societies will depend in part on the ability of citizens to read, hear, understand, and make sense of the scientific and technical issues of the day. Inherently, it is necessary for citizens to understand economic issues, foreign policy issues, and other complex matters to make informed judgments about the performance of government and to participate in political decision-making effectively. In this sense, scientific literacy is not separate from other citizenship skills, but a part of the fabric of knowledge and understanding necessary to sustain democratic participation in the 21st century.

The leaders of democratic societies in Asia, the European Union, and the United States agree that scientific literacy is a good thing and that having more of it would benefit their respective societies. In many countries, educational leaders and institutions focus substantial resources and organizational energy on the promotion of scientific literacy in student populations, sharing an assumption that the successful teaching of science and mathematics in the classroom will prepare students to function as scientifically literate citizens for the rest of their lives. Although there has been some growth in student science achievement scores during the secondary school years, cross-national studies find the overwhelming majority of adults unable to make sense of most of the science-related issues that now populate public policy agendas throughout the world.

This is a serious challenge to our long-standing commitment to participatory democratic governance. Too often, especially in Europe, the inability of adult citizens to make sense of important public policy issues that require some understanding of basic scientific constructs has been dismissed as a "deficit model" imposed by an arrogant scientific community (Wynne, 1991; 1996; Ziman, 1991). Ideology is no substitute for informed democratic debate. We need to set aside labels and slogans and think seriously about the implications of broad scientific illiteracy for our societies.

It may be useful to consider briefly two controversies that continue at various levels of intensity in all of our societies—the genetic modification of foods and the use of embryonic stem cells in biomedical research. Both of these issues have been the subject of serious public policy debates and, in some countries, these issues have become a partisan debate.

Greenpeace and other opponents of genetically modified foods have attempted to frame this debate as if genetic modification of plants and animals is a new phenomenon foisted upon an unsuspecting public by large corporations. This framing of the issue ignores or dismisses the centuries of distinguished British, European, and American work in plant breeding and animal husbandry. From the inception of agriculture in the Fertile Crescent, farmers have bred animals and propagated plants to improve their genetic characteristics. Today's methods of genetic modification are more precise in targeting single genes or specific sequences of genes, but it is the extension of a long and distinguished line of scientific inquiry. For those of us who value democratic political systems, it was discouraging to find at the end of the 20th century that only 29% of British and French citizens, 35% of German citizens, and 45% of American citizens could recognize that the statement "Ordinary tomatoes do not contain genes while genetically modified tomatoes do" was false (Pardo, Midden & Miller, 2002). Surely we cannot think that the majority of British, French, German, or American citizens were able to engage in an informed debate about genetically modified foods in 1999—the date of those surveys—or that we are significantly better off today.

More recently, a vigorous public policy debate has developed over the use of embryonic stem cells in biomedical research in Europe and in the United States. In the 2004 US presidential election, the subject of stem cell use was debated by the two major candidates in a nationally televised debate. Miller found that only 4% of US voters held a firm view on the stem cell issue after nearly a year of public debate, with a substantial majority reporting that it was too difficult for them to understand (Miller, Pardo & Kimmel, 2005). A 2005 Eurobarometer (64.3) found that only 4% of European adults claimed to be "very familiar" with the stem cell issue. These results cannot be encouraging to the friends of democracy on either side of the Atlantic.

In this context, this analysis will provide an empirical description of the current state of civic scientific literacy in thirty-four countries, examine some of the factors associated with the development of scientific literacy, and discuss the implications of these results for the future of democratic governance.

THE CONCEPT OF CIVIC SCIENTIFIC LITERACY

To understand the concept of civic scientific literacy, it is necessary to begin with an understanding of the concept of *literacy* itself. The basic idea of literacy is to define a minimum level of reading and writing skills that an individual must have to participate in written communication. Historically, an individual was thought of as literate if he or she could read and write their own name. In recent decades, there has been a redefinition of basic literacy skills to include the ability to read a bus schedule, a loan agreement, or the instructions on a bottle of medicine. Adult educators often use the term "functional literacy" to refer to this new definition of the minimal skills needed to function in a contemporary industrial society (Kaestle, 1985; Cook, 1977; Resnick and Resnick, 1977; Harman, 1970). The social science and educational literature indicates that about a quarter of Americans are not "functionally literate," and there is good reason to expect that roughly this proportion applies in most mature industrial nations and a slightly higher rate in emerging industrial nations (Ahmann, 1975; Cevero, 1985; Guthrie and Kirsch, 1984; Northcutt, 1975).

In this context, civic scientific literacy is conceptualized as the level of understanding of science and technology needed to function as citizens in a modern industrial society (Shen, 1975; Miller, 1983a, 1983b, 1987, 1995, 1998, 2000, 2004, 2010a, 2010b, 2010c). This conceptualization of scientific literacy does not imply an ideal level of understanding, but rather a minimal threshold level. It is neither a measure of job-related skills nor an index of economic competitiveness in a global economy.

THE MEASUREMENT OF CIVIC SCIENTIFIC LITERACY

In developing a measure of civic scientific literacy, it is important to construct a measure that will be useful over a period of years and that will be sufficiently sensitive to capture changes in the structure and composition of public understanding. If a time series indicator is revised too often or without consciously designed linkages, it may be impossible to separate the variation attributable to measurement changes from real change over time. The periodic debates over the composition of consumer price indices in the United States and other major industrial nations are a reminder of the importance of stable indicators over periods of time.

The durability problem can be seen in the early efforts to develop measures of the public understanding of science in the United States. In 1957, the National Association of Science Writers (NASW) commissioned a national survey of public understanding of and attitudes toward science and technology (Davis, 1958). Since the interviewing for the 1957 study was completed only a few months prior to the launch of Sputnik I, it is the only measure of public understanding and attitudes prior to the beginning of the space race. Unfortunately, the four major items of substantive knowledge were (1) radioactive fallout, (2) fluoridation in drinking water, (3) polio vaccine, and (4) space satellites. Fifty years later, at least three of these terms are no longer central to the measurement of public understanding.

Recognizing this problem, Miller attempted to identify a set of basic constructs, such as atomic structure or DNA, that are the intellectual foundation for reading and understanding contemporary issues, but which will have a longer durability than specific terms, such as the fallout of strontium 90 from atmospheric testing. In the late 1970s and the early 1980s, when the National Science Foundation began to support comprehensive national surveys of public understanding and attitudes in the United States, there was little experience beyond the 1957 NASW study in the measurement of adult understanding of scientific concepts. The first US studies relied heavily on each respondent's self-assessment of their level of understanding of various terms and concepts (Converse and Schuman, 1984; Sudman and Bradburn, 1982; Labaw, 1980; Dillman, 1978). This approach is still used in national studies in some countries, but provides a lower level of precision than direct substantive inquiries.

In a 1988 collaboration between Miller in the US and Thomas and Durant in the UK, an expanded set of knowledge items was developed that asked respondents direct questions about scientific concepts. These studies found that a combination of open-ended and closed-ended items provided significantly better estimates of public understanding. From this collaboration, a core set of knowledge items emerged that have been used in studies in Canada, China, Japan, Korea, India, New Zealand, and all 27 members of the European Union.

These core items have provided a durable set of measures of a vocabulary of scientific constructs, but it is important to continue to enrich the mix to reflect the growth of science and technology. For example, Miller's recent studies of the American public have included new open-ended measures of stem cell, nanotechnology, neuron, and genomic and new closed-ended knowledge items concerning the genetic modification of plants and animals, nanotechnology, ecology, and infectious diseases. Through the use of Item-Response-Theory (IRT), it is possible to incorporate marginal changes in the composition of a set of items and to produce comparable total scores across separate surveys or over time (Zimowski et al., 1996). An examination of the factor loadings in the 2005 US study illustrates the unidimensional nature of the construct (see Appendix Table 13.1). Open-ended items tend to have slightly higher loadings than closed-ended questions because they have a lower guessing component. Open-ended items, however, are more costly to collect and code and create more resistance from some respondents. A mixture of closed-ended and open-ended questions provides the optimal solution to the measurement of respondent knowledge.

THE CALIBRATION AND COMPUTATION OF INDIVIDUAL SCORES

To set the 2005 results from Europe and the United States in the context of previous studies, the responses from the 2005 studies were added to an IRT database that includes responses from previous US national studies conducted in 1988, 1990, 1995, 1997, 1999, 2001, and 2004; Eurobarometers from 1992 and 2001; and a 2001 national study in Japan. This multiyear IRT model includes more than 75,000 respondents. By computing a difficulty estimate, an efficiency estimate, and a guessing parameter, the

IRT technology can calibrate each of these items on a common metric over the full range of items, groups, and years and use various combinations of items to compute a total score for each individual (Zimowski et al., 1996).

The IRT scoring procedure produces a standardized score with a mean of zero and a standard deviation of 1.0. For reporting purposes, the mean (computed over the full database) was set to 50 with a standard deviation of 20. In terms of individual scores on the Index of Civic Scientific Literacy, the resulting scores range from approximately zero to 100.

A score of 70 or higher is indicative of a level of understanding sufficient to understand science and technology stories in the *New York Times* Science Times section or in *Science et Vie*. Like all threshold measures, the cut point is an approximation. An examination of the items that respondents are able to answer correctly indicates that individuals with a score below the middle 60s would have a difficult time making sense of the current debates over climate change or stem cell research. In other analyses, Miller (2010b, 2010c) has demonstrated that this threshold measure and a continuous measure of CSL produce similar results in structural equation models using US data. A similar comparison can be obtained by comparing the results in columns A and B on Table 13.1.

Using the threshold measure of civic scientific literacy, the results show a steady and significant improvement in the score of American adults over the last 18 years, in sharp contrast to the performance of American secondary school students. The proportion of American adults who are able to score 70 or higher in this Index grew from 10% in 1988 to 28% in 2005 (see Figure 13.1).

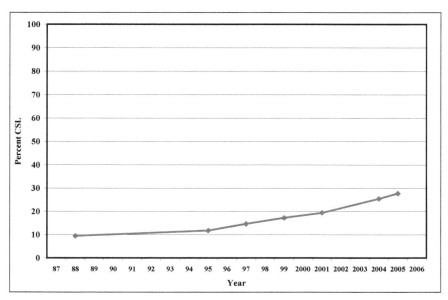

Figure 13.1 Civic scientific literacy in the United States, 1988–2005.

In cross-national terms, 35% of Swedish adults qualified as civic scientifically literate in 2005, followed by 28% of Americans (see Figure 13.2). On the same metric, 24% of Dutch adults and 22% of adults in Norway, Finland, and Denmark were classified as civic scientifically literate. In this ranking, differences of two or three percentage points do not reflect statistically significant differences. These results are consistent with earlier analyses of the European Union (then 15 members), Canada, Japan, and the US in the early 1990s (Miller, Pardo & Niwa, 1997).

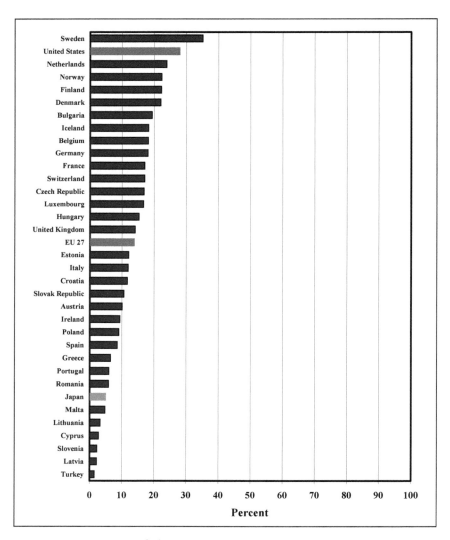

Figure 13.2 Civic scientific literacy in 34 countries, 2005.

The results in Figure 13.2 are shown on a zero-to-100 percent metric because there are no winners in this race. All of the countries included in this analysis are democratic in structure and assert a commitment to informed public policy debate on important issues. Yet, these results indicate that only one in three citizens in the highest ranking country are prepared to engage in and follow a public debate on important issues such as climate change, stem cell research, genetically modified foods, infectious diseases, and future energy sources. These will be the defining issues of the 21st century, and they all require some grasp of basic scientific constructs to make judgments about the seriousness of the problems and to decide among competing solutions.

THE SOURCES OF CIVIC SCIENTIFIC LITERACY

The issues that all democratic societies need to address are (1) what are the sources of civic scientific literacy for adults in each society, and (2) what can we do to expand and enhance the civic scientific literacy of citizens. Because of significant differences in the educational systems of various countries, the question of the sources of civic scientific literacy is not a simple inquiry. To understand the impact of the structure of science education in different societies, this analysis will look first at the United States and then at selected European countries.

THE SOURCES OF CIVIC SCIENTIFIC LITERACY IN THE UNITED STATES

Nearly three decades of international education studies have found that American secondary school students do relatively poorly on standardized tests in science and mathematics (Beaton et al., 1996; Schmidt, McKnight & Raizen, 1997). Recently, Miller, Scott & Okamoto (2006) found that American adults ranked 33rd among 34 countries in their acceptance of biological evolution. How, then, do American adults rank second in the world on these cross-national studies of civic scientific literacy?

Since Sputnik, a good deal of attention has been devoted to studying and revising elementary and secondary science and mathematics curricula in the United States. Despite substantial federal investments, the performance of American secondary school students has failed to improve. The results from the Second International Mathematics and Science Study (SIMSS) and the Third International Mathematics and Science Study (TIMSS) are compelling and disappointing (Beaton et al., 1996; Schmidt, McKnight & Raizen, 1997).

At the same time, many science and education policy leaders are unaware that the United States is the only major nation to require all of its college and university students to complete a year of college science as a part of their baccalaureate program. C. P. Snow's treatise on the two cultures is about the absence of general education requirements at Oxford and Cambridge and his plea that all educated persons should know both Shakespeare and the Second Law of Thermodynamics was a call for general education requirements in British higher education (Snow, 1959). Neither Snow nor subsequent advocates of general education have been able to change the system, and British students still read in only one field for a degree. Virtually all of the national education systems in Europe, Japan, China, and other countries use similar models. It is useful to examine existing US data to explore the influence of general education requirements—science courses for nonmajors—as one possible source of civic scientific literacy among adults.

It is also necessary to examine other possible sources of scientific literacy among American adults. In the six decades since the end of the Second World War, the number and range of informal adult science learning opportunities in the United States has increased markedly. In the last two decades, the availability of science on television and the Internet has increased exponentially. The sale of science books has grown sharply in recent years. The combination of these factors provides numerous opportunities for American adults to learn about science after the end of formal education, and available sales and use data indicate that many adults use these informal learning resources.

There are several factors that may encourage adults to seek science information after the end of formal schooling. One stimulus to continued science learning is individual and family health. Numerous studies show that adults who are diagnosed with cancer, for example, become active seekers of biological and biomedical information (Johnson, 1997; Miller & Kimmel, 2001). Another stimulus is occupational; a large number of adults work in firms and organizations that use science and technology, and some of these individuals may need to learn more about science for occupational reasons. The presence of minor children in a household is a third stimulus to continued adult science learning as parents seek to help with homework, science fairs, and questions from their children.

The growing number of public policy issues serves as another important stimulus to continued adult science learning. Although few adults could have studied stem cells, climate change, or nanotechnology during their formal schooling, all of these issues have become a part of current political discourse in the United States.

A STRUCTURAL EQUATION MODEL TO PREDICT CIVIC SCIENTIFIC LITERACY IN THE US

To explore these possible sources of influence on the development of civic scientific literacy, a structural equation analysis[1] of the 2005 US data set was conducted (Jöreskog & Sörbom, 1993). The analytic model included each individual's age; gender; highest level of education; number of college science courses completed; presence or absence of minor children in the household; interest in science, technology, medical, or environmental issues; personal religious beliefs; and use of informal science learning resources (see Figure 13.3).

A path model is useful for examining the relative influence of variables that have a known chronological or logical order. Each individual has a gender at birth and an age based on his or her birth date. An individual's gender may influence his or her education, although this influence appears to be diminishing in the United States and Europe. An individual's level of civic scientific literacy at any specific time may be thought of as the result of the combination of these and other factors (see Figure 13.3). In a path model, chronological or logical causation flows from left to right.

The product of the path coefficients is an estimation of the total effect of each variable on the outcome variable—CSL in this analysis. It is useful to look first at the total effect of each of the variables in this model, and then return to examine some of the specific path coefficients.

The number of college science courses[2] taken was the strongest predictor of civic scientific literacy, with a total effect of .74 (see Column B in Figure 13.4). Formal educational attainment[3] was the second best predictor of adult civic scientific literacy (.70). The third strongest predictor of adult civic scientific literacy was the use of informal science learning resources[4] (.25). The magnitude of this effect indicates that informal science learning is important for adults to maintain a functional level of

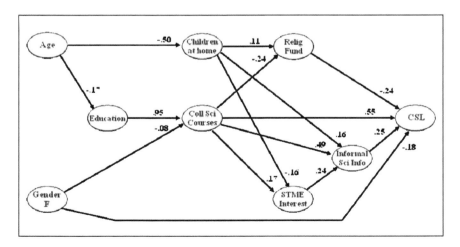

Figure 13.3 A path model to predict civic scientific literacy in adults, 2005.

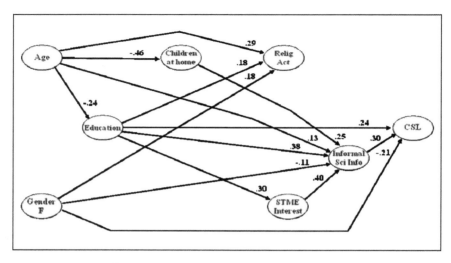

Figure 13.4 A model to predict CSL in Sweden, 2005.

civic scientific literacy and illustrates the linkage between formal schooling and adult learning later in life. The level of individual interest in scientific, technical, medical, or environmental issues influenced the level of use of informal science learning resources, but it had only a small net effect on civic scientific literacy (.06).

Personal religious belief was a relatively strong predictor of civic scientific literacy, with adults who hold fundamentalist religious beliefs[5] being significantly less likely to be scientifically literate than other adults (-.24). In this model, religious beliefs are current religious beliefs, and adults with more college science courses were significantly less likely to hold fundamentalist beliefs than other adults (-.24). Religious beliefs were not related to the use of informal science learning resources when age, gender, education, college science courses, and other variables were held constant.

Gender was a moderately strong predictor of adult civic scientific literacy, with a total effect of -.24 (see Column B in Table 13.1). The negative coefficient means that men were more likely to be scientifically literate than women among US adults. Older adults were less likely to be scientifically literate than younger adults (-.12), holding constant differences in education, gender, and other variables. The presence of preschool or school-aged children in the home had no net effect on adult civic scientific literacy in the United States.

This model explains 70% of the total variance in civic scientific literacy among US adults. This is a good fit for the model and other indicators confirm the fit of the model.

THE SOURCES OF CIVIC SCIENTIFIC LITERACY IN SELECTED EUROPEAN COUNTRIES

It is reasonable to assume that the major factors that promote civic scientific literacy in modern scientific industrial nations are similar. There are, however, some differences in the measurement of some of the variables derived from the 2005 Eurobarometer (63.1 science) that merit discussion prior to turning to the analysis of these data.

The first and most fundamental difference concerns education. Setting aside measurement issues temporarily, there are significant differences between postsecondary education in Europe and the United States. All college and university students in the US are required to complete one year of college-level science courses, as well as other general education courses in the humanities and the social sciences. In Europe and Asia, university students take courses in only the area of their academic concentration. As demonstrated in the preceding analysis, US performance on the Index of Civic Scientific Literacy was enhanced by the relatively high proportion of US students who attend a university and who enroll in one or more science courses. The Eurobarometer does not ask respondents about the number of postsecondary science courses that they have taken in their lifetime. This is a difference in educational systems and not a measurement problem per se.

There are, however, some important measurement issues related to education in the Eurobarometer. Throughout its history, the Eurobarometer has measured education in terms of the age of the respondent when he or she was last a full-time student. Although this provides a general indicator of formal education, it is imprecise in critical ways. The careful work of the OECD in mapping education demonstrates that there are substantial differences in educational tracks within and among European nations. Two individuals may have finished their full-time education at age 20, but one individual may have completed a university degree and the other may have completed a technical training program in a vocational training school. Clearly, these are not equivalent levels of education, and the OECD makes useful differentiations among the kinds of education and suggests some measures of equivalence across national educational systems. For these reasons, it is important to think of the Eurobarometer measure of education as a rough indicator of formal educational attainment.

A second area of a definitional and measurement difference concerns religion. The measures collected in the United States and analyzed in the preceding section characterize the content of religious beliefs held by respondents. Although the Protestant fundamentalism found in the United States has only faint echoes in Europe, it is clear from other studies that there is a wide range of theological beliefs among Europeans, ranging from strong deistic beliefs to more humanistic or agnostic beliefs. The only measures of religious activity available in Eurobarometer 63.1 assess denominational affiliation and frequency of religious service attendance. To differentiate

228 Jon D. Miller

these two kinds of measurements, the US variable is labeled as Religious Fundamentalism, and the European variable in labeled as Religious Activity. These are not identical variables.

Beyond these differences, there is substantial commonality in the two data sets used in this analysis. In both studies, age was collected in years and was entered into all of the models as a six-category ordinal variable. Virtually identical measures of gender and the presence of minor children in the home were used in both studies. The measure of issue interest is substantially the same. The measure of adult science learning resource use was similar in the two studies.

STRUCTURAL EQUATION MODELS TO PREDICT C. S. L. IN SELECTED EUROPEAN COUNTRIES

Using the same general model described above in regard to the US, it is possible to examine the relative importance of each of several variables on the prediction of civic scientific literacy in specific countries and in the European Union (using the data from the 27 member countries). For simplicity and brevity, this analysis will discuss the path diagram for Sweden and a summary table of total effects for the EU and for selected countries (see Table 13.1). For comparative purposes, the US model was recomputed using only a single measure of educational attainment (see Column C in Table 13.1).

Sweden

Looking at the model for Sweden (see Figure 13.4), there are several apparent structural differences. Religious activity is unrelated to civic scientific literacy (CSL) in Sweden, but it is not clear whether this is a reflection of the difference in the measurement of religion or simply a lower level of religiosity in Sweden. The presence of several paths from age to other intervening variables indicates stronger generational effects in Sweden than in the US.

It is more useful to compare the total effects of each of the independent variables in Sweden and the United States (see Table 13.1). The largest difference is associated with education, with a total effect of .66 in the US and .39 in Sweden. Given the top-ranked score of Swedish adults on the CSL Index, this result signals a higher level of homogeneity in educational experiences among Swedish adults than American adults (compare Columns C and E in Table 13.1). As in all countries, there is some disparity among Swedish adults in the number of years and the level of education that they attain in their lives, but the lower total effect does not indicate that the education is less effective in producing CSL, but rather that the higher level of homogeneity reduces the predictive power of the education

Table 13.1 Comparison of Total Effect of Selected Variables on Civic Scientific Literacy.

Total effect of...	U.S.-C	U.S.-T	U.S.-T*	E.U. 27	Sweden	Netherlds	Germany	Britain	Spain
	A	B	C	D	E	F	G	H	I
Gender (F)	-.17	-.24	-.22	-.33	-.25	-.43	-.35	-.43	-.20
Respondent age	-.11	-.12	-.07	-.23	-.11	-.16	-.11	-.24	-.28
Educational attainment	.59	.70	.66	.58	.39	.34	.41	.68	.40
College science courses	.62	.74	--	--	--	--	--	--	--
Kids at home	.00	.00	.00	.04	.00	.00	.00	.09	-.01
Religious fFund./aActivity	-.12	-.24	-.26	-.10	.00	-.10	.00	.00	-.11
STME issue interest	.06	.06	.10	.32	.12	.26	.13	.23	.08
Adult science learning	.26	.25	.34	.33	.30	.25	.30	.56	.20
R2 =	.48	.70	.64	.64	.28	.42	.35	.91	.29
Degrees of freedom	20	20	15	6	13	13	11	10	13
Chi-squares	117.2	114.0	33.0	23.1	31.6	35.7	37.6	30.1	27.3
RMSEA	.009	.007	.000	.017	.000	.000	.020	.019	.010
Upper confidence limit (RMSEA) 90%	.028	.027	.023	.032	.031	.028	.038	.039	.034
Number of cases	1,125	1,125	1,125	3,000a	965	972	1,324	1,212	988

U.S.-C is a model using the continuous CSL Index U.S.-T is a model using the threshold CSL Index (70 or higher) U.S.-T* is a model using the threshold CSL Index and omitting the college science course variable for consistency with E.U. countries a The Eurobarometer included 22,261 respondents, but the use of this number would destroy the comparability of the models with the United States or single European countries due to the sensitivity of LISREL fit statistics to large numbers. The reduction of the number of cases reported for significance testing is a standard practice with very large data sets.

variable in the total model. The influence of adult informal science learning was comparable in Sweden and the US (.30 and .34, respectively), suggesting that a substantial proportion of adults in both countries use informal resources to retain and enhance their understanding of emerging science and technology.

Educational attainment was the strongest predictor of civic scientific literacy in both Sweden and the US, but the relative effect was significantly stronger in the US than in Sweden (.64 and .41). Although this comparison uses the US model that excludes the number of college-level science courses, the strong correlation between postsecondary study and enrollment in a college-level science course means that the single US measure indirectly incorporates a good portion of the advantage conferred by these postsecondary science courses. Undoubtedly, the superiority of secondary school science education in Sweden (compared to the US) is one of the factors that accounts for the overall higher rating of Swedish adults on the CSL Index.

In Sweden, adult interest in issues about science, technology, medicine, or the environment was weakly related to adult scientific literacy (.12), which is comparable to the US (.10) but lower than the influence of issue interest in the EU (.32). An interest in science and technology issues was positively related to active adult science learning in Sweden (.30), the US (.25), and the EU (.46).

Overall, this model accounts for 28% of the total variance in the Swedish data in regard to the prediction of CSL. This is comparable to Spain but lower than the proportion of variance accounted for in the EU as a whole or in the US.

Netherlands

The Netherlands ranked third in the percentage of adults qualified as scientifically literate in 2005 and has a long tradition of distinguished secondary mathematics and science education. It has had a strong set of adult science learning resources, ranging from public science television broadcasting to science museums and "science shops."

In the Dutch model, women were significantly less likely to be scientifically literate than men (-.43), and this was one of the two strongest gender differences found in Europe or the United States (see Column F in Table 13.1). Older adults in the Netherlands were less likely to be scientifically literate than younger adults (-.16), which was slightly higher than the US (-.07) and slightly lower than the combined 27 members of the EU (-.23).

In the Netherlands, there was a strong positive relationship between educational attainment and CSL (.34). This relationship was slightly weaker than found in Sweden (.39), the US (.66), or in the combined 27 member states of the European Union (.58). Interest in science, technology, medical, or environmental issues was positively related to the use of informal science learning resources among Dutch adults (.44) and to scientific literacy

(.26). The level of use of informal science learning resources was positively related to adult scientific literacy in the Netherlands (.25).

Among Dutch adults, more frequent religious activity had a small negative relationship to CSL (-.10).

This model accounted for 42% of the variance in CSL in the Netherlands and displayed good fit parameters (see Table 13.1).

Germany

As the largest country in the European Union and a historic leader in European science, it is important to examine the structure of CSL in Germany. Although the quality of secondary education provided by the best German secondary schools—the gymnasiums—is legendary, in recent decades a substantial proportion of German secondary students have entered more vocational-oriented programs.

The model for Germany found a strong gender difference (-.35), a smaller age differential (-.11), and a strong positive relationship between educational attainment and scientific literacy (.41). The influence of education on CSL in Germany was similar to Sweden, slightly lower than the combined EU effect (.58), and substantially lower than the comparable US effect (.66). Neither the presence of minor children in the home nor the level of religious activity was related to CSL in Germany (see Column G in Table 13.1)

The level of interest in issues concerning science, technology, medicine, or the environment displayed a strong relationship to the use of adult informal science learning resources (.44) and a weaker positive relationship to scientific literacy (.13). The level of informal science learning activity was positively related to CSL in Germany (.30), which was similar to the impact found in the US and the EU (see Table 13.1).

The German model accounted for only 35% of the total variance in CSL, suggesting that there must be other variables that were not included in this model.

Great Britain

The long history of British leadership in science education demands a brief look at the structure of scientific literacy in the United Kingdom.

For almost all of the dimensions measured in this analysis, the direct effects of each of the independent variables found in Britain on CSL mirrors the European Union in the aggregate (see Column H in Table 13.1). British women are less likely to be scientifically literate than British men (-.43), and older adults in Britain were less like to be scientifically literate than younger residents in the U.K. (-.24). Educational attainment had a strong positive relationship to CSL in Britain (.68), and the presence of minor children at home has a small positive impact on adult scientific literacy (.09). Religious activity in Britain was unrelated to CSL.

There was a strong positive relationship between interest in scientific, technological, medical, and environmental issues and the use of informal science learning resources—newspapers, magazines, books, the Internet, other people (.41). This level of science issue interest had a positive impact on civic scientific literacy (.23), and the level of informal science learning resources had a substantial positive influence on CSL (.56).

Overall, this model accounts for 91% of the variance in scientific literacy among British adults and was the best fitting model in this analysis.

Spain

Finally, it is useful to look at Spain because it is one of the five largest European countries, it represents some of the traditions of southern Europe, and it has a long conservative Catholic tradition.

There is a strong positive relationship between education and CSL in Spain, holding constant differences in age and gender (.40). This effect is similar to the level of educational influence in Sweden and Germany. In a pattern similar to that found in the US and the EU, the level of interest in science, technology, medical, and environmental issues was positively related to the use of informal science learning resources (.42). Both the level of issue interest and the level of informal science resource use were positively related to the level of CSL (.08 and .20 respectively), but less strongly than found in the US or the EU.

Spanish women were less likely to be scientifically literate than Spanish men (-.20), and older Spanish adults were less likely to be scientifically literate than younger Spanish adults (-.28). The presence of minor children at home had no impact on adult scientific literacy (-.01), and the level of religious activity was negatively related to scientific literacy among Spanish adults (-.11).

DISCUSSION

The issue of scientific literacy is inextricably linked to democratic government. The long and distinguished history of the British people in the development of democratic institutions and practices—from the Magna Carta to today's vigorous democratic processes—should remind us of the centrality of the idea that effective participation in the affairs of government requires some understanding of both democratic practices and the substance of the issues in dispute. Only George Orwell could imagine a "democracy" in which citizens would not need to understand the issues, and he was not advocating that model for our adoption.

The work described above outlines the situation as it exists today. Most adults in Europe and the United States are currently unable to understand the substance of some of the most important issues facing modern societies on planet Earth. These results have often provoked two opposite and equally unacceptable responses.

Some individuals in the scientific community have read these results to justify the exclusion of scientific and technological issues from the normal political process and have argued for the creation of "science courts" to resolve contested decisions. Some, like Morris Shamos, have argued that science is inherently too difficult for ordinary people and that democratic processes will be necessarily ill informed (Shamos, 1995). This approach is an affront to the basic premises of democracy and has been largely set aside in most countries.

A milder version of this same reaction is that it is impossible to fully educate all adults, but that it is possible to select representative samples of adults, give them a cram course on the science of a specific issue or issues, and then guide them through a structured discussion of the issue. Although these "consensus conferences" are more democratic in spirit than the science court proposals of earlier decades, the results to date have attempted to utilize relatively short experiences (one to three days) to elucidate issues that parliamentary committees have struggled with for years.

At the other end of the spectrum, some individuals have tried to deny the results through derisive terms such as the "deficit model" and to suggest that citizens can determine the truth of issues without knowing the science behind them (Wynne, 1991, 1996; Ziman, 1991). This reaction is built on the notion that all science is relative and that any group's take on a scientific issue is as good as any other because the process is fundamentally political. Wynne asserts:

> This approach to public understanding of science therefore underlines the point reflected in other sociological analyses of scientific knowledge, that the boundaries of the scientific and the social are social conventions, predefining relative authority in ways that may be inappropriate, and which are open to renegotiation. (Wynne, 1996, p. 39)

This approach represents a fundamental misunderstanding of both science and politics.

Three decades ago, Shen (1975) argued that the public understanding of science might be usefully divided into practical scientific literacy, cultural scientific literacy, and civic scientific literacy. In this context, civic scientific literacy refers to a level of scientific terms and constructs to read a daily newspaper or magazine and to understand the essence of competing arguments on a given dispute or controversy. Shen argued:

> Familiarity with science and awareness of its implications are not the same as the acquisition of scientific information for the solution

of practical problems. In this respect civic scientific literacy differs fundamentally from practical scientific literacy, although there are areas where the two inevitably overlap. Compared to practical scientific literacy, the achievement of a functional level of civic scientific literacy is a more protracted endeavor, Yet, it is a job that sooner or later must be done, for as time goes on human events will become even more entwined in science, and science-related public issues in the future can only increase in number and in importance. Civic scientific literacy is a cornerstone of informed public policy. (Shen, 1975, p. 49)

For the long-term health of democracy in the modern world, we need to reject both the idea that scientific literacy is impossible and that it is irrelevant. Scientific literacy is not a myth, but has been achieved by millions of adults in Europe and the United States. But, as Shen suggests, the development of civic scientific literacy is a "protracted endeavor" that begins with a solid footing during formal schooling and is sustained through continuing attention to quality science reporting throughout one's adult years. It is not a one-time inoculation that lasts for a lifetime, but it is more like the acquisition of language—we begin with basic vocabulary and grammar in school but continue to acquire new words, concepts, and ideas as long as we live.

The development of scientific literacy by a majority of adults is widely endorsed by political and educational leaders, but it has proven to be more difficult to achieve than to endorse. It is appropriate to conclude this analysis with a discussion of what we know about the development of civic scientific literacy in modern societies and what policies might be adopted to expand the proportion of adults who attain civic scientific literacy.

There is broad agreement that the roots of scientific literacy must begin in the elementary and secondary schools. Most countries have invested heavily in advancing the study of science and mathematics, and we have extensive cross-national testing programs in place to monitor our progress in meeting this objective. Science and mathematics educators are divided over the optimal approaches to help students acquire a firm grasp of basic scientific constructs, but there is a general recognition among both educators and political leaders that progress in this area has been unsatisfactory. The United States has been notably ineffective in increasing the scientific literacy of its secondary school graduates, although it may well lead the world in spending for this purpose.

The unique advantage of the American educational system has been its insistence on general education requirements for all college and university students. As the preceding analyses have shown, it is these college-level science courses that account for the relatively positive ranking of US adults. In the context of the generally dismal performance of secondary school science in the United States, this result is even more impressive in that it appears to overcome a serious educational shortfall and provide a minimal level of scientific literacy for nearly 30% of American adults.

At the 50th anniversary of C. P. Snow's Rede lecture in Cambridge, it may be useful for European and Asian educational and political leaders to revisit his arguments. Recognizing the difficulties associated with an American offering advice about European science education, it is time for Europe and the prosperous nations of Asia to think about the arguments of Snow and others that the present system of specialized higher education is not serving the best interests of Europe in the 21st century. When this specialization is combined with the early stratification of students into vocational and academic tracks, the result is an adult population whose last formal exposure to scientific ideas and constructs may have been in the 9th or 10th year of school. This is inadequate for adults that are expected to make sense of climate change, embryonic stem cells, genetic modifications, or nanotechnology.

The introduction of general education requirements into higher education systems in Asia or Europe would involve making university studies into a full four-year curriculum, but today's university students will be in the work force and will function as citizens for 40 or 50 years after the completion of their formal studies. Surely, the addition of one year would be a justifiable investment for prosperous societies who depend on the engine of science and technology for their economic future. The introduction of general education—including a year of university-level science—for all university students should not be expected to increase the number of new patents or Nobel Prizes, but it would produce a citizenry that would be better prepared to participate in democratic decision-making about scientific and technological issues in our societies.

Finally, it is important to recognize the importance of adult learning about science throughout the life cycle and the contribution of informal science learning resources to this process (Miller, 2010a). The results from the US indicated that exposure to college-level science courses significantly enhanced the public use and comprehension of informal science learning resources, and the use of these resources was strongly associated with the maintenance of civic scientific literacy among adults in all countries. Governments, corporations, and universities need to recognize the vital role played by these institutions and support their work.

The task of creating a scientifically literate public is not simple and will not be accomplished in the immediate future. But it is an essential part of building and maintaining our democratic institutions. We need to acknowledge the centrality of this objective and to think seriously and programmatically about how to obtain this end.

APPENDIX

Table 13.2 Confirmatory Factor Analysis of Science Knowledge Items, 2005

	Loading
Provide a correct open-ended definition of a "stem cell."	.84
Provide a correct open-ended definition of a "molecule."	.80
Disagree: "A nanometer is equal to 1/100th of an inch."	.77
Provide a correct open-ended definition of a "neuron."	.76
Disagree: "Lasers work by focusing sound waves."	.76
Provide a correct open-ended definition of "DNA."	.75
Disagree: "Ordinary tomatoes ... do not have genes but genetically modified tomatoes do."	.73
Disagree: "Stem cells occur only in plants."	.73
Disagree: "Antibiotics kill viruses as well as bacteria."	.69
Disagree: "Nuclear power plants destroy the ozone layer."	.68
Provide a correct open-ended definition of "what it means to study something scientifically."	.67
Agree: "Electrons are smaller than atoms."	.67
Disagree: "Global warming is increasing primarily because the level of direct radiation from the Sun is increasing."	.65
Agree: "The center of the Earth is very hot."	.65
Agree: "The continents on which we live have been moving their location for millions of years and will continue to move in the future."	.61
Provide a correct open-ended definition of an "experiment."	.61
Agree: "More than half of human genes are identical to those of mice."	.60
Disagree: "For the first time in recorded history, some species of plants and animals are dying out and becoming extinct."	.59
Indicate that the Earth goes around the Sun once each year through two closed-ended questions.	.58
Indicate that light travels faster than sound.	.57
Disagree: "The earliest humans lived at the same time as the dinosaurs."	.56
Agree: "Over periods of millions of years, some species of plants and animals adjust and survive while other species die and become extinct."	.56
Agree: "All plants and animals have DNA."	.54
Agree that astrology is not at all scientific.	.52
Disagree: "Humans have somewhat less than half of their DNA in common with chimpanzees."	.50
Agree: "The universe began with a huge explosion."	.49

(continued)

Table 13.2 (continued)

	Loading
Agree: "The greenhouse effect causes the Earth's temperature to rise."	.48
Agree: "The primary human activity that causes global warming is the burning of fossil fuels such as coal and oil."	.47
Indicate a correct understanding of the meaning of the probability of one in four.	.44
Agree: "One of the effects of global warming will be that some species of plants and animals will thrive and other species will become extinct."	.41
Agree: "Human beings, as we know them today, developed from earlier species of animals."	.37

Chi-squares = 9580; degrees of freedom = 456; Root Mean Square Error of Approximation (RMSEA) = .040; 90% confidence interval for RMSEA= .038, .043.

NOTES

1. In general terms, a structural equation model is a set of regression equations that provides the best estimate for a set of relationships among several independent variables and one or more dependent variables. For the structural analysis presented in this paper, the program LISREL was used, which allows the simultaneous examination of structural relationships and the modeling of measurement errors. For a more comprehensive discussion of structural equation models, see Hayduk (1987) and Jöreskog and Sörbom (1993).
2. This variable is a measure of the number of college science courses, including courses in both community colleges and four-year colleges and universities. The number of courses was divided into three levels: (1) no college-level science courses, (2) one to three courses, and (3) four or more courses. Individuals with one to three courses are the students who took college science courses as a part of a general education requirement rather than as a part of a major or a supplement to a major. The use of an integer measure would have given undue weight to majors and minimized the impact of general education science courses in the analysis.
3. Educational attainment was measured with a five-category ordinal variable. The lowest level included all individuals who did not complete secondary school or obtain a GED. The second category included high school graduates and GED holders. The third category included respondents with an associate degree, and the fourth category included individuals who earned a baccalaureate but not a graduate or professional degree. The highest category included all individuals who completed a graduate or professional degree.
4. The Index of Informal Science Education (ISE) includes seven separate measures of science information acquisition: each individual's use of science magazines, news magazines, science books, science television, science museums, science and health web sites, and a public library. All seven items loaded at the .5 level or higher in a confirmatory factor analysis. A factor score was computed for each respondent and converted into a

zero-to-100 scale. For the purpose of this model, a five-category ordinal variable was constructed.
5. The Index of Religious Beliefs is a count of the number of times that a respondent indicated strong agreement with (1) The Bible is the actual word of God and is to be taken literally and (2) There is a personal God who hears the prayers of individual men and women, and strong disagreement with (3) Humans beings as we know them developed from earlier forms of life. Individuals who scored three on this Index were classified as Fundamentalist (28%); individuals who scored two on the index were classified as Conservative (28%); individuals who scored one on the index were classified as Moderate (26%), and individuals who scored zero on the scale were classified as Liberal (19%).

REFERENCES

Ahmann, S. 1975. The Exploration of Survival Levels of Achievement by Means of Assessment Techniques, in Duane M. Nielsen (Ed.), *Reading and Career Education*. Newark, DE: International Reading Association. Pp. 38–42.

Beaton, A. E., M. O. Martin, I.V.S., Mullis, E.J. Gonzalez, T.A. Smith, & D. L. Kelly. 1996. *Science Achievement in the Middle School Years: IEA's Third International Mathematics and Science Study (TIMSS)*. Chestnut Hill, MA: Boston College.

Cevero, R. M. 1985. Is a Common Definition of Adult Literacy Possible? *Adult Education Quarterly* 36:50–54.

Converse, J. M. & H. Schuman. 1984. The Manner of Inquiry: An analysis of question form across organizations and over time, in C. F. Turner & E. Martin (Eds.), *Surveying Subjective Phenomena*. New York: Russell Sage Foundation.

Cook, W. D. 1977. *Adult Literacy Education in the United States*. Newark, DE: International Reading Association.

Davis, R. C. 1958. *The Public Impact of Science in the Mass Media*. Ann Arbor, MI: University of Michigan Survey Research Center, Monograph No. 25.

Dillman, D. 1978. *Mail and Telephone Surveys: The Total Design Method*. New York: Wiley.

Guthrie, J. T. & I. S. Kirsch. 1984. The Emergent Perspective on Literacy. *Phi Delta Kappan* 65:351–355.

Harman, D. 1970. Illiteracy: An Overview. *Harvard Educational Review* 40:226–230.

Hayduk, L. A. 1987. *Structural Equation Modeling with LISREL*. Baltimore: The Johns Hopkins University Press.

Jöreskog, K. & D. Sörbom. 1993. *LISREL 8*. Chicago: Scientific Software International.

Johnson, J. D. 1997. *Cancer-Related Information Seeking*. Creskill, NJ: Hampton Press.

Kaestle, C. F. 1985. The History of Literacy and the History of Readers, in E. W. Gordon (Ed.) *Review of Research in Education* 12:11–54. Washington: American Educational Research Association.

Labaw, P. J. 1980. *Advanced Questionnaire Design*. Cambridge, MA: Abt Books.

Miller, J. D. 1983a. *The American People and Science Policy*. New York: Pergamon Press.

Miller, J. D. 1983b. Scientific Literacy: A Conceptual and Empirical Review. *Daedalus* 112(2):29–48.

Miller, J. D. 1987. Scientific Literacy in the United States, in D. Evered & M. O'Connor (Eds.), *Communicating Science to the Public.* London: Wiley.
Miller, J. D. 1995. Scientific Literacy for Effective Citizenship, in R.E. Yager (Ed.), *Science/Technology/Society as Reform in Science Education.* New York: State University Press of New York.
Miller, J. D. 1998. The Measurement of Civic Scientific Literacy. *Public Understanding of Science* 7:1–21.
Miller, J. D. 2000. The Development of Civic Scientific Literacy in the United States, in D. D. Kumar & D. Chubin (Eds.), *Science, Technology, and Society: A Sourcebook on Research and Practice.* New York: Plenum Press. Pp. 21–47.
Miller, J. D. 2004. Public Understanding of, and Attitudes toward Scientific Research: What we know and what we need to know. *Public Understanding of Science* 13:273–294.
Miller, J. D. 2010a. Adult Science Learning in the Internet Era. *Curator* 53:2:191–208.
Miller, J. D. 2010b. The Conceptualization and Measurement of Civic Scientific Literacy for the 21st Century, in J. G. Hildebrand & J. Meinwald (Eds.), *Science in the Liberal Arts Curriculum.* Cambridge, MA: American Academy of Arts and Sciences.
Miller, J. D. 2010c. Civic Scientific Literacy: The role of the media in the electronic era, in D. Kennedy & G. Overholser (Eds.), *Science and the Media.* Cambridge, MA: American Academy of Arts and Sciences.
Miller, J. D. & Kimmel, L. G. 2001. *Biomedical Communications: Purposes, Audiences, and Strategies.* New York: Academic Press.
Miller, J. D., R. Pardo, & L. Kimmel. 2005. The Impact of Attitudes toward Stem Cell Research on the Vote for President. A paper presented to the annual meeting of the American Association for Public Opinion Research, Miami Beach, Florida, May 12, 2005.
Miller, J. D., R. Pardo, & F. Niwa. 1997. *Public Perceptions of Science and Technology: A Comparative Study of the European Union, the United States, Japan, and Canada.* Madrid: BBV Foundation Press.
Miller, J. D., E. Scott, & S. Okamoto. 2006. Public Acceptance of Evolution. *Science* 313:765–766.
Northcutt, N. W. 1975. Functional Literacy for Adults, in D. M. Nielsen & H. F. Hjelm (Eds.), *Reading and Career Education.* Newark, DE: International Reading Association. Pp. 43–49.
Papacostas, A. *Eurobarometer 64.3: Foreign Languages, Biotechnology, Organized Crime, and Health Items, November-December 2005* [Computer file]. ICPSR04590-v3. Cologne, Germany: GESIS/Ann Arbor, MI: Inter-university Consortium for Political and Social Research [distributors], 2010-06-23. doi:10.3886/ICPSR04590.
Pardo, R., C. Midden, & J. D. Miller 2002. Attitudes Toward Biotechnology in the European Union. *Journal of Biotechnology,* 98:9–24.
Resnick, D. P. & L. B. Resnick. 1977. The Nature of Literacy: An Historical Exploration. *Harvard Educational Review* 47:370–385.
Schimdt, W. H., C. C. McKnight, & S. Raizen. 1997. *A Splintered Vision: An Investigation of U.S. Science and Mathematics Education.* Boston: Kluwer Academic Press.
Shamos, M. H. 1995. *The Myth of Scientific Literacy.* New Brunswick, NJ: Rutgers University Press.
Shen, B. J. 1975. Scientific Literacy and the Public Understanding of Science, in S. Day (Ed.), *Communication of Scientific Information.* Basel: Karger.

Snow, C. P. 1959. *The Two Cultures and the Scientific Revolution.* New York: Cambridge University Press.
Sudman, S. & N. M. Bradburn. 1982. *Asking Questions.* San Francisco: Jossey-Bass.
Wynne, B. 1991. Knowledges in Context. *Science, Technology & Human Values* 16:111–121.
Wynne, B. 1996. Misunderstood Misunderstandings: Social identities and public uptake of science, in A. Irwin & B. Wynne (Eds.). 1996. *Misunderstanding Science? The Public Reconstruction of Science and Technology.* Cambridge, UK: Cambridge University Press.
Ziman, J. 1991. Public Understanding of Science. *Science, Technology & Human Values* 16:99–105.
Zimowski, M. F., E. Muraki, R. J. Mislevy, & R. D. Bock. 1996. *BILOG-MG: Multiple-group IRT Analysis and Test Maintenance for Binary Items.* Chicago: Scientific Software International.

14 Using Latent Trait Models to Assess Cross-National Scales of the Public's Knowledge about Science and Technology[1]

Sally Stares

SURVEY MEASURES OF SCIENTIFIC KNOWLEDGE

The formulation of the construct 'knowledge', or 'literacy' as it is sometimes called, has been a subject of considerable debate amongst those studying the relationship between science and the public (Sturgis and Allum, 2004). Different types of knowledge have been proposed (e.g. Shen, 1975), with 'civic' scientific knowledge attracting the greatest degree of attention. In their turn, different types of civic scientific knowledge have been distinguished. Of these, knowledge of the content or vocabulary of science has received most consideration—perhaps partly because it is arguably more straightforward to operationalise, using survey questionnaires, than elements such as knowledge of the institutional aspects of science.

A standard set of closed-ended items for capturing general science knowledge has been developed and used by Jon Miller in the US (Miller, 1998), and John Durant and colleagues in the UK (Durant, Evans, & Thomas, 1989). Known as the 'science literacy scale' in the US, this comprises a set of around ten statements that respondents are asked to identify as true or false. Such statements include, for example, 'The centre of the Earth is very hot', and 'All radioactivity is man-made'. The statements are not intended to comprise a comprehensive interrogation of respondents' knowledge; rather they are to be viewed as a sample of facts from a wider domain—an approach commonly used in surveys of political knowledge (e.g. Converse, 2000). A similar set of items has been developed to capture knowledge specifically about biotechnology and genetics, in particular, for inclusion in studies of public perceptions of biotechnology. This has been applied in Eurobarometer surveys on the topic since 1993. A number of important objections have been raised regarding the content of such knowledge items, questioning for example whether they capture scientific knowledge that is relevant for the public (e.g. Irwin and Wynne 1996) and whether it is meaningful at all to gauge people's science 'literacy' by means of their ability to recall a set of isolated facts (Jasanoff 2000). Such concerns are apposite in any individual country context. The theme of this paper, however, is on the considerations that come to bear in using the same items, whatever their content, to assess science knowledge across different social settings or groups. In this chapter the focus is on cross-national comparisons. Country comparisons are by no means the only contrasts that could be made. For example, Pardo and Calvo (2004) analyse the measurement properties of science knowledge items for groups of different ages and different levels of education, and Raza et al. (2002) investigate cultural comparisons within India. However, there are a number of reasons why cross-national comparisons are relevant for Public Understanding of Science (PUS) studies. One is substantive, and relates quite simply to the fact that for many science actors it is important to know how the climate of opinion varies from one polity to another—research in the field of PUS is inevitably situated firmly in its political and economic context. Methodologically, cross-national comparisons need to be scrutinised in order to investigate whether meanings vary systematically with the different languages in which the surveys are administered, and whether measurement errors vary systematically with styles of survey administration, which is organised on a country-by-country basis. This is important both for informing survey analyses and for informing future questionnaire design. Some publications have already suggested that knowledge scales might be quite differently composed within Europe (e.g. Pardo & Calvo, 2004; Peters, 2000). In the broadest sense, this chapter is concerned with the cross-national comparability of scales of science knowledge that are built using what has become the standard question design. More specifically, it centres on whether the items used 'work' in similar enough ways in different national contexts for us to be able to make fair cross-national comparisons using them. Even more precisely, it works with a definition of 'comparability' that is purely statistical. This is of course a very narrow approach to a very broad concern. In this chapter I hope to show, however, that notwithstanding its limited remit, information about statistical comparability can contribute useful diagnostic information to the large and multifaceted task of assessing the validity of survey measures of knowledge of science and technology.

DATA

Two sets of items will be analysed, both from the Eurobarometer survey series. The 'biotechnology' items, as they will be called, are intended to capture knowledge about biology and genetics. The 'science literacy' items are intended to capture knowledge about science in a broader sense. Since published critique has been directed much more often to the general science literacy items than to the similar set on biotechnology, I devote most attention to the latter in this chapter. My motive for doing so is not only to redress the imbalance of attention in the literature, but also because Allum et al. (2008) suggest that although measures of general attitudes towards science and general knowledge of science are weakly correlated, measures of more specific attitudes and knowledge (e.g. of biotechnology) are more strongly related to each other. So it may be worthwhile to spend some time developing a tool for capturing knowledge about biotechnology in particular.

The two sets of data analysed are summarised in Table 14.1, with grey highlighting indicating the correct answers. In the first half of the table are the ten biotechnology items from the 2002 Eurobarometer on public perceptions of biotechnology. These are followed by the thirteen general science literacy items posed in the 2005 Eurobarometer on Public Understanding of Science (PUS). In both surveys the questions were posed to respondents in the then fifteen EU member states. Two key characteristics of the data should be noted at this point. One is that some items are answered correctly by only a few respondents, while others are answered correctly by many. The other is that 'don't know' (DK) response rates are very high for many of the items.

Table 14.1 Distribution of Responses to Knowledge Questions across Fifteen EU Countries

			% responses[2]		
No.	Label	Statement	True	False	DK
Knowledge about biotechnology, 2002					
1	bacteria	There are bacteria which live from waste water.	84	3	12
2	tomato	Ordinary tomatoes do not contain genes, while genetically modified tomatoes do.	35	36	29
3	clone	The cloning of living things produces genetically identical copies.	66	16	18
4	fruit	By eating a genetically modified fruit, a person's genes could also become modified.	20	49	31
5	mother	It is the mother's genes that determine whether a child is a girl.	23	53	24
6	yeast	Yeast for brewing beer consists of living organisms.	63	14	23
7	test	It is possible to find out in the first few months of pregnancy whether a child will have Down's Syndrome.	79	7	14
8	animal	Genetically modified animals are always bigger than ordinary ones.	27	38	35
9	chimpanzee	More than half of human genes are identical to those of a chimpanzee.	52	15	33
10	transfer	It is not possible to transfer animal genes into plants.	29	26	45

$n=16,040$

(continued)

Table 14.1 (continued)

No.	Label	Statement	% responses[2]		
			True	False	DK
Knowledge about science in general, 2005					
1	sun	The Sun goes around the Earth.	31	65	4
2	hot	The centre of the Earth is very hot.	87	6	6
3	oxygen	The oxygen we breathe comes from plants.	80	15	4
4	milk	Radioactive milk can be made safe by boiling it.	10	75	16
5	electrons	Electrons are smaller than atoms.	45	30	25
6	plates	The continents on which we live have been moving for millions of years and will continue to move in the future.	88	5	7
7	mother	It is the mother's genes that decide whether the baby is a boy or a girl.	20	65	15
8	dinosaurs	The earliest humans lived at the same time as the dinosaurs.	22	67	11
9	antibiotics	Antibiotics kill viruses as well as bacteria.	40	49	11
10	lasers	Lasers work by focusing sound waves.	26	47	27
11	radioactivity	All radioactivity is man-made.	27	60	13
12	human	Human beings, as we know them today, developed from earlier species of animals.	72	19	9
13	month	It takes one month for the Earth to go around the Sun.	19	65	16

n=15,518

METHODS FOR SCALING

The typical approach to creating a scale from these items is to add up the number of correct responses. This simple sum-score approach is potentially problematic in a number of ways, however. First, it does not allow for any measurement error in the items. Second, it does not allow the possibility of distinguishing between a substantively incorrect response and a DK response. Third, it usually means assigning equal weights to all items in the scale, even though some are more diagnostic of people's knowledge than others.

In these sets of items, it would be beneficial to address the three concerns. Firstly, it is well understood, both in educational testing and in attitude research that items are imperfect indicators of the concepts that we try to capture using them. Secondly, it can be informative to distinguish between a DK response and substantive response, in order to investigate possible response effects in the data, that is, particular types of measurement error. DK response rates are very high in both item sets analysed in this chapter, and particularly for the biotechnology items. Thirdly, it would make sense to award credit to items differently, depending on their content, just as would be done in a typical examination or test.

In the analyses below I use latent trait models to address these three concerns (for an introduction, see Bartholomew et al. 2008). So I hypothesise the existence of a construct, 'science knowledge', which cannot be directly observed, and take answers to the individual survey questions as imperfect manifestations of that latent variable. If the items all tap into the construct, then they should be statistically associated with each other; these associations should be explained by the latent variable 'knowledge'.

Latent variable models in general can be thought of most simply as regression models with multiple observed response variables and a smaller number of unobserved explanatory variables. It may be the case, with these knowledge items, that more than one latent variable, more than one unobservable attribute, is required to adequately explain the patterns of responses to these questions. Perhaps, by way of illustration, the responses that are given to these items are best explained by two latent variables, one of which captures levels of knowledge, and the other, which picks up a certain response style that influences people's answers but separately from their level of knowledge. An expected response style might be, for example, acquiescence bias—the tendency to agree or give a positive response to questions—that is often found in survey data. Latent variable models enable us to address the three current concerns in the following ways. First, they allow for measurement error by specifying a probabilistic rather than a deterministic relationship between item responses and the latent variables that explain them. Second, particular latent variable models can be employed that enable us to retain the distinction between substantive and DK responses. In survey research latent variable models are most commonly applied in the form of factor analyses based on linear regression models, which assume that latent variables and observed items are continuous, interval-level variables. However, these are inappropriate when observed items are categorical, which is the case with the items to be analysed, and indeed is often the case throughout attitudinal surveys such as the Eurobarometer. Latent trait models are then more suitable: they define logistic regressions for the relationship between categorical item responses and continuous latent traits. Latent trait models are better known in some fields as Item Response Theory (IRT) models (van der Linden and Hambleton 1997). In the models presented here, my first approach is to treat the observed knowledge items as three-category nominal items, that is, essentially to use a multinomial logit model for the relationship between each item and the latent trait(s).

Last, the question of determining whether some survey items should receive more or less weight in a scale of knowledge, and exactly what weights, can be addressed using the model's parameter estimates. Latent trait models provide a number of useful pieces of information about a set of items. For each item (or item category, if there are more than two categories) the slope coefficient or *loading* tells us to what extent the item (or category) enables us to discriminate between respondents on the trait; the higher the loading, the greater the discrimination, the more pertinent the item is to the substance of the trait, the greater its weight should be in determining a person's position or 'score' on the trait. Items with high discrimination power will help us to capture 'knowledge' and to calibrate respondents in terms of their levels of knowledge. Items that have low discrimination power do little work for us in characterising respondents and may be candidates for deletion from future survey waves. Alongside discrimination parameters, the model specifies constants or intercepts. Whereas these are not of great interest in the linear factor model, in latent trait models they have heuristic value when expressed as difficulty parameters for the items. The difficulty of a particular response is defined as the probability of giving that response for the median individual on a trait. It is desirable to have items with a range of difficulty levels in any one scale. The combined information from item loadings and intercepts can be represented graphically, by calculating a selection of fitted probabilities of item responses for a range of values for a latent trait (fixing the other trait(s) at some values, if there is more than one trait—in this chapter, the other trait(s) are fixed at their means when this is done). From such fitted probabilities we can draw Item Characteristic Curves (ICCs) or trace lines, which show at a glance the changing probabilities of choosing each of the response categories at any point along the latent trait. ICCs show for each item its discrimination power, via the steepness of the slopes of its response curves (the steeper the slope, the greater the discrimination), and its difficulty, by way of the location of the curve in the plot (the higher on the latent trait, the greater the difficulty).

Jon Miller's various analyses of the closed form knowledge items—both of the science literacy scale, and of the Eurobarometer-style questions focusing on biotechnology-related facts—employ latent variable models, and constitute the most advanced approach in the PUS literature, to my knowledge. He takes binary items (where there are DK responses these are recoded as 'incorrect'), uses a preliminary factor analysis to identify items that form a unidimensional scale, then applies a three-parameter logistic model to these items, that is with difficulty and discrimination parameters, and an extra parameter to correct for guessing. Miller, Pardo, and Niwa (1997) and Miller and Pardo (2000) employ these models for comparative studies of the EU, US, Japan, and Canada.

In this chapter I adopt a slightly different approach from Miller and colleagues to deal with possible response effects. As the literature suggests that a third, 'guessing' parameter is only weakly identified (Skrondal & Rabe-Hesketh, 2004; Thissen & Wainer, 1982), and my samples are of standard sizes (roughly a thousand in each country), I keep to a two-parameter logistic model, but allow more than one trait to try to account for response effects. This is potentially advantageous not only because of statistical identification reasons, but also because I suspect that there is a more complex response style in the Eurobarometer items than guessing between two options. The data contain high rates of DK responses, and in the biotechnology items the 'true=correct' items tend to be answered correctly more often than the 'false=correct' items. These data may therefore contain a mixture of acquiescence bias, propensity to guess, and propensity to profess ignorance—in addition to the knowledge we are trying to capture. So I retain the distinction between DK and an incorrect response, use two-parameter logistic models, and allow the possibility of needing more than one trait to represent variation in the data. Further, I model the latent traits as discrete, with seven levels, rather than as continuous (see Heinen 1996). This has the particular advantage that it does not impose the assumption of a normal distribution for the latent trait, so we can allow for the possibility that knowledge is not symmetrically distributed among the population.[3]

In order to address the question of cross-national comparability, I focus on differences and similarities between European countries in the construct 'knowledge' that can be derived from the items using latent trait models. The logic is that if the estimated item parameters (loadings and difficulties) for a trait representing 'knowledge' are very different from country to country, then the relationships between the survey items and the construct 'knowledge' is different from country to country. This would imply that in a statistical sense the items do not capture knowledge in the same way from place to place. We should then be cautious when drawing comparisons between countries using these items, and consider modifying them in future surveys towards the goal of creating broadly comparable scales. If, however, the item loadings and difficulties can be constrained to be the same between countries without compromising model fit too much, then we can reasonably speak of a common construct 'knowledge', and it makes statistical sense to use it as a means of comparing levels of knowledge from country to country. Whether it also makes substantive sense is the next question—and not for this chapter. The statistical approach here is roughly analogous to those applied to continuous observed items in factor analysis, for which there are many more demonstrations and discussions of 'non-equivalence' or 'invariance' of item parameters in the literature (for a recent example and overview see Allum, Read and Sturgis, forthcoming).

Procedurally, in this chapter I first of all find adequate models for knowledge for each country separately. Then for each of these models I compare their parameter estimates informally. I then attempt to find a single model across the whole data set, with country as a covariate, where the loadings and intercepts are fixed to be the same for all countries. The criteria for 'finding a model' are those of model fit and model interpretation. In order to assess model fit I focus on two-way marginal residuals, following Bartholomew et al. (2008), Bartholomew and Knott (1999), and Jöreskog and Moustaki (2001).[4] As well as fitting adequately, a model must make substantive sense. In this context, at the most basic level it means that the odds of answering questions correctly should increase as one moves up the latent trait 'knowledge'.

RESULTS OF SCALING THE BIOTECHNOLOGY KNOWLEDGE ITEMS

By way of orientation, I present here the results from a two-trait model for the ten items in the British sample, which fits the data well (2.7 per cent of standardised marginal residuals > 4). Figure 14.1 shows the ICCs of what I label the 'knowledge' trait for this model. Each plot refers to one of the items, with the value of the latent trait on the horizontal axis and the probability of giving the responses 'true', 'false', or 'DK' on the vertical axis. It can be seen that as one moves from the lower to the higher end of this trait (from lower to higher levels of knowledge), broadly speaking, the probability of giving a correct response increases, and the probability of giving an incorrect or a DK response decreases. The other trait (which is less interesting and not presented here) seems to summarise the tendency to give a DK rather than a substantive response—this is the response effect trait that I expected to find.

This broad interpretation of the knowledge trait comes with two qualifications, however. First, all those items listed in the left-hand side of the figure have very shallow slopes or low discrimination power: it hardly matters whether one is 'low' or 'high' on knowledge; the probability of giving a correct response is very similar at every point. Moreover that probability is high: these items are easy to get right; *chimpanzee* is the most difficult, but even for this item, a person at the lower end of the trait has around a 60 per cent chance of answering it correctly. So these items do not do a lot of 'work' for us in helping to distinguish between people who have high versus low levels of knowledge. Notably, these items share the characteristic that 'true' is the correct answer. The items displayed on the right-hand side, for which, incidentally, 'false' is the correct answer, generally have much greater discrimination power: a person at the lowest end of the knowledge scale has about a 20 per cent chance of answering the item *tomato* correctly, whereas a person at the highest end has about an 80 per cent chance of giving the right answer. Note that *tomato* is a more difficult item than *fruit* (just below it); a person with the median level of knowledge is more likely to answer *fruit* than *tomato* correctly. The second qualification to the general interpretation is that a few of the slope estimates suggest a problematic interpretation. The signs of the true and false slopes for *clone* and *chimpanzee* are actually in the wrong order: the probability of giving the incorrect response (for both items, 'false') *increases* as the level of the trait increases. The slope estimates for 'true' and 'false' responses on these two items are actually not significantly different from each other (at p<0.05). However, the implication of the model for the calculation of posterior scores ('factor scores') for the trait is that a person answering all other items correctly but

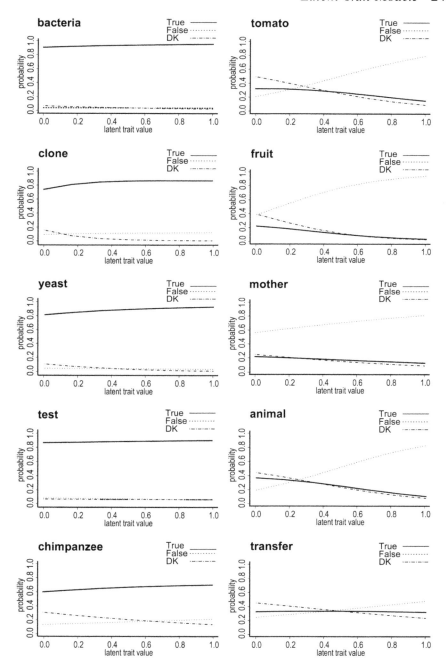

Figure 14.1 Item characteristic curves for the 'knowledge' trait from a 2-trait discrete trait model for 3-category nominal items, British sample.

incorrectly saying 'false' to these two items would be assigned a slightly higher score than a person answering all items correctly. Under this model, the former has a score of 0.943 and the latter, 0.917.

The next step in fitting a model to the British data would be to drop the problematic items from the scale, in order to find a model in which the slope coefficients are aligned in the directions that fit logically with a scale defining high knowledge at one end. This exercise is in fact quite problematic. Removing items from the scale noticeably destabilises the items remaining—most notably, the 'true=correct' items, making it difficult to find a model in which the slope coefficients take the required signs. Within the scope of the several models that I attempted, I could not find one that represented any improvement over the ten-item scale, especially in terms of producing factor scores that were logically ordered according to numbers of correct answers. So the ten-item trait remains the final model for this section on the British sample, but with a caveat attached.

Moving on to explore a joint model of knowledge items for the fifteen countries, separate country-by-country analyses suggest that two-trait models are a feasible starting point. Using all ten items in the set, two traits fit well for all country samples: the percentage of large two-way standardised marginal residuals ranges from 0.2 in Finland to 7.2 in Germany, with a mean of 2.4 per cent across the fifteen countries. In all countries, one trait can be reasonably labelled 'knowledge', whereas the interpretation of the other trait varies a little more between samples—of the range of interpretations, the most common is a response effect trait, with DK responses at one end, and 'false' at the other. Focusing on the 'knowledge' trait, Table 14.2 gives a qualitative summary of the few items and few countries for which slope coefficients deviate from the pattern to be expected in a trait capturing knowledge. It reflects the model of the British data, in that many of the 'true=correct' items lack discrimination power in some countries, and in a number of cases, whilst the overall probability of a correct response is highest at the highest point of the trait, the slope for the incorrect response is increasing—suggesting the problems with factor scores encountered in the British data. However, these are not such serious problems compared with the last item, *transfer*, for which in five countries, at the 'high knowledge' end of the trait the probability of giving the incorrect response is greater than the probability of giving the correct response. In these cases it is very clear from the ICCs that the item does not fit logically with the others in the scale. From this point it is dropped from the item set. Repeating these exploratory analyses with nine items leaves the qualitative summary of them in Table 14.2 essentially unchanged.

From the example of ICCs given in Figure 14.1 it is clear that some items have greater discrimination power than others. In a joint trait model, the relative discrimination of the items would be fixed to be the

Table 14.2 Qualitative Summaries of Unusual Item Characteristic Curves (ICCs) on Biotechnology 'Knowledge' Traits, from 2-trait Models, 15 Countries

	'True=correct' items					'False=correct' items				
	bacteria	clone	yeast	test	chimp-anzee	tomato	fruit	mother	animal	transfer
Austria										
Belgium	c-, i+				d					d, c-, i+
Denmark					i+					I, c-, i+
Finland				c-, i+				i+		
France	c-, i+		c-, i+	c-, i+	c-, i+					
Germany		c-, i+	c-, i+	c-, i+	i+					I, i+
Greece	i+			i+						
Ireland	i+				i+					c-, i+
Italy	c-, i+	c-, i+	c-, i+	c-, i+						
Luxembourg	d	d		d, i+						I, d
Netherlands	d	d		d	d					I, d
Portugal				i+		I, i+				
Spain	c-, i+	c-, i+		c-, i+						I, i+
Sweden	d			d						
UK	d			d	i+					d

Key
d   Low discrimination: very flat ICCs
c-   Slope for correct response decreasing slightly with higher levels of 'knowledge'
i+   Slope for incorrect response increasing slightly with higher levels of 'knowledge'
I   'Incorrect' most likely response at top end of trait
regular font   Slight effect
bold font Strong effect: more seriously problematic

same between countries. Unfortunately, it seems that the differences in ICCs between countries are too large for a joint trait model, with the same measurement model for each country, to fit well.[5] In the joint version of the two-trait models for nine items, 42 per cent of standardised two-way marginal residuals are large. The number of large residuals is notably high for four items: *tomato, yeast, animal,* and *chimpanzee,* in terms of two-way item-by-item margins as well as country-by-item margins, and three-way item-by-item margins. Dropping these from the scale almost halves the proportion of high residuals, but the rate is still 26.3 per cent. Increasing the number of traits also helps model fit: a three-trait model with these four problematic items removed reduces the proportion of high residuals to 15.9 per cent. This is still arguably too high (compare for example with the latent class models in Mejlgaard and Stares, this volume). Moreover, the measurement of knowledge from these items is considerably unstable. In the models with equal measurement models between countries for all traits, I found that different numbers and combinations of items produced quite different solutions—echoing the findings from the two-trait model of British data presented as an example. Some combinations of items failed to return a trait successfully representing correct 'knowledge' at one end, even when the model contained three traits. So although these items are intended to constitute a sample from a wide universe of knowledge items, the interpretation of the construct 'knowledge' seems to depend, more than is desirable, on the combination of items contributing to it.

Because the objective for this model is to derive a measurement of knowledge, there should be no compromise to the model by allowing the second trait, which fills the role of accounting for response styles, to differ between groups. That is, a feasible joint model might be one with a fixed trait representing 'knowledge', and a country-specific trait for response effects. However, for this set of items, it does not provide a solution. Both with nine items, and with a reduced set of five items (chosen by means of inspecting large two-way and three-way marginal residuals, as above), the fixed trait cannot be interpreted as 'knowledge'. It seems to be closer to a response effect, with DK at one end and 'false' at the other, regardless of whether this is the correct response. Albeit these models represent a great improvement in fit (13.2 per cent large two-way residuals for the nine-item model for example), they do not return a viable representation of 'knowledge'. From these analyses, then, it seems that finding a viable joint model for these items is a difficult task. The models attempted here either fit badly or do not identify a trait that could feasibly be interpreted as 'knowledge', and seem to be numerically unstable. Out of the models presented in this section, the three-trait model with five items is the best representation of 'knowledge', cross-nationally. The ICCs for the 'knowledge' trait from this model are presented in Figure 14.2. Note that for this trait low scores denote high levels of knowledge.

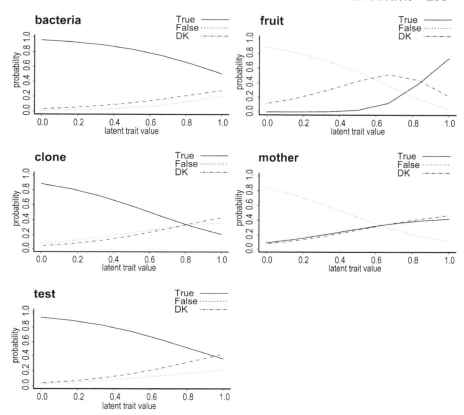

Figure 14.2 Item characteristic curves from a 3-trait discrete trait model for 3-category nominal items, with measurement models equal for all traits, for 15 countries, biotechnology 'knowledge' trait.

Having decided on the best cross-national scale that can be derived from these items, the question to ask is how knowledge is distributed country-by-country. Table 14.3 shows the distribution of levels of knowledge according to this trait (reversed from the original model so that high levels of the trait denote high levels of knowledge). Specifically it shows the percentage of the population estimated to belong at each of the seven levels of the trait, by country, and for the fifteen countries together, weighted by their respective populations. Countries are ordered from highest to lowest mean knowledge score. The distribution of the trait among countries is consistent with expectations from the PUS literature: high levels of knowledge are found among the Northern European countries and, with some exceptions, lower levels among those in the South. Overall, Europeans

254 Sally Stares

Table 14.3 Percentages of Respondents in Each Level of the Final Joint Model of Biotechnology Knowledge Items

% within country[6]	Low knowledge			Level		High knowledge		Mean knowledge
Sweden	0	0	0	7	10	47	36	0.85
Denmark	0	0	0	16	15	45	23	0.79
Netherlands	0	0	0	17	15	44	22	0.78
UK	1	0	1	22	17	42	18	0.75
Finland	1	0	1	30	19	36	12	0.70
France	2	0	1	30	19	36	12	0.70
Luxembourg	2	0	1	31	19	35	11	0.69
Italy	4	1	1	34	19	32	9	0.66
Germany	4	1	2	36	19	30	9	0.65
Ireland	5	1	2	36	18	30	9	0.65
Spain	5	1	2	37	19	29	8	0.64
Belgium	6	1	2	38	18	28	8	0.63
Greece	8	1	2	43	18	23	5	0.59
Austria	11	2	3	46	17	19	4	0.55
Portugal	18	2	3	45	15	15	3	0.49
Europe total (pop. weighted)	4	1	1	32	18	33	12	0.68

score quite highly on this scale, with very few people falling into the lower three levels of the trait, and with an EU-wide average level of 0.68.

Given the lack of success in finding a well-fitting joint model for the polytomous items, the question naturally arises whether the difficulties with the models above are due to retaining the distinction between DK and substantively incorrect responses. As it is much more common to model binary versions (correct versus incorrect, with DK counted as an incorrect answer) in the PUS literature, I briefly present the results of analyses of binary items here.

Country-by-country analyses suggest that one trait is sufficient to represent the data: percentages of large two-way marginal residuals range from 0 in five countries to 3.9 in Spain, and with a mean of 2.10, and all items take loadings of the same sign. The country-by-country models may be qualitatively similar, but their parameters are different enough to make a joint model, fixing the measurement model between countries, fit very poorly (43.1 per cent large two-way marginal residuals). As before, both deleting problematic items from the scale, and increasing the number of traits, improves fit dramatically. With

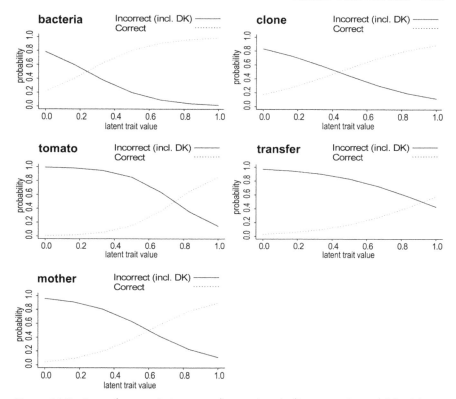

Figure 14.3 Item characteristic curves from a 1-trait discrete trait model for binary items, with measurement models equal for 15 countries.

five items, 26.3 per cent of two-way marginal residuals are large. Notably, a somewhat different set of items are retained here in comparison with the model for polytomous items—in particular, *transfer* is included in the scale. A two-trait model for this set of items improves the fit further (14.2 per cent large residuals), but with flat response curves for three of the five items, it is of questionable value, and might be interpreted as a case of over-fitting. The one-trait model for five items is therefore arguably the preferred model, all things considered. ICCs for it are shown in Figure 14.3. It is interesting that correct responses to the item *transfer* are predicted to belong to only those at the very top of the scale.

RESULTS OF SCALING THE GENERAL SCIENCE KNOWLEDGE ITEMS

Speculating on possible causes of the problems encountered with the biotechnology items, their content might be something to consider. Because in different models it was different items that caused the problems with fit, it seems not to be the case that odd question wording in a few places is to blame. Carrying out a similar set of analyses on the science literacy items therefore provides a useful point of comparison.

The analyses here are somewhat truncated, for reasons of space. Modelling items as polytomous, and running two-trait discrete trait models, country-by-country analyses identify a few items as problematic, as with the biotechnology items: the signs of some item loadings on a country's 'knowledge' trait return a counterintuitive interpretation. Deleting some items from the scale is therefore a necessary step. Based on informal analyses of the kind summarised in Table 14.2, nine items are retained for a first step towards finding a joint model. A two-trait model for these nine items, with equal measurement models between countries on both traits, fits poorly—although notably, not as poorly as the two-trait nine-item model for the biotechnology items (cf. 28.1 per cent large two-way marginal residuals for the former, versus 41.8 per cent for the latter). Reducing the number of items in the scale does very little to reduce the proportion of large two-way marginal residuals. However, a great improvement in fit is obtained by relaxing part of the measurement model: namely, allowing both slopes and intercepts to vary between countries on one trait, while constraining slopes to be equal across countries (with intercepts free to vary between countries) on the other. In this model only 4.1 per cent of two-way marginal residuals are large, and the fixed trait can feasibly be interpreted as representing low to high knowledge. Figure 14.4 shows ICCs for this model, for UK respondents—that is, with the fixed, Europe-wide slopes but UK-specific intercepts. There are a few points to note here. First, the slopes for correct responses are relatively steep, for all items, compared with those in the biotechnology set (cf. Figure 14.1). That is, the science items broadly speaking have greater discrimination power than the biotechnology items. Second, it is *not* the case for the science items that all of the 'false=correct' items are more difficult than the 'true=correct' items. The first three items in the diagram, in the left-hand column, are those for which 'true' is the correct response. The second, 'Electrons are smaller than atoms', is a relatively difficult item, whose difficulty is not attributable to response style. These are two attractive features of the item set—features that would be very desirable in the biotechnology items.

DISCUSSION: IMPLICATIONS FOR FUTURE SURVEY DESIGN

The results of these analyses suggest that deriving a statistically comparable scale of knowledge across the fifteen European countries, using the items available, is a difficult task—and more so for the biotechnology than for the science literacy items. However, the models also point quite clearly to a number of modifications that might be considered in designing future surveys that include these kinds of items.

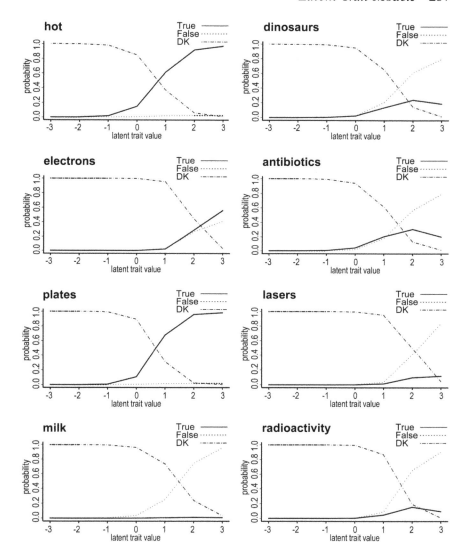

Figure 14.4 Item characteristic curves from final science literacy scale model, curves for UK.

Considering first of all the internal workings of the scale within countries, the most striking feature of the knowledge items is that many of them are relatively easy, and therefore not diagnostic, as the majority of respondents answer them correctly. However, it may not be that the items straightforwardly test facts that are very widely known. Those items for which 'true' is the correct response are easier (that is, more people answer them correctly) than those for which 'false' is correct. This could be an indication that the high rates of correct answers are attributable to response effects such as guessing, or acquiescence bias. The biotechnology items and the science literacy items are similar in this regard, although the problem is more acute for the biotechnology items. In their methodological analysis of the science items, Pardo and Calvo (2004) suggest that the scale could be improved by adding or substituting more difficult items in the set. They specifically recommend using more 'false=correct' items to increase the difficulty level of the test. They also suggest offering a four-point Likert answer scale, to allow respondents to differentiate between whether they think each statement is 'definitely' or 'probably' true or false, to alleviate guessing or other response effects. However, I would take a different approach. Response styles such as acquiescence bias and guessing are known to be more likely among certain demographics, including cultural groups (Smith, 2003). Increasing the number of 'false=correct' items might therefore lead the scale to favour a particular type of respondent, making it even more open to charges of bias. The very odd scaling behaviour of the item *transfer*, which requires a double negative for a correct response, is a good warning against over-using this strategy. It would undoubtedly be useful to try to increase the number of difficult items, but this would be more effective if it could be ensured that it was the content of the item, not the required response, which was difficult. So it would be advantageous if these more difficult questions were mixed, with some requiring 'true' and some 'false' as correct responses. The four-point Likert response scale may to some extent reduce the possible effects of guessing and acquiescence bias, but a more likely successful strategy might be to remove the true versus false dichotomy from the exercise altogether, instead asking respondents to choose between the two. Many of the original science literacy and biotechnology knowledge scale items could easily be reformulated in this style. For example, *It is the mother's genes that determine the sex of the child* would become a task of choosing between the statements *It is the mother's genes...* and *It is the father's genes...* Multiple choice items might also be considered. For example, the statement, *Ordinary tomatoes do not contain genes, while genetically modified ones do*, could be reformulated as a question, such as:

Which of the following contains genes?

A. Human beings
B. Fruits and vegetables
C. GM fruits and vegetables
D. A and B
E. B and C
F. All of the above.

Ideally the order of the first three response options would also be rotated. Multiple choice questions are more complex to analyse, and more costly to field, but should be seriously considered as a possible way of alleviating the response effects in the data.

Changing the item format altogether is a relatively drastic move: new items may always work less well than established ones, and new items prevent the analysis of trends over time. A more moderate strategy would be to add more difficult items to the existing set for the next survey, and evaluate the effectiveness of this before considering changing the question format altogether in future waves. Given the greater success in finding an adequate model for the science literacy items than the biotechnology items, steering the content of these questions towards general science and away from specific applications, such as biotechnology, might be advisable. Adding items to an existing set seems to carry a smaller risk of failure than moving wholesale to a different question format, but even with this approach, due consideration needs to be given to respondent fatigue and the relative importance of this construct versus the other topics that need to be covered in a given questionnaire. The matter requires a strategic decision by the survey designers. If a model could be found to fit the items in their current format, this may be the best solution to creating a measure of knowledge. The two-trait model (with one fixed 'knowledge' trait and one free 'response effect' trait) that fits the science literacy items is an example of such a model. However, using more than one trait, in order to account for response effects, such a model might not be as intuitive to a lay audience as a unidimensional knowledge scale. On a more positive final note, remember that in within-country analyses, the existing two-trait models deliver a good deal of useful information about the items. For cross-country comparisons, given the importance of knowledge to the PUS research field, more efforts to improve the scale would be valuable. Methodological critiques of biotechnology and science knowledge scales emphasise this. Pardo and Calvo (2004) point out that the weak association often found between measures of knowledge and attitudes might be partly attributable to the quality of the scales used. In their meta-analysis of the relationship between knowledge and attitudes, Allum et al. (2008) find the greatest variance in their model attributable to the measures used, and very little to cross-national differences. This in particular provides motivation to work towards a better fitting cross-national measure.

NOTES

1. This chapter is derived from research funded by a scholarship from the UK Economic and Social Research Council.
2. Weighted frequencies, weighting each country's contribution to the total according to their respective population sizes: totals do not always sum to 100 per cent due to rounding.
3. Models are estimated using Latent GOLD (Vermunt & Magidson 2005). Item characteristic curves and some fit statistics are calculated using functions in S-Plus kindly written by Dr Jouni Kuha.

4. For responses to each pair of items, I create a two-way marginal table, by collapsing over responses to the other variables. I then compare O, the observed frequency in a single cell of such a table, with E, the expected frequency for that same cell. The residual for each cell is calculated in standardised version, as $(O-E)^2/E$, where values greater than 4 are taken to indicate poor fit (Bartholomew et al., 2002). The greater the number of large residuals, the worse the model is, and I take as the fit statistic for each model the percentage of standardised marginal residuals greater than 4, overall as well as conditional on country, for cross-national models. For fuller details, see the same approach applied in the context of latent class models, in Stares (2009).
5. As a brief preliminary analysis for such a joint model, an indicative analysis was carried out, focusing on item discrimination power, defined as the discrimination parameters of correct in comparison with incorrect responses—that is, ignoring the slope estimates for DK responses, for the moment. If the discrimination parameters of certain pairs of items were in a significantly different *order* in different countries—say, if *tomato* were more highly discriminating than *bacteria* in some countries, but less highly discriminating than *bacteria* in others—it would be a clear sign that finding a well-fitting joint model would be difficult. The analysis was carried out with S-Plus software, using 95 per cent confidence intervals around the differences between slope estimates, and applying the Bonferroni correction to allow for multiple comparisons. In fact, only two pairs of items (*bacteria* with *yeast* and *tomato*) appear to have significantly different relative discrimination powers, and then, only between Portugal and Spain, for the first pair, and Portugal versus Spain and Denmark, for the second. This gives grounds for optimism that a cross-country model of knowledge, with fixed measurement models between countries, might be feasible.
6. Within countries percentages are unweighted, due to computation problems.

REFERENCES

Allum, N., Read, S., and Sturgis, P. (2010). Evaluating Change in Social and Political Trust in Europe Using Multiple Group Confirmatory Factor Analysis with Structured Means. In E. Davidov, J. Billiet, and P. Schmidt (Eds.) *Methods for Cross-cultural Analysis: Basic Strategies and Applications*. London: Routledge Academic.

Allum, N., Sturgis, P., Tabourazi, D., & Brunton-Smith, I. (2008). Science knowledge and attitudes across cultures: A meta-analysis. *Public Understanding of Science, 17*(1), 35–54.

Bartholomew, D. J., Steele, F., Moustaki, I. and Galbraith, J. G. (2008). *Analysis of multivariate social science data (Second edition)*. Boca Raton, FL: Chapman & Hall/CRC.

Converse, P. E. (2000). Assessing the capacity of mass electorates. *Annual Review of Political Science*, (3), 331–353.

Durant, J., Evans, G., & Thomas, G. (1989). The public understanding of science. *Nature, 340*, 11–14.

Heinen, T. (1996). *Latent Class and Discrete Latent Trait Models: Similarities and Differences*. Thousand Oaks, CA: Sage.

Irwin, A., & Wynne, B. (Eds.). (1996). *Misunderstanding Science? The Public Reconstruction of Science and Technology*. Cambridge: Cambridge University Press.

Jasanoff, S. (2000). The "Science Wars" and American Politics. In M. Dierkes & C. von Grote (Eds.), *Between Understanding and Trust: the Public, Science and Technology*. Amsterdam: Harwood Academic Publishers.

Jöreskog, K. G., & Moustaki, I. (2001). Factor analysis of ordinal variables: a comparison of three approaches. *Multivariate Behavioral Research*, 36, 347–387.
van der Linden, W. J., & Hambleton, R. K. (Eds.) (1997). *Handbook of Modern Item Response Theory*. New York: Springer-Verlag.
Miller, J. D. (1998). The measurement of civic scientific literacy. *Public Understanding of Science*, 7, 203–223.
Miller, J. D., & Pardo, R. (2000). Civic Scientific Literacy and Attitude to Science and Technology: A Comparative Analysis of the European Union, the United States, Japan, and Canada. In M. Dierkes & C. von Grote (Eds.), *Between Understanding and Trust: The Public, Science and Technology*. Amsterdam: Harwood Academic Publishers.
Miller, J. D., Pardo, R., & Niwa, F. (1997). *Public Perceptions of Science and Technology: A Comparative Study of the European Union, the United States, Japan, and Canada*. Madrid: BBV Foundation.
Pardo, R., & Calvo, F. (2004). The cognitive dimension of public perceptions of science: Methodological issues. *Public Understanding of Science*, 13, 203–227.
Peters, H. P. (2000). From Information to Attitudes? Thoughts on the Relationship between Knowledge about Science and Technology and Attitudes toward Technologies. In M. Dierkes & C. von Grote (Eds.), *Between Understanding and Trust—the Public, Science and Technology*. Amsterdam: Harwood Academic Publishers.
Raza, G., Singh, S., & Dutt, B. (2002). Public, Science, and Cultural Distance. *Science Communication*, 23(3), 292–309.
Shen, B. S. P. (1975). Scientific Literacy and the Public Understanding of Science. In S. Day (Ed.), *Communication of Scientific Information*. Basel: Karger.
Skrondal, A., & Rabe-Hesketh, S. (2004). *Generalized Latent Variable Modeling: Multilevel, Longitudinal and Structural Equation Models*. Boca Raton, FL: Chapman & Hall/CRC.
Smith, T. W. (2003). Developing Comparable Questions in Cross-National Surveys. In J. Harkness, F. van de Vijver, & P. P. Mohler (Eds.), *Cross-Cultural Survey Methods*. Hoboken, NJ: Wiley.
Stares, S. (2009) Using latent class models to explore cross-national typologies of public engagement with science and technology in Europe. *Science, Technology and Society*, 14(2), 289–329.
Sturgis, P. and Allum, N. (2004) Science in society: Re-evaluating the deficit model of public attitudes. *Public Understanding of Science*, 13, 55–74.
Thissen, D., & Wainer, H. (1982). Some standard errors in item response theory. *Psychometrika*, 47, 397–412.
Vermunt, J. K., & Magidson, J. (2005). *Technical Guide for Latent GOLD 4.0: Basic and Advanced*. Belmont, MA: Statistical Innovations Inc.

15 Statistical Modelling of Public Understanding of Science

Anil Rai and Rajesh Shukla

Recent advances in the field of computer technology have made it easier to apply complex statistical models/techniques to analyse survey data more meaningfully, which was not possible in the past. A number of computer statistical techniques have been developed for intensive data exploration and aggregations to draw meaningful conclusions.

In large-scale surveys the data on various quantitative as well as qualitative characteristics of a population are collected from sampled units. These surveys can be broadly divided in two categories on the basis of the type of analysis of the data. The first category of surveys is known as *descriptive*, in which the main concern of the researchers is drawing inferences through estimation of parametric functions of the target population. The second category of survey is known as *analytical*. In this category the main concern of the researchers lies in structural analysis of the population so that inferences drawn on the basis of this analysis are applicable to other similar populations as well. Categorical data (such as level of education, gender, occupation group etc.) analysis falls in the second category, especially in case of analysis of qualitative variables of survey.

In Public Understanding Surveys (PUS) respondents are asked a large number of questions, and most of the information collected is qualitative in nature. Variables considered in PUS are usually either ordinal or nominal in nature. In case of nominal variables order of level has no meaning or, in other words, it is non-informative, gender, i.e. male or female. But in the case of ordinal qualitative variables order and name of the level is quite informative and relevant, such as level of education (categorised as primary, secondary, high school, graduate and postgraduate). Therefore, aggregation of these variables using different perspectives and then drawing conclusions regarding different domains of human understanding with respect to science and technology has its challenges. Further, survey design for data collection is mostly complex in nature and sampling weights play an important role in unbiased estimation of parameters of interest.

The PUS surveys can be classified as analytical surveys, where the basic interest of researchers is to explore structural relationships among variables in the population. Here, level of scientific understanding and knowledge is dependent variables whereas other qualitative variables and predictors are explanatory variables. These relationships can be further used in future predictions. Therefore, statistical modelling of qualitative and quantitative variables from survey data is one of the major issues. In this article an attempt will be made to describe some of these techniques in the context of public understanding of science surveys. Further illustrative examples for these techniques will be taken from data collected in a National Science Survey conducted by the National Council of Applied Economic Research, New Delhi.

There are mainly three types of problems encountered in categorical data analysis. First is the measure of association, by which the degree of relationship between any two qualitative variables can be measured. Second is the testing of hypotheses, under which various hypotheses of interest, like goodness-of-fit, independence of attributes and homogeneity of proportions can be tested using chi-square test statistics. Third is the development of models using qualitative and quantitative variables. In the case of PUS, data testing of hypotheses for independent of attributes such as dependence of level of understanding and knowledge of science and technology on gender, level of education, social groups etc. is very important when drawing inferences related to population behaviour. Similarly, level of understanding about science and technology across the country/regions/social groups can be statistically tested using chi-square tests for estimating homogeneity of proportions. Hypothesis-testing of two-dimensional contingency tables is quite common, but applications of these tests in the case of multidimensional contingency tables have only been made possible recently, with the help of log-linear models. Bishop, Fienberg and Holland (1975) provided one of the earliest books in this rapidly expanding field. Books by Agresti (1996), Christensen (1997) etc. also provide details regarding these models. An application of these models in case of survey data is a little more complex.

Assessment of knowledge and understanding of an individual based on sets of scientific questions is a cumbersome and costly process as it requires aggregation of large number of qualitative variables to draw a meaningful conclusion. It is possible, instead, to develop a stochastic model to assess the level of understanding and knowledge of science and technology based on PUS survey data. Using this model, one can classify an individual with certain probability into different ordinal classes of knowledge (such as high, medium and low) and understanding and attitude toward science and technology.

In this article, an attempt has been made to develop a logistic regression model using PUS data so that any individual from a similar population can be classified into different knowledge category levels in probabilistic terms, based on his personal characteristics, i.e. without responding to a set of scientific questions. Data for development of this model has been taken from the National Science Survey conducted by the National Council of Applied Economic Research, New Delhi. Further, in the case of PUS surveys, researchers may be interested in exploring the relationship among different qualitative variables such as level of education, level of income, level of media exposure, occupational status, and gender along with level of understanding and knowledge about science and technology. Quite useful for this purpose are log-linear models, closely related to logistic regression modelling for hypothesis-testing of independence of attributes, in the case of a multidimensional contingency table. A brief description of these models is provided in the next section.

LOG-LINEAR MODELS

Categorical variables can be classified into two broad categories: (i) nominal and (ii) ordinal. Levels of nominal variables have no natural ordering, whereas ordinal variables have ordered levels. Interval variables are another class of variable; they have no distance between two levels of scale. In the measurement hierarchy, interval variables are at the highest level followed by ordinal and nominal variables. What this means in practice is that statistical techniques suitable for a lower level of measurement hierarchy can be applied to an upper level, but not at lower level. Interval variables are also called quantitative variables. The measure of association in 2x2 tables can be calculated by odds ratio but for I x J dimensional contingency tables, association is measured by taking the weighted average of various odd ratios. In order to measure ordinal association, subjects are classified into pairs as concordant or discordant. The pair is concordant if a subject ranked higher with respect to variable X is also ranked higher with respect to variable Y. The pair is discordant if the subject is ranked higher with respect to X and lower with respect to Y.

Statistical properties of log-linear models lie somewhere between analysis of variance (ANOVA) and the regression model. In this model, observed cell frequencies of contingency tables are modelled in the form of logarithms. There are a number of justifications for this form of modelling. First, it supports large sample theory and arises as a natural extension of mathematics from Poisson sampling. Further, in case of multinomial cell frequencies, bounds are quite awkward, i.e. between 0 and N, due to which estimation of parameters may be difficult, but in case of log-linear models, estimation and interpretation of parameters are quite straightforward. Extendability and scalability allow log-linear models to be extended for the analysis of multidimensional contingency tables.

Log-linear models are special cases of generalised linear models. All variables of interest in log-linear models are treated as response variables, as it does not discriminate between dependent and independent variables. If there is a need to treat variables of interest as dependent and independent variables, then logit or logistic regression technique is preferred. In case of logit or logistic regression analysis, quantitative variables can also be used as independent variables along with qualitative variables. The technique of log-linear modelling is based on fitting models to observed frequencies of the contingency tables, and then the model is capable of estimating an expected frequency under various null hypotheses under consideration. Different patterns of association among categorical variables can also be estimated with the help of odds ratios. In order to ensure simplicity of model fitting, it is desirable to fit a hierarchical log-linear model. For a hierarchical log-linear model, if a higher order interaction term is present in the model, then a corresponding lower order term should also be considered (see statistical details in Appendix 1).

There are a number of important and interesting interpretations regarding the inferential aspects of these models in terms of odds ratios. There are two approaches in model fitting (i) top-down approach and (ii) bottom-up approach. In a top-down approach the model is fitted by testing a saturated model and dropping lower order terms until the best fitted model is obtained, whereas in the case of a bottom-up approach, one starts with the simplest model (model of independence) and higher order terms are added until the best fitted model is obtained. Generally, parameters of the model are obtained using an iterative proportional fitting algorithm (Deming–Stephan algorithm). This algorithm generates maximum likelihood estimates of expected cell frequencies for a hierarchical model. Testing of model fitting is done using a chi square test or likelihood ratio test. The quality of model fit can be evaluated with the help of residual analysis. In this process standardised residuals are calculated by dividing the difference between observed and expected frequency by the square root of the expected frequency. A large difference in residual indicates the appropriateness of the fitted model to the data set. The following points need to be kept in mind before fitting log-linear models:

1. The number of variables is as few as possible.
2. The number of sample observations is five times more than number of cells.
3. The number of expected frequencies is greater than 1, and not more than 20% of cell expected frequencies are less than 5.

Applications of log-linear models for analysis of complex survey data require strong computational power. A brief review is given in subsequent section.

LOG-LINEAR MODELS IN SURVEY DATA

Cohen (1976) investigated a very special case of general problems with testing goodness-of-fit when data are collected with the help of a complex sampling design, particularly the special case of cluster sampling under the model of constant design effect. The design effect as defined by Kish and Frankel (1974) is the ratio of the variance under the sampling design to the variance under simple random sampling with replacement. The most sustained work on tests of independence from complex samples has been carried out by Nathan (1969, 1972, 1973, 1975). He also reviewed the work of several authors, such as Bhapkar and Koch (1968) and Chapman (1966). These authors concentrated their efforts on statistics that are closely related to the null hypothesis of testing independence of attributes. Several unbiased statistics have been proposed based on sample reuse techniques, notably the balanced repeated replication (BRR) method. The variance-covariance estimates for different cells of two-way contingency tables for statistics were estimated with the help of the same sample reuse method. The test statistics proposed by the above authors are in the form of a quadratic function of the above statistics. Unfortunately, test statistics proposed by this technique perform very badly with respect to their achieved significance levels, due to high correlation between the numerators and denominator of the proposed test statistics. The simulation results reported by Nathan (1973) are flawed, as pointed out by the author in his subsequent paper in 1975.

After careful examination of the problem of estimating the variance of non-linear statistics from complex samples, Fellegi (1980) proposed two test statistics in the light of existing literature. The first test statistic is based on Taylor's approximation of the statistics considered by Nathan (1973). The second set of test statistics is based on the approach of eliminating the effect of complex survey design from the usual Pearson's chi-square statistics, by dividing the statistic by the average of cell design effects. Rao and Scott (1979, 1981) have shown that for complex sampling designs, customary test procedures are not valid even asymptotically. It has been proved that Pearson's chi-square test statistic (X^2) and the likelihood ratio test statistic (G^2) are asymptotically distributed as a weighted sum of independent chi-square random variables, each with one degree of freedom. The weights attached to each chi-square random variable are eigenvalues of the design effect matrix, which is based on the concept of the usual design effect matrix, which is in turn based on the concept of the usual design effect of individual cells. Further, first-order corrections to the above statistics, i.e. X^2 and G^2, have been proposed so that their first moments become equal to their degrees of freedom as in the case of applications for multinomial sampling. Also, a second-order correction based on the Satherthwaite approximation (1946) has been applied to modify the X^2 and G^2 when data are obtained with the help of complex survey designs. Rao and Scott (1984, 1987) extended these modifications for multidimensional contingency tables with the help of log-linear models. Fay (1985) extended such modifications for testing for independence and various forms of conditional independence in contingency tables, by using resampling techniques such as jackknifing and BRR.

Some relatively less important work in the field of categorical data analysis in survey sampling is by Shuster and Downing (1976) who proposed methods for testing independence, quasi-independence and marginal symmetry in contingency tables derived for a variety of sampling schemes. Brier (1980) used Dirichlet multinomial distribution as a model for contingency tables generated by cluster sampling schemes. Koch et al. (1975) discussed certain aspects of multivariate analysis of the data from possibly complex survey designs in terms of large sample methodology involving weighted least-squares algorithms for the computation of Wald statistics. Holt et al. (1980) empirically studied the survey design effect on tests of goodness-of-fit, tests of homogeneity and tests of independence for British Economic Survey (BES) and General Health Survey (GHS) data. Similarly, other important empirical studies in this area are by Rao and Hidiroglou (1981), Kumar and Rao (1984), Fay (1984), Thomas and Rao (1984, 1985), Singh and Kumar (1986), Fay (1989) etc. Since, procedure of analysis of categorical data from survey sampling has been incorporated in SAS© (a data analysis software); it applies both first- and second-order corrections proposed by Rao and Scott (1981).

LOGISTIC REGRESSION ANALYSIS

Structural relationships among various qualitative variables in the population can be quantified using a number of techniques. In these techniques, primary interest lies on a dependent factor, which is dependent on other independent factors. This dependent factor is known as response factor. In this model building process, various log odds related to response factors are modelled. As a special case, if the response factor has only two categories with probabilities ϖ and $(1-\varpi)$ respectively then the odds of getting category one is $\varpi / (1-\varpi)$. If $\log(\varpi / (1-\varpi))$ is modelled using an analysis of variance (ANOVA) type of model, it is called a logit model. If the same model is being treated as a regression type model then it is called a logistic regression model. In a real sense, logit and logistic are names of transformations. For logit transformation, a number p between values 0 and 1 is transformed with $\log\{\varpi / (1-\varpi)\}$, whereas for logistic transformation a number x between $-\infty$ to $+\infty$ is transformed with $e^x /((1 + e^x)$ function. It can be seen that these two transformations are the reverse of each other, i.e. if logit transformation is applied on a logistic transformation function, it provides value x and similarly, if logistic transformation is applied to logit transformation function it provides value ϖ. (See statistical details in Appendix 2.)

General regression analysis is a method for investigating functional relationships among variables. Multiple regression linear models are based on one dependent variable and a number of independent or explanatory variables. Most of the variables in this model are quantitative in nature. Estimation of parameters in this regression model is based on four basic assumptions. First, the response or dependent variable is linearly related to the explanatory variables. Second, model errors are independently and identically distributed as normal variables with a mean zero and common variance. Third, independent or explanatory variables are measured without errors. The last assumption concerns equal reliability of observation. If our response variable in the model is qualitative in nature, then the probability of this response variable falling into given categories can be modelled in place of the response variable itself, using the same model, but with a number of constraints in terms of assumptions of the multiple regression model. First, the range of probability is between 0 and 1, whereas, right-hand side functions in the case of multiple regression models is unbounded. Second, the error term of the model can take only limited values, and error variance is not constant but depends on the probability of a response variable falling into a particular category.

THE PRESENT DATA SOURCE: THE INDIAN NATIONAL SCIENCE SURVEY OF 2004

Results presented in this chapter are primarily based on information collected through an all-India field survey called the "National Science Survey 2004" undertaken by the National Council of Applied Economic Research. Sample respondents, individuals over 10 years of age, were selected by adopting a multistage stratified random sampling design from a wide cross-section of people (age, education and sex) in the country. In view of India's diversity in terms of languages and locations, the sample size and selection procedure were designed to provide state-level estimates. Respondents were selected from the entire country by covering both rural and urban areas, with the objective of enhancing estimate precision.

The rural sample was selected from a representative number of districts across the country, and the urban sample was sampled from big metropolitan cities to small towns with populations below 5,000. A total of 346,000 individuals (115,000 rural and 231,000 urban) were listed, covering 553 villages in 152 districts as rural and 1128 urban blocks in 213 towns as urban. Over 30,000 individuals were selected from the listed individuals to collect detailed information through a questionnaire approach involving face-to-face interviews.

DATA ANALYSIS:

A major objective of this data analysis is to demonstrate the development of a stochastic model to find the probability of an unknown individual falling into different levels of knowledge categories, based on measurable and verifiable characterisation. After this model is developed, any individual belonging to similar population can be classified reliably into different knowledge categories such as high, medium and low based on his profile (i.e. without testing his knowledge through a questionnaire). This model can be developed using quantitative, qualitative or combination of quantitative and qualitative explanatory variables. A total 9,954 rural respondents covering major Indian states were selected for this analysis. The knowledge of individual respondents was categorised as low, medium and high based on index (score) obtained through responses to different questions from various sectors of science and technology. These responses were appropriately aggregated after normalisation, rescaling and weighting. The cut-off for categories is based on the average score and its standard error. Level of knowledge was taken as a response variable with three levels. The explanatory variables are level of education, occupation, age, sex, social group and mass media exposure. Of these, five are qualitative variables and exposure to mass media is quantitative variable. The extent of exposure to mass media of the respondent has been obtained based on frequency of exposure and perceived importance of the media. Table 15.1 provides details of the categorical variables.

It is difficult to assess the knowledge, understanding and attitude towards science and technology of a respondent as these must be based on a number of questions from a well-designed questionnaire. Further, analysis and aggregation of responses is also challenging and time-consuming as most responses to questions are qualitative in nature. Therefore, it is desirable to develop a model, based on a large-scale survey, which may be able to predict the probability of categorising respondents in a particular knowledge class based on simple and verifiable personal attributes. In order to take account of qualitative and quantitative variables for prediction of probabilities for categorising respondents in each knowledge category, logistic regression models have been fitted to the data collected.

In order to determine categories of dependent or response variable (level of understanding and knowledge about science and technology,) logistic regression model scores, obtained by aggregating correct responses to each question related to science and technology from the questionnaire, are calculated for each respondent. The scores of 9,954 rural respondents were arranged in ascending order and boundary cut-off points for each knowledge category, i.e. low, medium and high level of knowledge, were obtained based on percentiles. Scores below the 25th percentile, between the 25th percentile and 75th percentile and above the 75th percentile were assigned as low, medium and high levels of understanding and knowledge about science and technology, respectively.

Further, sampling design has been taken into account through sampling weights in this analysis. Stepwise logistic regression procedure has been adopted in fitting this model. In this procedure, knowledge categories have been taken as the response variable and stepwise explanatory variables

Table 15.1 Definition of Variables Used in the Analysis

Characteristics/Variable	Detail	Level Code
Response Variable		
	Level of knowledge, i.e. High (1), Medium (2) and Low (3), obtained through development of index (score) based on responses obtained for the science questions in the National Science Survey questionnaire.	
Explanatory Variables		
Education Categories	Below matriculation	1
	Matriculation (10th) up to 12th	2
	Graduate in any other subject	3
	Graduate in science	4
	Postgraduate in any other subject	5
	Postgraduate in science	6
	Diploma	7
Occupation Categories	Accountants and auditors	1
	Scientists and related workers	2
	Administrative, executive and managerial workers	3
	Clerical and related workers	4
	Service workers	5
	Production related workers	6
	Farmers	7
	Agricultural workers, non-agricultural workers and housewives	8
	Student	9
	Unemployed	10
	Others	11
Age Categories	Child (<18 years)	1
	Young (18–30 years)	2
	Middle Age (30–45 years)	3
	Upper Middle Age (45–60 years)	4
	Old Age (60> years)	5
Social Group	Scheduled Caste (SC)	1
	Scheduled Tribe (ST)	2
	Other Backward Caste (OBC)	3
	General	4

have been added to the model. At every step, the importance of the variable is assessed based on the chi-square score. The most important explanatory variable is selected in the first step, then all remaining variables are assessed and the next variable is entered into the model; the prediction performance of the models is assessed after removing the first variable. If performance improves then the first variable is dropped from the model. Otherwise both variables were considered, and the next most important of the remaining variables enters into the model. This process continues until all important explanatory variables are included in the model. Stepwise variables considered in this model building process have been given in Table 15.2.

This table also gives measures of association, Aikaki Information Criterion (AIC), percentage concordance, percentage discordance, percentage ties and Tau-C statistics. Here, the percentage concordance is a measure of average agreement of the actual category with the predicted category of level of understanding and knowledge about science and technology over all respondents. The Tau-C statistic is a similar measure.

It can be seen that level of education is the most important factor, followed by sex, age and social group, in enhancing level of knowledge in the field of science and technology. The contribution of other factors in increasing level of knowledge is either already explained by these factors or has little significance. The value of Tau-C lies between 0 and 1. Values closer to 1 indicate strong cumulative association of explanatory variables.

Table 15.3 shows estimates of various parameters for different levels of various factors along with percent of standard errors and level of significance. It may be noted that estimates of these parameters show the effect of various levels of parameters on knowledge, subject to constraints that the sum of all these effects of various levels for a given parameter is zero. Therefore, estimates are provided for K–1 levels if a factor has K levels. It can be seen from this table that all estimates are significant. Therefore all parameters need to be included in this logistic model.

Table 15.2 Results of Stepwise Logistic Regression: All India

Step	Variable Entered	AIC	Percentage Concordance	Percentage Discordance	Percentage Ties	Tau-C
1	Intercept	1.25	-	-	-	-
2	Education level	1.18	45.30	10.70	44.00	0.67
3	Sex	1.17	59.30	19.00	21.80	0.70
4	Age	1.16	68.10	26.30	5.60	0.71
5	Social Group	1.15	70.60	27.40	2.00	0.72
6	Occupation	1.15	71.50	27.40	1.00	0.72
7	Mass Media Exposure	1.15	72.30	27.40	0.30	0.73

Table 15.3 Logistic Regression Estimation Results: All India

Parameter	Level	Estimate	S.E. (%)	Pr>Chi-Square
Intercept 1	-	-1.7114	0.04	<.0001
Intercept 2	-	0.00814	7.33	<.0001
Sex (Male)	1	-0.2406	0.04	<.0001
Age categories				
Child (<18 years)	1	0.5756	0.05	<.0001
Young (18–30 years)	2	-0.3762	0.04	<.0001
Middle Age (30–45 years)	3	-0.2307	0.07	<.0001
Upper Middle Age (45–60 years)	4	-0.0149	1.22	<.0001
Education categories				
Below matriculation	1	1.4076	0.04	<.0001
Matriculation 10th up to 12th	2	0.4207	0.13	<.0001
Graduate in any other subject	3	0.0284	1.98	<.0001
Graduate in science	4	-0.0287	3.06	<.0001
Postgraduate in any other subject	5	-0.8744	0.11	<.0001
Postgraduate in science	6	-0.4342	0.49	<.0001
Occupation categories				
Accountants and auditors	1	0.2539	1.23	<.0001
Scientists and related workers	2	-0.1648	0.34	<.0001
Administrative, executive and managerial workers	3	-0.8474	0.15	<.0001
Clerical and related workers	4	-0.2809	0.27	<.0001
Service workers	5	0.0358	1.89	<.0001
Production related workers	6	0.2767	0.19	<.0001
Farmers	7	-0.0274	1.45	<.0001
Agricultural workers, non-agricultural workers and housewives	8	0.3194	0.12	<.0001
Student	9	-0.0587	0.78	<.0001
Unemployed	10	0.2816	0.17	<.0001
Social Group				
Scheduled Caste (SC)	1	-0.0821	0.21	<.0001
Scheduled Tribe (ST)	2	0.3397	0.06	<.0001
Other Backward Caste (OBC)	3	0.0300	0.04	<.0001
MME	-	-0.3049	0.05	<.0001

Using estimated parameters of the logistic regression models for different levels of response and explanatory variables, models can be developed based on respondent profile, and the probability of classification into different response levels can be obtained.

Table 15.4 provides details about measure of association when separate logistic models were fitted for different states. In this case percentage concordance ranges from 68.1 to 86.9. It is least for state 2 and highest for state 22. Conversely, percentage concordance is highest (31.6) for state 2 and minimum (12.9) for state 22. There is not much difference in terms of percentage ties. Tau-C statistics follow the expected trend according to percentage concordance or percentage discordance. Because states are likely to be more homogenous, accuracy of prediction for most of the states is quite high.

Table 15.4 Measure of Association for Different Statistics

S. No	State	Percentage Concordance	Percentage Discordance	Percentage Ties	Tau-C
1	Andhra Pradesh	75.1	24.5	0.3	0.753
2	Assam	68.1	31.6	0.3	0.683
3	Bihar	70.1	29.6	0.3	0.703
4	Haryana	77.5	22.3	0.2	0.776
5	Himachal Pradesh	83	16.8	0.2	0.831
6	Karnataka	78.6	21.1	0.4	0.788
7	Kerala	68.4	31.3	0.3	0.686
8	Madhya Pradesh	75.3	24.5	0.3	0.754
9	Maharashtra	79.7	20.1	0.2	0.798
10	Meghalaya	77.7	22	0.3	0.778
11	Orissa	82.7	13.8	3.5	0.844
12	Punjab	86.4	13.4	0.2	0.865
13	Rajasthan	83.3	16.4	0.3	0.835
14	Tamil Nadu	75.2	24.5	0.3	0.753
15	Uttar Pradesh	81.1	18.7	0.2	0.812
16	West Bengal	76.6	23.1	0.3	0.767
17	Pondicherry	72	27.7	0.3	0.722
18	Chhattisgarh	86.9	12.9	0.1	0.87
19	Uttaranchal	83.3	16.6	0.1	0.834
20	Jharkhand	84.9	14.9	0.2	0.85

It can be seen from this analysis that logistic regression analysis can be a very powerful tool for integration of qualitative and quantitative variables through modelling. It may be useful to make reliable predictions about an individual's characteristics, such as knowledge, understanding and attitude, on the basis of his measurable and verifiable attributes. Further, by using these models, one can predict social behaviour relating to science and technology without actually conducting large-scale surveys, as most predictor variables used in the model are available in many surveys conducted for some other purpose. These models can also be useful for development of a decision support system related to public understanding of science and technology.

APPENDIX 1: LOG-LINEAR MODELS

Let a two-dimensional contingency table be formed with the I and J levels of qualitative variables X and Y respectively. Let m_{ij}, i=1, 2, 3,————,I and j= 1,2,3, ... J denote the observed cell frequency in ij-th cell of the contingency table. These observed cell frequencies of ij-th cell can be modelled with the help of saturated log-linear model as

$$\log m_{ij} = \lambda_0 + \lambda_i^x + \lambda_j^y + \lambda_{ij}^{xy} \qquad \ldots \ldots (1)$$

where $\sum_{i=1}^{I} \lambda_i^x = \sum_{j=1}^{J} \lambda_j^y = \sum_{i=1}^{I} \lambda_{ij}^{xy} = \sum_{j=1}^{J} \lambda_{ij}^{xy} = \sum_{i=1}^{I}\sum_{j=1}^{J} \lambda_{ij}^{xy} = 0$

Here

l_0 = General mean

λ_i^x = effect of i-th level of variable X

λ_j^y = effect of j-th level of variable Y

λ_{ij}^{xy} = effect of interaction between i-th levels of variable X and j-th level of variable Y.

Model (1) is called a saturated log-linear model due to the fact that the number of parameters to be estimated in the model are equal to the number of degrees of freedom of observation in contingency table (i.e. I*J-1), so there is no degree of freedom for the error component. As a consequence of this, the estimated cell frequencies obtained from this model after fitting are always equal to observed cell frequencies. Under the null hypothesis of independence of row and column attributes (i.e. $\lambda_{ij}^{xy} = 0$) this model can be written

$$\log m_{ij} = \lambda_0 + \lambda_i^x + \lambda_j^y \qquad \ldots (2)$$

where $\sum_{i=1}^{I} \lambda_i^x = \sum_{j=1}^{J} \lambda_j^y = 0$

After fitting reduced model (2) we get expected frequencies under null hypothesis as

$$\log \hat{m}_{ij} = \hat{\lambda}_0 + \hat{\lambda}_i^x + \hat{\lambda}_j^y \qquad \ldots (3)$$

The usual chi-square test for statistics for independence of attributes can be calculated as

$$X^2 = \sum_{i}^{I} \sum_{j}^{J} (m_{ij} - \hat{m}_{ij})^2 / \hat{m}_{ij} \qquad \ldots (4)$$

The X^2 will follow chi-square distribution with (I-1)(J-1) degrees of freedom. The above log-linear model for two-dimensional contingency tables can further be extended to three-dimensional contingency tables for three variables (factors) X, Y and Z. This three-factor saturated model can be written as:

$$\log m_{ijk} = \lambda_0 + \hat{\lambda}_i^x + \lambda_j^y + \lambda_K^Z + \lambda_{ij}^{xy} + \lambda_{ik}^{xz} + \lambda_{jk}^{yz} + \lambda_{ijk}^{xyz} \qquad \ldots (5)$$

where, $\sum_{i=1}^{I} \lambda_i^x = \sum_{j=1}^{J} \lambda_j^y = \sum_{k=1}^{K} \lambda_k^z = 0$

$\sum_{i=1}^{I} \lambda_{ij}^{xy} = \sum_{i=1}^{I} \sum_{j=1}^{J} \lambda_{ij}^{xy} = \sum_{j=1}^{J} \lambda_{ij}^{xy} = 0$

$\sum_{i=1}^{I} \lambda_{ik}^{xz} = \sum_{k=1}^{K} \lambda_{ik}^{xz} = \sum_{i=1}^{I} \sum_{k=1}^{K} \lambda_{ik}^{xz} = 0$

$\sum_{j=1}^{J} \lambda_{jk}^{yz} = \sum_{k=1}^{K} \lambda_{jk}^{yz} = \sum_{j=1}^{J} \sum_{k=1}^{K} \lambda_{jk} = 0$

$\sum_{j=1}^{J} \lambda_{ijk}^{xyz} = \sum_{j=1}^{J} \lambda_{ijk}^{xyz} = \sum_{k=1}^{K} \lambda_{ijk}^{xyz} = 0$

$\sum_{i=1}^{I} \sum_{j=1}^{J} \lambda_{ijk}^{xyz} = \sum_{i=1}^{I} \sum_{k=1}^{K} \lambda_{ijk}^{xyz} = \sum_{j=1}^{J} \sum_{k=1}^{K} \lambda_{ijk}^{xyz} = 0$

$\sum_{i=1}^{I} \sum_{j=1}^{J} \sum_{k=1}^{K} \lambda_{ijk}^{xyz} = 0$

Interpretation of the terms in the three-factor model is analogous to that in the two-factor model. This model provides an opportunity to test the number of null hypothesis. Under the hypothesis of complete independence, i.e. every factor is independent of every other factor, model (5) can be written as

$$\log m_{ijk} = \lambda_0 + \lambda_i^x + \lambda_j^y + \lambda_k^z \qquad \ldots (6)$$

Further, under the null hypothesis that one factor is independent of other two factors when the relationship between the other two factors has no role, the model can be reduced in following form depending upon structure of null hypothesis under testing.

Case 1: X is independent of Y and Z

$$\log m_{ijk} = \lambda_0 + \lambda_i^x + \lambda_j^y + \lambda_k^z + \lambda_{jk}^{yz} \qquad \ldots (7)$$

Case 2: Y is independent of X and Z

$$\log m_{ijk} = \lambda_0 + \lambda_i^x + \lambda_j^y + \lambda_k^z + \lambda_{ik}^{xz} \qquad \ldots (8)$$

Case 3: Z is independent of X and . Y

$$\log m_{ijk} = \lambda_0 + \lambda_i^x + \lambda_j^y + \lambda_k^z + \lambda_{ij}^{xz} \qquad \ldots (9)$$

The null hypothesis of conditional independence, i.e. given a particular level of a factor, the other two factors are independent. There are three ways in which these conditioning factors can be chosen and saturated Log-linear model (9) can be written according to null hypothesis under consideration, i.e.

$$\log m_{ijk} = \lambda_0 + \lambda_i^x + \lambda_j^y + \lambda_k^z + \lambda_{ik}^{xz} + \lambda_{ik}^{yz} \qquad \ldots (10)$$

$$\log m_{ijk} = \lambda_0 + \lambda_i^x + \lambda_j^y + \lambda_k^z + \lambda_{ij}^{xy} + \lambda_{jk}^{yz} \qquad \ldots (11)$$

$$\log m_{ijk} = \lambda_0 + \lambda_i^x + \lambda_j^y + \lambda_k^z + \lambda_{ij}^{xy} + \lambda_{ik}^{xz} \qquad \ldots (12)$$

APPENDIX 2: LOGISTIC REGRESSION ANALYSIS

Let X be a vector of explanatory variables and denote the probability of a binary response variable. Logistic model is then given by

$$\log it(\pi) = \log\left(\frac{\pi}{1-\pi}\right) = \alpha + \mathbf{X}\boldsymbol{\beta} = g(\pi)$$

where α is the intercept parameter and β is a vector of slope parameters. In case the response variable has ordinal categories, say 1,2,3,------, I, I+1, then generally the logistic model fits with the common slope, based on cumulative probabilities of response categories instead of individual probabilities. This provides parallel lines of regression model. The mathematical form of this model can be written as:

$$g[\text{Prob}(y \leq i(x))] = \alpha_i + x\beta, \quad 1 \leq i \leq I$$

where $\alpha_1, \alpha_2, \text{------}, \alpha_k$ are k intercept parameters and β is the vector of slope parameters. However, in case of nominal response logistic model, logit model can be extended as

$$\log\left(\frac{\Pr(y=i/X)}{\Pr(y=i+1/X)}\right) = \alpha_i + \beta_i' x_i, \quad i = 1, 2, \text{------}, I$$

where $\alpha_1, \alpha_2, \text{------}, \alpha_I$ are intercept and $\beta_1, \beta_2, \text{------}, \beta_I$ are vector slope parameters. This is also known as a multinomial model. These models can be fitted through the maximum likelihood procedure, and effect of complex survey design can be included in this model fitting procedure using the SURVEYLOSTIC procedure in SAS. This procedure uses either the Fisher-scoring algorithm or Newton Raphson algorithm. As already mentioned, it can be seen that logit function $g(\pi) = \log\left(\frac{\pi}{1-\pi}\right)$ is inverse of logistic function, i.e.

$$\pi = \frac{e^z}{(1+e^z)} = \frac{1}{(1+e^{-z})}$$

where $z = X\beta$

The generalised logit function can be written by extending binary logit to a vector of levels $(\pi_1, \pi \text{-----}, \pi_{z+1})$ by contrasting each level with a fixed level

$$g(\pi_i) = \log\left(\frac{\pi_i}{\pi_{I+1}}\right), \quad i = 1, 2, \text{---}, I$$

Model fitting statistics can be obtained with the help of the following three criteria.

* - 2 Log—Likelihood :

$$-2 \log L = -2 \sum_{i=1}^{I} w_i f_i \log(\hat{\pi}_i)$$

where w_i and f_i denotes weight and frequency values of the ith observation.

* *Akaiki Information Criterion:*

$$AIC = -2 \log L + 2\,[(I-1) + (k+1)g]$$

*where I is total number of response levels and k is number of explanatory variables.

* *Generalised Coefficient of Determination:*

Cox and Snell (1989) proposed the following generation of the coefficient of determination.

$$R^2 = 1 - \left\{\frac{L(0)}{L(\hat{\theta})}\right\}^{2/n}$$

where $L(0)$ is likelihood of intercept model, and $L(\theta^{\wedge})$ is likelihood of specified model. Maximum value of R^2 can be obtained as

$$R^2_{\max} = 1 - \{L(o)\}^{2/n}$$

Nagelkerke (1991) proposes the following adjusted coefficient, which can achieve a maximum value of 1.

$$R^{*2} = \frac{R^2}{R^2_{\max}}$$

In order to incorporate the effect of complex sampling design in this process of model fitting and estimation, let be the response variable with $1,2,3,\ldots,I, I+1$ categories. Let there be K covariates of Y, which were denoted by vector **X**. Let population be stratified into H strata, and hth stratum consists of n_h, $h=1,2,---,H$ clusters. Let

$$n = \sum_{h=1}^{H} n_h,$$

i.e. total number of clusters in the population. Further, m_{hc} denotes number of sampling units in i^{th} cluster, $c=1,2,\ldots, n_h$ of ith stratum. Let Y_{hcj} be I-dimensional vector of indicator variables for the first I categories from ith cluster of hth stratum. In this case, if jth sampling unit of ith cluster in hth

Statistical Modelling of Public Understanding of Science

stratum falls in i^{th} category then i^{th} row of the vector is 1, and remaining elements are zero. Also, $Y_{hcj(I+1)}$ denotes a similar vector for $(I+1)^{th}$ category. Similarly, x_{hij} denotes R dimensional row vector of explanatory variables for j^{th} sampling unit from i^{th} cluster in h^{th} stratum. If there is intercept then $x_{hij}=1$. Let $\underline{\pi}_{hij}$ be the expected vector of response variables, i.e.

$$\pi_{hij} = E\left(\frac{Y_{hij}}{\underline{x}_{hij}}\right)$$

and $\pi_{hij(I+1)} = E\left(\dfrac{Y_{hij(I+1)}}{\underline{x}_{hij}}\right)$

The generalised logistic model can be defined as

$$\pi_{hij(i)} = \frac{e^{\underline{x}hij\underline{\beta}_i}}{1+\sum_{i=1}^{I} e^{xhij\beta_i}},$$

where $\underline{\beta}_i = (\beta_{i1}, \beta_{i2}, \ldots, \beta_{ik})$

The estimate of variance for the parameter of the model can be obtained using the Taylor approximation, i.e.

$$\hat{v}(\hat{\theta}) = \hat{\theta}^{-1}\hat{G}\hat{\theta}^{-1}$$

where $\theta = (\beta'_1, \beta'_2, \ldots, \beta'_I)$

$$\hat{\theta} = \sum_{h=1}^{H}\sum_{i=1}^{nh}\sum_{j=1}^{mch} W_{hij}\, \hat{\mathbf{D}}_{hij} \left[diag(\hat{\underline{\pi}}_{hij}) - \hat{\underline{\pi}}_{hij}\hat{\underline{\pi}}'_{hij}\right]^{-1} \hat{\mathbf{D}}^{1}_{hij}$$

$$\hat{G} = \frac{n-1}{n-k}\sum_{h=1}^{H} \frac{n_h(1-f_h)}{n_h-1} \sum_{i=1}^{n_h} \left(\underline{\mathbf{e}}_{hi} - \overline{\underline{\mathbf{e}}}_{h..}\right)\left(\underline{\mathbf{e}}_{hi} - \overline{\underline{\mathbf{e}}}_{h..}\right)'$$

$$\underline{e}_{hi} = \sum_{j=1}^{mhc} W_{hij}\, \underline{\hat{\mathbf{D}}}_{hij}\left(diag(\hat{\underline{\pi}}_{hij}) - \hat{\underline{\pi}}_{hij}\hat{\underline{\pi}}'_{hij}\right)\left(\mathbf{Y}_{hij} - \hat{\underline{\pi}}_{hij}\right)$$

$$\overline{e}_{h..} = \frac{1}{n_h}\sum_{i=1}^{nh} e_{hi}$$

Hence, $\hat{D}_{hij}$ is the matrix of the partial derivative of $\underline{\pi}$, and f_h is a sampling fraction of hth stratum.

NOTES

1. Dr. Anil Rai, Principal Scientist, Indian Agricultural Statistics Research Institute, Library Avenue: New Delhi (India); Dr. Rajesh Shukla, Director, NCAER-CMCR, New Delhi (India).

REFERENCES

Agresti, A.C. (1996). *Categorical Data Analysis*. New York: Wiley & Sons.
Bhapkar, V.P. and Koch, C.G. (1968). Hypothesis of no interaction in multidimensional contingency tables. *Technometrics*, 10, 107–123.
Bishop, Y.M.M., Fienberg, S.E. and Holland, P.W. (1975). *Discrete Multivariate Analysis: Theory and Practice*. Cambridge: Massachusetts Institute of Technology Press.
Brier S. S. (1979). Categorical data models for complex sampling schemes. Ph.D. thesis University of Minnesota USA.
Chapman, D.W. (1966). An approximate test of independence based on replication of complex survey design. Unpublished master's thesis, Cornell University.
Christensen, R. (1997). *Log-Linear Models and Logistic Regression*. New York: Springer-Verlag.
Cohen, J.E. (1976). The distribution of chi-square statistics under cluster sampling from contingency tables, *J. Amer. Stat. Assoc.*, 71 (355), 665–670.
Cox, D. R. and Snell, E. J. (1989). *The Analysis of Binary Data, Second Edition*, London: Chapman and Hall.
Fay, R.E. (1985). A jackknife chi-square test for complex samples. *J. Am. Statist Assoc.*, 80, 148–157.
Fay, R.E. (1989). Additional evaluation of chi-square methods for complex samples. *Proc. Am. Statist. Assoc. Sur. Meth. Sec.*, 680–685.
Fellegi, I.P. (1980). Approximate tests of independence and goodness of fit based on stratified multistage samples. *J. Am. Statist. Assoc.*, 75, 261–268.
Kish L. and Frankel, M.R. (1974). Inference from complex samples. *J. Roy. Stat. Soc., Series B*, 36, 1–37.
Koch, C.G. Freeman, D.H. and .Freeman, J. L. (1975). Strategies in Multivariate Analysis of Data from Complex Surveys. *Int. Statist. Rev.*, 43, 59–78.
Kumar, S. and Rao, J.N.K. (1984). Logistic regression analysis of labour force survey data. *Sur. Meth.*, 10(1), 62–81.
Nagelkerke, N. J. D. (1991),. "A Note on a General Definition of the Coefficient of Determination," *Biometrika*, 78, 691–692.
Nathan, G. (1969). Test of independence in contingency tables from stratified samples. In N.L. Johnson and H. Smith, eds. *New Developments in Survey Sampling*. New York: Wiley, 578–600.
Nathan, G. (1972). On the asymptotic power of tests for independence in contingency tables from stratified samples. *J. Am. Statist. Assoc.*, 6, 917–920.

Nathan, G. (1973). Approximate tests of independence in contingency tables from stratified samples. *N.C.H.S., Vital and Health Statistics. Series-2, No. 53.* Washington D.C.

Nathan, G. (1975). Test of independence in contingency tables from stratified proportional samples. *Sankhya*, 37 (C) Part I, 77–87.

Rao, J.N.K. and Scott, A.J. (1979). Chi-squared tests for analysis of categorical data from complex surveys. Proceedings of American Statistical association, Section on Survey Research Methods, 58–66.

Rao, J.N.K. and Scott, A.J. (1981). The analysis of categorical data from complex sample surveys. Chi-square tests for goodness-of-fit and independence of two-way tables. *J. Am. Statist. Assoc.*, 76, 221–230.

Rao, J.N.K. and Scott, A.J. (1984). On chi-square tests for multiway contingency tables with proportions estimated from survey data. *Ann. Statist.*, 12, 46–60.

Rao, J.N.K. and Scott, A.J. (1987). On. Simple adjustment of chi-squared tests with sample survey data, *Ann. Stat.* 15, 385–397

Satterthwaite, F. F.(1946).An Approximate Distribution of Estimates of Variance Components. *Biometrics*, 2, 110–114.

Shuster, J.J, and Downing (1976). Two-Way Contingency from from complex sampling schemes. *Biometrika*, 63, 271–276.

Skinner, C.J., Holt, D. and Smith, T.M.F. (1989). *Analysis of Complex Surveys.* New York: Wiley & Sons.

Thomas, D.R. and Rao, J.N.K. (1984). A Monte Carlo study of exact levels of chi-squared goodness-of-fit statistics under cluster sampling. *Technical Report, 66,* Ottawa: Carlton Univ.

Thomas, D.R. and Rao, J.N.K. (1985). On the power of some goodness-of-fit tests under cluster sampling. *Technical Report, 66,* Ottawa: Carlton Univ.

Thomas, D.R. and Rao, J.N.K. (1987). Small sample comparisons of level and power for simple goodness of fit statistics under cluster sampling. *J. Am. Statist. Assoc.*, 82, 630–636.

Thomas, D.R., Singh, A.C. and Roberts, G. (1989). Size and power of independence tests for R*C tables from complex surveys. *Proc. Am. Statist. Assoc.*, 763–768.

16 The Cultures of Public Understanding of Science
Defining Cultural Distance

Gauhar Raza and Surjit Singh

Scientific ideas, laws, information, data and methods generated by specialists through what is termed scientific practice follow a complex trajectory to reach the masses (Raza et al. 2002). Even after revalidation and acceptance within the scientific community, a scientific idea takes a fairly long time to become part of the cognitive structure of a given cultural set of common citizens. The notions that are today considered commonsense by a majority of citizens all over the world have taken a few centuries to become an integral component of the worldview of a sizeable population segment. The revolution of the earth, absence of celestial spheres to hold heavenly bodies, bacteria and viruses causing disease and lightning caused by charged clouds are but a few examples of the lag between discovery and its percolation among the lay public. In recent times the speed, spread, efficacy and efficiency of communication channels has increased the velocity of propagation of scientific information, and acceptance of new technologies among the masses has also increased (Stamm, Clark, and Eblacas, 2000).

Despite considerable reduction in time lag, the resistance offered to scientific information or ideas varies a great deal across the board.[1] What are the operand factors that determine the relative resistivity? Do ideas compete with each other to remain or become part of what Bauer calls 'commonsense' (Bauer, 2009)? Is this 'commonsense' a social, cultural or ethnic constructions? Therefore as these parameters change does 'commonsense' also change? In the context of public and science, is there one and only one definable 'commonsense' or are there many 'commonsenses' that could be placed at varying cultural distances from the scientific knowledge system. If there is a natural divide between the scientific method of configuring reality and peoples' worldviews then is the divide measurable? These are some of the pertinent questions with which scholars have been grappling for quite some time.

A BRIEF HISTORICAL OVERVIEW

The debate on science and its relationship with the public is almost as old as modern science. It has been argued that the nature of the discourse has undergone radical changes (Raza et al. 2002: 293)[2]. In the eighteenth century scholars realised that the scientific community had developed distinct 'linguistic peculiarities' that posed a hurdle in science reaching ordinary citizens. By the mid-eighteenth century 'the term 'science' came to be used in a narrow sense to refer just to the 'physical' or 'natural' sciences' (Collini, 1993: xi). Stefan Collini informs us that in the 1930s the nomenclature proposed, by analogy of *artists,* for 'students of knowledge of material world' was *scientist*. For the next 30 to 40 years science had to struggle to gain anything like parity in national curricula in Europe. Gradually, as the community gained acceptance it also developed a formal language, which depended heavily on mathematical expressions.

Mathematics offered precision, reducing any ambiguity that could be introduced while sharing scientific tenets in ordinary language. In order to communicate methodology and results of scientific research, scientists increasingly used mathematical symbols, notations and equations. In 1937, J.D. Bernal, one of the most outstanding minds of his time, recognised these inherent characteristics of science and asserted that 'science itself has proceeded almost entirely by the method of isolation and precise definition of categories independent of time', and this 'isolation in science, however, can only be achieved by a rigorous control of the circumstances of the experiment or application (Bernal, 1937). The lay public, even in the twenty-first century, is not expected to achieve 'rigorous control of circumstances', and therefore, the distance between common citizen and science, if left without intervention, is bound to increase.

In the early stages of the discourse it was realised that the divide—besides being caused by a natural lag in knowledge-generation-validation and its dissemination—is also cultural (Snow, 1993). Although Snow acknowledged the heterogeneity that exists among subgroups within the large set of scientists, for him the hallmarks of scientific culture were common attitudes, common standards and patterns of behaviour, common approaches and assumptions prevalent among the community of scientists. Thus he concluded, " . . . the scientific culture is really a culture, not only in an intellectual but also in an anthropological sense" (Snow, 1993: 9).

SCIENCE, SCIENTIFIC METHOD AND CULTURE

The scholarly debate that followed Snow's lectures has questioned almost all the building blocks on which his argument rested, yet the conclusion—that science as a social activity and scientists as agents of knowledge generation have a distinctive culture—still remains the focus of the debate (Leavis, 1962).

Science, as a human activity with its own culture, develops filters to sift science from non-science on a continual basis. After passing through a set of membranes the residue is accepted as valid scientific knowledge. It should be noted that as science progresses, the filters also undergo a process of refinement, and old-valid-knowledge is subjected to re-filtration. Different scientists and disciplines within science use distinct sets of filters to arrive at conclusions, which, after acceptance among the experts, constitute valid scientific knowledge. A set of filters put together could be broadly categorised as the 'scientific method'. Durant argues that scientists during their investigations use 'a great variety of exploratory stratagems' for arriving at a set of results that could be accepted, first, by the community of experts, then granted passage into the 'corpus of existing knowledge' and eventually assimilating into the common citizens' worldview (Durant, 1993).

On the other hand, 'the ambiguity of the concept of culture is notorious' (Sardar and Loon, 1997). Anthropologists have made attempts to define culture in many ways, yet none of them are universally acceptable. One end of the definition-spectrum is occupied by statements such as 'culture is that whole complex which includes knowledge, belief, arts, morals, law, customs, and other capabilities and habits acquired by man as a member of society' (Tylor, 1924: 1). On the other end, cultural anthropologists define culture as a mental construct or an ensemble of stories we tell ourselves about ourselves. Geertz notes that 'it is an essentially contested concept, like democracy, religion, simplicity, or social justice, it is a multiply defined one, multiply employed, incredibly imprecise' (Geertz, 1999: 45).

PUBLIC AND PUBLIC UNDERSTANDING OF SCIENCE

The notion of the gap between science and the public and the notion of public have undergone radical changes over the past three centuries (Bensaude-Vincent, 2001). In the first phase, i.e. in the eighteenth century, the wedge was positioned in the 'style of argumentation'; later with the 'formalisation and mathematization of science it became linguistic one' and in mid-twentieth century 'the scientists and ordinary people live(d) in two different worlds' (108). Bensaude-Vincent notes that during the first two phases of discourse, the public was not presupposed to be just a mass consumer of science; the notion of an ignorant public only emerged in mid-twentieth century.

In the recent past, indicators for judging the level of scientific literacy were developed, and large-scale surveys were administered, first in the United States, followed by Europe and elsewhere (S. Miller, 2001: 116; Zhongliang, 1991: 314).[3] Without explicitly stating it, these pronounced a large segment of the population 'scientifically illiterate'.[4] Serious doubts were raised about the purpose and intent of this approach and finally effected the 'end of the deficit model', at least in United Kingdom, if not in other parts of the academic world engaged in large-scale survey studies (Raza, Singh and Dutt, 1995). However, Steve Miller points out that it 'does not mean there in no knowledge deficit' among lay people (Miller, 2001: 118). These criticisms played a crucial role in the emergence of a more realistic perspective usually termed the 'contextual approach'.

Godin and Gingras, while analysing the evolutionary trajectory of various approaches, suggest that scientific knowledge generation and its appropriation is essentially a form of social organisation of culture (Godin and Gingras, 2000). It is crucial to point out here that the contextual model they suggested needs further refinement (Godin and Gingras, 2000: 53). The model places the scientific community at the centre of the broad set of culture. There is a need to recognise that the large universal set of human culture consists of smaller subsets, each one protected by a boundary. Though these membranes are permeable, they offer varied resistance to the transfer or even appreciation of knowledge generated and attitudes and perceptions practiced in one cultural subset to another. It is also important to note that these boundaries often cut across each other and could only be defined in terms of various multidimensional socialisation processes prevalent in a given society. For example social determinants such as linguistic barriers (Lee, Fradd and Sutman, 1995) or caste differentiation or a combination of both could influence the permeability of these boundaries to a significant extent.

DEFINING CULTURAL DISTANCE

For the purpose of the present investigation cultural distance is defined as the distance that a worldview, an attitude, a perception or an idea, generated within one cultural context, travels on a time scale, for its democratisation within the thought structure of the other cultural group/s. Let us now make an attempt to test the hypothesis that various ideas generated within the scientific realm of thought could be placed at varying cultural distances from the quotidian life of a set of common citizens. A few social determinants of the permeability of cultural membranes have been listed; however they do not constitute the focus of the argument.

The authors have argued elsewhere that factors inherent within the scientific knowledge system, such as complexity involved in explaining the phenomenon, duration of its life cycle, the control that an individual or a collective can exercise and the intensity with which the phenomenon could influence life of a common citizen, are the most important factors that determine this cultural distance. It was also observed earlier that as the extent of inherent mathematical obscurity and conceptual contortion required for explaining a phenomenon increases, the pace of its propagation reduces progressively (Raza, Singh and Dutt, 1996).

A common citizen who has not been exposed to higher levels of mathematics and abstract scientific ideas is likely to invoke intuitive, cultural or even religious explanations when confronted with a complex natural phenomenon (Raza, Dutt, Singh, and Wahid, 1991). But such explanations offered by the populace cannot be construed as a measure of their scientific-illiteracy or irrationality (Durant and Bauer, 1992). Non-experts, including the illiterate segments of society, to a high degree rely on scientific explanations, as was observed during a survey conducted in Delhi among plague-affected citizens (Raza, Dutt and Singh, 1997).

INDEX OF DEMOCRATISATION AND MEASURING THE CULTURAL DISTANCE

At this stage we propose a simple method by which one can determine this cultural distance. In order to explain the point we consider a dichotomous response variable where the first category is constituted by valid scientific answers and all scientifically invalid responses are clubbed together to form the second category (see Figure 16.1). The percentage response variable is plotted on a scale constituted by years of formal schooling that a respondent has received. For convenience sake let us call it the *education variable* and plot it on the x-axis. The y-axis characterises the dichotomous percentage response variable and is represented by two curves: i.e. scientifically valid and scientifically invalid responses offered by the populace interviewed.[5] Because the response variable is dichotomous the two curves will always intersect each other at a point where fifty per cent of those who were interviewed offered valid scientific explanations.[6] The perpendicular drawn from this point on the x-axis would show that the concept or idea or information under scrutiny at this point has become part of the cognitive structure of fifty per cent of population under discussion. In other words in the region on the right side of the perpendicular more than fifty per cent of the populace subscribe to scientific explanation, and on the left is the region where more than fifty per cent subscribe to it. Let us call this point *index of democratisation (id)* of a concept beyond which an incremental increase in the value of X would mean that more than fifty per cent of the population subscribe to valid scientific explanation.

Conversely, we could say that a piece of scientific information or concept has to travel on the x-axis for a given number of years in order to achieve the threshold level *id* at a given point of economic and sociocultural development (Miller, Pardo and Niwa, 1997).[7]

The idea here is not to establish that socialisation in modern education is the only determinant that influences the worldview of a common citizen, and a change in educational level will bring about an identical change in the worldview, irrespective of other factors.[8] It has been established repeatedly

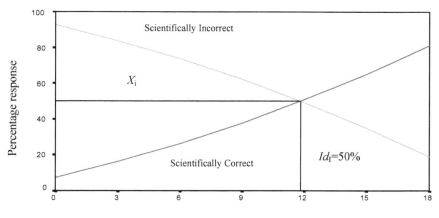

Figure 16.1 Measuring cultural distance: the average number of years of schooling when 50% of a population attain correct answers on the item.

that a whole host of factors, external to the nature of scientific information, such as gender, occupation, access to non-formal channels of information, economic status, predisposition to cultural and religious activities and age, have a bearing on the worldview of cultural formations and subgroups (Raza et al., 1991). We here propose to develop a scale on which the comparative cultural distance of various scientific concepts and information from quotidian life can be mapped. The method helps us measure the distance of the index of democratisation *(id)* on the x-axis from the origin. In most survey studies, besides other factors, respondents' years of schooling are recorded (NSF Report, 1998).

Using curve-fitting techniques any empirical data set collected from the field can be used to determine id_i and the distance x_i for each of the questions c_i posed to the populace. The curve plotted on a two-dimensional graph, i.e. response variable verses education, can now be reduced to a one-dimensional plot without loss of information. We could on the education scale plot *id* for each concept at a corresponding cultural distance x and also measure their comparative distance from the quotidian life of the interviewed populace. The larger the distance x_i for a given natural phenomenon or episode in the first quadrant, the farther would be the phenomenon from the quotidian life of the population segment under scrutiny.

BROAD PROPERTIES OF THE DATA SETS USED

For validation three data sets were used. The first and the second data sets were gathered during two surveys conducted at Allahabad, a town in northern India where a large section of the Indian population had gathered to participate in a religio-cultural event (Bhattacharya, 1983). These samples were collected in the years 2001 and 2007.[9] The third data set was generated as part of a national survey carried out in 2004.[10]

The first two data sets collected in 2001 and 2007 predominantly represent the north-central part of India. Although almost all the states of India were represented in the data sets, the sample was skewed in favour of three states, Uttar Pradesh, Madhya Pradesh and Bihar[11] (see Tables 16.1 and 16.2). It could also be argued that people who visited Allahabad during the Kumbh Mela[12] were favourably predisposed towards religious and cultural structures of configuring natural phenomena.

The third data set[13] (CG_2) represents the national population. Sample respondents, individuals over 10 years of age, were selected by adopting a multistage stratified random sampling design from a wide cross-section of people (age, education and sex) in the country by covering both rural and urban areas. In view of India's diversity in terms of languages and locations, the sample size and selection procedure were designed to provide state level estimates. A total of about 347,000 individuals (115,000 rural and 232,000 urban) were listed, covering 553 villages in 152 rural districts and 1,128 urban blocks in 213 towns. Over 30,000 individuals were selected from those listed; from them, detailed information was collected using a questionnaire involving face-to-face interviews.

For convenience, the data sets have been placed here chronologically. CG_2 covers 23 states of the country. In order to perform statistical operations, representative weighted data was used. Therefore, it could be safely concluded that this data set represents the economic, social and cultural diversities and imbalances that exist across the Indian provinces.

EFFICACY OF CULTURAL MODEL OF PUBLIC UNDERSTANDING OF SCIENCE

At the outset it should be stated that out of all the questions posed to the respondents, thirty-seven figured in the schedules used for administering the three survey studies. Out of these, five questions were chosen for testing the efficacy of the suggested method (see Table 16.1). These questions were related to the area of physics and cosmology.

All three data sets show a fair degree of statistical consistency (Table 16.2).[14] In order to determine the precise values of cultural distance Xc_i, a dummy variable was created. The aggregate values of $X_{c\ mean}$ were also computed, through SPSS, using this dummy variable. In 2002, after plotting the dichotomous response graphs, the values of Xc_i were computed manually and thus were approximate. This explains the difference between the present values and the earlier reported values Xc_i for 2001 data.

Table 16.1 Questions Posed to the Respondents

Idea	Question	Concept	Scientifically Valid Responses	Scientifically Invalid Responses
C_1	What is the shape of Earth?	Rotundity of earth	Earth is round/ spherical	Flat, Long, Beautiful
C_2	How do day and night form?	Rotation of earth	Earth rotates on its axis	Revolution of Sun, Sun goes behind mountains at night, God's will
C_3	What causes eclipse?	Revolution of earth	Shadow of earth or moon/earth / moon come between Sun and moon/earth	Shadow of stars, Rahu-Ketu, God's will
C_4	What is Akaash Ganga (Milky Way)?	Formation of galaxy	Collection of stars	Mist, heavenly river, Bhagwati Ganga, path for ghosts, path for heavenly Gods
C_5	How did humans come to being?	Theory of evolution	Humans evolved from other life forms	Monkeys, Baba Adam and Eve, Brahma, the creator

The values of cultural distance clearly show that the notion 'rotundity of earth' could be placed closest to the quotidian life of all three cultural groups. The values were -1.0, -1.4 and -2.0 for 2001, 2004 and 2007. The data analysis also revealed that the theory of evolution occupies the farthest end of cultural distance scale. The values of cultural distance (X_c) for C_5 were 18.8(2001), 16.5(2004) and 18.0(2007) for the three data sets. The other four concepts occupy positions between these two extremes. Measured from the point of origin, X_{c3}, revolution of earth, occupied third position on the scale, its cultural distance from the three sampled populations was 9.1(CG_1), 8.5(CG_2) and 9.1(CG_3). In India, formation of eclipse (C_4) is a phenomenon that is deeply associated with myths, superstitions and religious practices. Therefore, as expected the comparative cultural distance of C_4 from the quotidian life of three sampled sets is large when compared to the previous concepts. The computed values of X_{c4} for CG_1, CG_2 and CG_3 were 12.0, 11.5 and 11.5 respectively.

The $X_{c\ mean}$ reported in the table are not the simple average of cultural distances but computed from individual response values recorded in the database.

Table 16.2 Statistical Properties and Cultural Distance

Estimate	Years		
	2001 (CG1)	2004 (CG2)	2007 (CG3)
$Xc1$ shape	-1.0	-1.4	-2.0
$Xc2$ rotation	8.5	5.2	7.1
$Xc3$ revolution	9.1	8.5	9.1
$Xc4$ galaxy	12.0	11.5	11.5
$Xc5$ evolution	18.8	16.5	18.0
Xc mean	9.5	8	9.2
Cronbach's alpha	0.721	0.805	0.749
Standard deviation	5.47	5.47	5.15
Skewness	-1.094	-0.816	0.755
Kurtosis	2.188	1.746	2.10
Standard error (Mean)	0.094	0.032	0.084
Standard error (Skewness)	0.042	0.040	0.040
Standard error (Kurtosis)	0.084	0.079	0.080
Coefficient of variation	0.58	0.68	0.56

CULTURAL DISTANCE OF VARIOUS GROUPS FROM SCIENTIFIC CONCEPTS

We also proposed earlier that for a given scientific concept, notion, piece of information or tenet, various cultural groups could be placed at different cultural distances. Table 16.2 also shows that the cultural distances of CG_2 from all scientific concepts were consistently lower compared to the other two sets. The reason could be attributed to the change in the nature of the sampled population. It was pointed out earlier that CG_1 and CG_3 predominantly represented the three north Indian states. The provinces located in the east, west and south of India are, economically, educationally and culturally far more developed. Some of these provinces have 100 per cent literacy level. Thus, the national data set CG_2 represents a universe that is different from CG_1 and CG_3. In fact, the former two are subsets of CG_2. Several scholars designate the three states that constitute about 90 per cent of the first two cultural groups as backward states of India (**Question for authors—this point would be enhanced by a reference**). They were sampled

The Cultures of Public Understanding of Science   291

on time scale at two different points, 2001 and 2007. The analysis also confirms the hypothesis that the cultural distance of a scientific phenomenon is also determined by a gamut of extrinsic factors, such as socio-economic conditions, literacy levels, access to media channels, geographical location and caste compositions of the sampled population.

The mean cultural distance computed for all three data sets X_{CG} varied in magnitude. The value of mean cultural distance X_{CG1} was 9.5, X_{CG3} was 9.2 and it was the lowest for CG_2, i.e. 8.0. It clearly suggests that among the three, the national sample (CG_2) is at the lowest cultural distance from the set of five chosen scientific concepts. The sample collected in 2007 at Kumbh Mela takes the second position, and the population interviewed at the same site in 2001 occupies the furthermost end on the cultural distance scale.

The method also suggests that none of the groups could be categorised as scientifically literate or illiterate. Instead they could be placed at varying degrees of cultural distance from a scientific idea or set of ideas that we may decide to probe during a survey study.

CULTURAL DISTANCE OF VARIOUS PROVINCES

Using the national representative data collected in 2004, values of $X_{ci=1\ to\ 5}$ were computed for all twenty-three Indian states. For the present discussion two states from each broad geographic region were selected. Kerala and Andhra Pradesh represent southern parts of the country, Uttar Pradesh and Bihar are located in the north of India, West Bengal and Assam are the eastern states, Delhi and Haryana are situated in the centre and Rajasthan and Maharashtra represent the eastern geographical region. Values of mean Xc_i for each of these states were also computed (see Table 16.3). Subsequently, the ten provinces were ranked on the basis of their cultural distance from each of the five selected scientific concepts. They were also ranked in ascending order according to the mean value of Xc_i. Value 1 was assigned for the lowest mean cultural distance, and 10 occupied the outermost end.

The relative position of cultural distance for all the concepts remained the same across each state. For all ten provinces C_1 could be placed at the shortest cultural distance, and C_5 could be placed at the furthermost end. In between, the value of cultural distance for each concept, i.e. C_2, C_3 and C_4, increased progressively. For example, Kerala ranked first, with a mean value of $Xc_i=4.5$; the computed values of Xc_1, Xc_2, Xc_3, Xc_4 and Xc_5, for Kerala, were -0.5, 3.0, 6.2, 6.6 and 7.4, respectively. For West Bengal, which scored tenth position, the respective values were 1.2, 6.3, 8.9, 15.5 and 17.1. This shows, quite clearly, that parameters intrinsic to a scientific notion play a predominantly significant role in not only determining the absolute value but also in ascertaining its relative position on the cultural distance scale.

Table 16.3 Relative Cultural Distance and Ranking of Selected Indian States

State	Xc_1 Shape of Earth	Xc_2 Rotation of Earth	Xc_3 Revolution of Earth	Xc_4 Formation of Galaxy	Xc_5 Evolution of Mankind	Xc_{mean}
Kerala	-0.5(4)	3.0(1)	6.2(4)	6.6(1)	7.4(1)	3.5(1)
Rajasthan	0.0(5)	5.0(4)	6.0(1)	7.2(2)	11.0(2)	5.8(2)
Delhi	-2.0(1)	5.2(5)	8.2(6)	9.8(5)	14.2(4)	7.2(4)
Bihar	2.4(9)	5.7(6)	6.8(3)	9.1(4)	16.1(6)	7.2(5)
Haryana	0.5(6)	5.0(3)	6.2(2)	8.5(3)	21.0(10)	6.8(3)
Andhra Pradesh	-0.5(3)	6.8(9)	9.3(8)	10.0(6)	16.9(7)	8.0(6)
Uttar Pradesh	-0.9(2)	4.8(2)	8.0(5)	12.0(7)	19.0(9)	8.4(7)
Maharashtra	2.8(10)	6.2(7)	9.9(9)	12.5(8)	15.9(5)	9.0(8)
Assam	1.8(8)	8.5(10)	11.3(10)	13.5(9)	13.8(3)	9.8(9)
West Bengal	1.2(7)	6.3(8)	8.9(7)	15.5(10)	17.1(8)	9.8(10)

Note: Ranking given in parentheses

However, it should be noted that the absolute value of cultural distance, for various physical concepts, varies a great deal across provinces. Delhi scored the lowest on the cultural distance scale for the concept C_1, shape of earth: the value of Xc_1 was -2.0, and Maharashtra scored 2.8 for the same scientific notion. Haryana was placed at the largest cultural distance for the theory of evolution; the value of Xc_5 was 21.0, whereas for Kerala the value of Xc_5 was quite low at 7.4. It is evident that, using this empirical model, if a cultural subgroup is taken as the reference point, the cultural distance of each phenomenon could be mapped, and strategies to bridge the cultural distance for each of the scientific concepts could be devised.

When states were ranked according to the mean cultural distance and arranged in ascending order Kerala scored the lowest followed by Rajasthan, Delhi, Bihar, Haryana, Andhra Pradesh, Uttar Pradesh, Maharashtra, Assam and West Bengal. However, individual ranking according to different concepts portrayed a detailed picture of reality.

Kerala, which scored the first position on the mean cultural distance distribution scale, occupied fourth position when it was ranked according to its cultural distance from C_1 and C_4. Conversely, if we take a scientific concept as the reference point, then various states, taken here as cultural subgroups, could be placed at varying degrees of cultural distance. For example, in order to democratise the notion of revolution of the earth (C_3), the cultural subgroup represented here as Kerala (R_{c3} =4) will have to travel a longer cultural distance compared to the population of Rajasthan (R_{c3} =1), Haryana (R_{c3} =2) and Bihar (R_{c3} =3).

If we take C_5, theory of evolution, as the reference point, the relative positions of various states change significantly. Kerala occupies the first rank, a position that represents shortest cultural distance, followed by Rajasthan ($R_{c\ mean}$ =2). Haryana and Bihar move to the tenth and sixth positions respectively. Thus, we conclude that the strategy to communicate 'revolution of earth' or 'theory of evolution' to the people of Kerala may not work in Haryana or Bihar. In other words, if a scientific notion is to be democratised among a cultural subgroup, specificities of their cultural-cognitive-structure will have to be taken into account.

COMPARATIVE SHIFT IN CULTURAL DISTANCE

We now move on to discuss the shift in cultural distance observed over a period of six years. For this we have used the samples gathered at two different points on the time scale, i.e. in 2001 and 2007, at Kumbh Mela.

In order to compute the shift in cultural distance the following equation was used.

$$\Delta Xc_i = \sum Xc_{it2} - \sum Xc_{it1}$$

Where

ΔXc_i: denotes the shift in cultural distance

t_2: is the latest point of observation on time scale

t_1: is the earliest point of observation on time scale.

It should be noted that both the polarity and magnitude of ΔXc_i are important. The magnitude denotes degree or extent of shift that has taken place over a time period Δt, and the polarity signifies the direction of this change.

Analysis reveals that in six years values of cultural distance for all the scientific concepts under scrutiny have reduced. Rotundity of earth, which was at a cultural distance of -1.0, in six years has moved closer to the cultural-cognitive-structure of the north-central Indian populace; the value of Xc_i computed for CG_3 was -2.0. The magnitude of ΔX_{c1} was -1.0, and it was a significant change. The negative polarity shows that the perception of 'a round earth' has moved closer to the cultural cognitive structure of the sampled populace (see Table 16.4). The computed values of ΔXc_2, ΔXc_3, ΔXc_4 and ΔXc_5 were -1.4, -0.0, -0.5 and -0.8 respectively. The negative polarity of shift in cultural distance across the chosen scientific concepts shows that all these notions, though to a varying degree, have moved closer to the people's cultural thought structure.

Table 16.4 Magnitude and Polarity of Shift in Cultural Distance

Concepts	Shift
ΔXc mean	-0.3
ΔXc_1 shape	-1.0
ΔXc_2 rotation	-1.4
ΔXc_3 revolution	0.0
ΔXc_4 galaxy	-0.5
ΔXc_5 evolution	-0.8

Here we would like to introduce a word of caution: if the observed value of ΔXc_i is negative, to draw a conclusion that the scientific awareness level of the populace has gone up could be erroneous. Even after arriving at a conclusion based on aggregate or mean values of the shift in cultural distance (ΔXc_{mean}=-0.3), the absolute value and the polarity of each ΔXc_i also needs to be scrutinised.

The above analysis shows that for all the scientific concepts being examined in this article the absolute value of ΔXc_i varies significantly, but the polarity of shift in cultural distance consistently remains negative across C_1-C_5. In other words, notwithstanding the apprehensions we have about the phrase '*scientific awareness level*', it could be concluded that during the past six years, in the area of physics and cosmology, the scientific awareness level of the north Indian population has increased. We report with a high degree of confidence that in India, over the past six years, a definite reduction in cultural distance has been observed.

CONCLUSIONS

It has been repeatedly shown that spaces available to communicate science are sparse and narrow (Cees, Mark and Ivar, 2006). Therefore, the efficacy of communication needs to be increased. The method presented here is a step in that direction. It clearly marks the magnitude of cultural distance of scientific ideas from quotidian life. Mapping the cultural distances could be used to formulate effectual strategies of science communication.

On the basis of our analysis, it could be argued that the cultural model of public understanding of science suggested in this article is well grounded in the conceptual framework that rejects categorisation of 'publics' into 'scientifically literate' and 'scientifically illiterate'.

In order to travel through the permeable membranes that define boundaries of a cultural grouping or a cultural subgroup, ideas face varying degrees of resistance. Empirically, this resistance could be measured in terms of cultural distance.

On the basis of the analysis, an assertion can be made that policy makers, educators and communicators of science, keeping in mind the nature of parameters intrinsic to these scientific phenomena, will have to devise and implement effectual strategies so that these explanations become integral components of the cognitive structure of a specific cultural group.

The method lends itself to meaningful statistical tests and can be used for determining the relative distance between 'scientific structures of configuring reality' and 'peoples' cultural worldviews'.

The analysis carried out on different cultural data sets leads us to the observation that the cultural model suggested in this article can be effectively used for computing the cultural distance of each of these groups from a given set of scientific notions.

Where time series data is available for a given cultural group, the method can also be used to observe shifts in cultural distance over a period of time. If the public has been observed for a sufficiently long duration, the observations may also lead to prediction of future trends.

Reduction in cultural distance across all five scientific concepts indicates that during the past six years there has been a noticeable increase in the public understanding of science in India.

NOTES

1. Almost all survey studies show that the percentage of respondents who offer correct explanation to apparently similar questions in an area of scientific investigation varies a great deal. For example see Lightman, A. P. and Miller J.D. (1989), Contemporary cosmological beliefs. *Social Studies of Science* 19: 127–136. Surveys administered and reported in later years in OECD countries, USA, China, Japan, South Africa and India also indicate a similar pattern.
2. A shorter version of this argument was published in Raza et al., 2002.
3. See National Science Foundation (NSF 1993, 1996, 1998a, 2000). Similar surveys were administered in European countries, Japan, China, Canada, South Africa and Korea during the last decade of the previous century.
4. The authors find the notion of *scientific literacy* arrogant. It presumes that there is a segment of the population that is scientifically illiterate. Literature on public understanding of science, especially the deficit model approach, liberally uses the phrase. See various *Science and Engineering Indicators* reports of NSF or Miller, Pardo and Niwa (1996).
5. All the three categories of responses, other than scientifically correct were clubbed together to form 'scientifically invalid' category.
6. If we plot all the four categories of responses separately the value of *id* will be less than fifty per cent and will also vary across the concepts. The point at which the curve of the invalid response intersects the scientifically valid response curve will determine value of *id*. However, the method proposed is still valid because at this point the percentage of respondents who offer an invalid response will be equal to the percentage of respondents who offer the scientifically correct explanation.

7. In this volume, Miller et al., while discussing the problem of indicators of scientific attitude and knowledge, raise an interesting issue—that the indicators quickly become obsolete. They do not elaborate on the reasons for the fast pace of obsolescence. In our opinion, the likely impact of the phenomenon in the life of the common citizen reduces as the cultural distance increases, thus the indicator associated with the occurrence loses its efficacy for the prevalent standard methods of analysing public understanding of science.
8. The rise in level of education will necessarily result in a rise in percentage of those respondents who offer scientifically valid explanations. See 'Promoting Public Understanding of Science. Proceedings of OECD Symposium on Promoting Public Understanding of Science', Tokyo 5–6, November 1996, OCDE/GD(97)52. The chapter on recommendations concludes: 'primary education is fundamental to increasing scientific literacy'. The recommendations are based on more than 20 years of periodically administered surveys of public understanding of science.
9. The two survey studies, conducted in 2001 and 2007, were administered by Gauhar Raza and Surjit Singh, National Institute of Science Technology and Development Studies, India. Partial funds for both studies were provided by the National Council for Science & Technology Communications, Department of Science & Technology, India.
10. This study was carried out by the National Council of Applied Economic Research, India.
11. National Census reports have repeatedly shown that in these states, the majority of people live in abysmal poverty, they are afflicted with low literacy levels and access to health facilities is appalling (Census of India, 2001). It should be pointed out here that these provinces are geographically quite large and also densely populated, representing about 29 per cent of the total population of India.
12. Kumbh Mela is a religio-cultural conglomeration of people at the confluence of two holy rivers of the country at Allahabad (Prayag), held every twelve years; half Kumbh is held every six years, between Kumbh.
13. For the detailed survey methodology, please refer to Appendix III of India Science Report (Rajesh Shukla, 2005), available at www.insaindia.org/India%20Science%20report-Main.pdf
14. Cronbach's alpha value for all three years is more than 0.72. Estimates of skewness, kurtosis and their standard errors are presented in Table 16.2. These estimates show that the distribution of characteristics in the population is asymmetric. For instance, the skewness and the standard error (SE) of the distribution for 2001, 2004 and 2007 is (-1.09, 0.042); (-0.82, 0.039) and (0.75, 0.040), respectively. This shows the asymmetry of the distribution of the 2001 and 2007 data. Also a positive value indicates a positively skewed distribution (that is, with scores bunched up on the low end of the scale). In 2004, as twice the standard error of the skewness is 0.080 (0.040 x 2=0.080), lower than the absolute value of the skewness statistic (-0.82), this implies that the distribution is significantly skewed. Also the ratio of skewness to SE for the three years is 26.02, -20.64 and 18.84. Having examined the statistical characteristics of the data set, dichotomous curves were plotted for each of the five scientific concepts.

REFERENCES

Bauer, M. (July 2009), Editorial. *Public Understanding of Science*, 18(4): 378–382.

Bensaude-Vincent, B. (2001), A genealogy of the increasing gap between science and the public. *Public Understanding of Science*, 10: 99–113.
Bernal, J. D. (Winter 1937), Dialectical Materialism and Modern Science, *Science and Society*, Vol II, No. 1, (available at: http://www.marxists.org/archive/bernal/works/1930s/dsams.htm) accessed July 11, 2009.
Bhattacharya, N. (1983), *The Cultural Heritage of India, Vol IV: The Religions*. The Ramakrishna Mission, Institute of Culture, Calcutta.
Cees, M. K., Mark J. W. B. and Ivar, E. V. (September 2006), Through which medium should science information professionals communicate with public: Television or the internet. *Journal of Science Communication*, 5(3): 1–8.
Census of India (2001), Registrar General & Census Commissioner, India.
Collini, S. Introduction in Snow, C. P., (1993 [first published in 1959]), *The Two Cultures*. Cambridge, UK: Cambridge University Press. p xi.
Durant, J. (1993), What is Scientific Literacy? In *Science and Culture in Europe*, edited by Durant, J. and Gregory, J. London: Science Museum.
Durant, J. and Bauer, M. (Feb. 1992), British public perception of astrology: An approach from the sociology of knowledge. Paper presented at Annual Meeting of the American Association of Advancement of Science, Chicago, USA.
Geertz, C. (1999), A Life of Learning, Charles Homer Haskins Lecture for 1999, American Council of Learned Societies. Occasional Paper No. 45.
Godin, B. and Gingras, Y. (2000), What is scientific technological culture and how is it measured? A multidimensional model. *Public Understanding of Science*, 9: 43–58.
Leavis, F. R. and Yudkin, M. (1962) *Two cultures?: The significance of C. P. Snow*, London.
Lee, O., Fradd, S. H. and Sutman, F. X. (1995), Science knowledge and cognitive strategy use among culturally and linguistically diverse students. *Journal of Research in Teaching*, 32(8): 797–816.
Miller, J. D., Pardo, R. and Niwa, F. (1997), *Public Perception of Science: A Comparative Study of European Union, the United States, Japan and Canada*. Chicago: Chicago Academy of Science.
Miller, S. (2001), Public understanding of science at the crossroads. *Public Understanding of Science*, 10: 115–120.
National Science Foundation (NSF) (1998), A Report on Science and Engineering Indicators. Chapter 7. *Science and Technology: Public Attitude and Public Understanding*. Virginia, USA: National Science Board, Virginia.
Raza, G., Dutt, B. and Singh, S. (1997), Kaleidoscoping public understanding of science on hygiene, health and plague: A survey in the aftermath of a plague epidemic in India. *Public Understanding of Science*, 6: 247–267.
Raza, G., Dutt, B., Singh, S. and Wahid, A. (1991), Prototype of the forms of scientific cognition: A survey of cultural attitude to natural phenomena. NISTADS Reports I and II.
Raza, G. and Singh, S. (Dec. 2004), Cultural distance between peoples' worldview and scientific knowledge in the area of public health. *Journal of Science Communication*, 3(4): 1–8.
Raza, G., Singh, S. and Dutt, B. (March 2002), Public, science and cultural distance. *Science Communication*, 23(3): 293–308.
Raza, G., Singh, S. and Dutt B. (Feb. 1995), Peoples' attitudes to scientific knowledge: The context of culture. *Journal of Scientific and Industrial Research*, 54: 108–121.
Raza, G., Singh, S., Dutt, B. and Chander, J. (1996), *Confluence of Science and People's Knowledge at the Sangam*. New Delhi: Ishtihaar.

Sardar, Z. and Loon, B. V. (1997), *Introducing Cultural Studies*. NY: Totem Books.

Snow, C. P. (1993 [first published in 1959]), *The Two Cultures*. Cambridge: Cambridge University Press, UK.

Stamm, K. R., Clark, F. and Eblacas, P. R. (2000), Mass communication and public understanding of environmental problems: The case of global warming. *Public Understanding of Science,* 9: 219–237.

Tylor, B. 1924 (original 1871), *Primitive Culture*. 7th edition (2). New York: Brentano's.

Zhongliang, Z. (1991), People and science: Public attitude in China towards science and technology. *Science and Public Opinion,* 18: 311–317.

Part IV
Cultural Aspects of Sensitive Topics

17 Beliefs about Astrology across Europe

Nick Allum and Paul Stoneman

Anyone reading this chapter is likely at some point to have read their horoscope. Astrology columns are widespread in print media, and have been a staple for a surprisingly long time. The first columnist was 17th-century astrologer William Lily, who famously predicted the Great Fire of London, albeit 14 years early. Merriam-Webster's dictionary defines astrology as 'divination of the supposed influences of the stars and planets on human affairs and terrestrial events by their positions and aspects.' A horoscope on the other hand is defined as a 'diagram of the relative positions of planets and signs of the zodiac at a specific time (as at one's birth) for use by astrologers in inferring individual character and personality traits and in foretelling events of a person's life.' The more common understanding of horoscopes is that they are astrological forecasts, such as those that appear in newspapers. It is this definition that we use for the rest of the chapter. Ten years ago, just under half of Americans read their horoscope at least occasionally (National Science Board, 2000), and there is little reason to think that the numbers have declined since then.

In this chapter, we examine the relationship between Europeans' beliefs about astrology and about science and how the two bear on each other. After presenting some background and previous research, we focus on two empirical, analytic perspectives. The first contrasts belief about astrology with beliefs about science and asks: what are the bases on which European publics differentiate science and para-science? The second perspective examines the extent to which beliefs about science and para-science are psychologically mutually opposed and asks instead: what are the bases of simultaneous belief in science and para-science, and what are the consequences for engagement in science?

ASTROLOGY OR SCIENCE?

It is one thing to read an astrology column for amusement or entertainment, but quite another to believe that astrological predictions about events or personality will come true. A surprisingly large quantity of scientific research has been carried out to evaluate the claims of astrology. Less surprisingly, there is really no evidence to support such claims (Blackmore & Seebold, 2001; Carlson, 1985; Eysenck & Nias, 1982). But perhaps people do not set any store by astrological predictions, and a health warning is really not necessary. After all, one does not need to believe something is true to be entertained by reading it.

Even if people do believe in astrology, or ghosts and alien abductions for that matter, does this have a bearing on people's understanding of and engagement with science? Again, the evidence is that it probably does. For not only do sizeable proportions of the American and European publics believe in the efficacy of astrology, they also believe that it is scientific. Surveys in Europe and America have tracked public beliefs about astrology and science since 1988. In America, respondents have been asked whether astrology is 'very scientific, sort of scientific or not at all scientific.' Around 60 per cent said astrology is not at all scientific with around 30 per cent saying it is 'sort of scientific' in seven surveys between 1988 and 2001. In 2004, the proportion rejecting astrology rose slightly, to 66 per cent (National Science Board, 2006). In Europe, there appears to be even more widespread belief that astrology is scientific. In 2001, 53 per cent thought astrology was 'rather scientific' (European Commission, 2001a).

What might account for these widespread beliefs? The first part of this chapter sets out to evaluate several potential explanations for variation in the credibility given to astrology qua science by European citizens, using a recent survey. In doing so, we also examine how astrology is viewed alongside other knowledge-generating practices, scientific or otherwise, in order to understand where astrology is located in the European public's representational field. In the following sections, we briefly outline some putative factors that we consider might account for variation in citizens' beliefs before describing in more detail the data and methods used for the empirical analysis.

THE 'IMMUNISATION' HYPOTHESIS

From a traditional science communication perspective, it is scientific knowledge, particularly knowledge of the methods of science, that would be expected to 'immunise' citizens against false belief in pseudoscience. Hence those who are more scientifically literate, who understand principles of experimentation, the combining of empirical evidence with logical inference, and so forth, should be more likely to realise that astrology, for all its formalistic presentation, is not consistent with the tenets of scientific method. The survey evidence broadly supports this hypothesis, albeit indirectly. In both Europe and America, correlates of scepticism about astrology's 'scientificness' tend to be higher levels of education and higher social class and income, although there is some inconsistency between surveys. In 1992 Eurobarometer surveys showed that more highly educated Europeans were less likely to think that astrology is scientific, whereas in 2001, this was not the case (European Commission, 2001a). In America, education has been a consistent predictor. For example, the most recent NSF data show that whereas 84 per cent of college graduates think that astrology is not at all scientific, just 62 per cent of those who only graduated from high school share this belief (National Science Board, 2008). Education is not, of course, coterminous with scientific knowledge, but it is strongly correlated (Allum, Sturgis, Tabourazi, & Brunton-Smith, 2008; Miller, 2004). Income and social class are themselves invariably correlated with education. Examining the relationship between scientific knowledge and belief in astrology as a science net of education, income, and social class, would be a stronger test of the 'immunising' hypothesis.

WHAT'S IN A NAME?

One of the better established findings in survey measurement is that responses can be extremely sensitive to the particular form of words used in the asking of the question (Schuman & Presser, 1996). In the European surveys reviewed earlier, the English version of the questionnaire uses the word 'astrology' as the stimulus object of the item: "People can have different opinions about what is scientific and what is not. I am going to read out a list of subjects. For each one tell me how scientific you think it is by the scale on this card . . . [other subjects] . . . Astrology." It is possible that people are unfamiliar with this term but may be more familiar with terms like 'horoscopes,' 'star signs,' 'sun signs,' which, if asked how scientific they are, might elicit different results. In fact, there is some evidence on this from the 1992 Eurobarometer survey. Respondents were randomised to two different versions of the list of subjects. One simply contained the single word, as above, whereas in the other condition, a brief explanation of each subject was given. For astrology, the explanation was: ". . . that is the study of occult influence of stars, planets etc. on human affairs." There was no significant difference in responses between the two conditions (INRA, 1993). However, in most European languages the suffixes 'ology,' 'ologie,' 'ologia' connote an academic field of study. This may be enough to encourage respondents to think that astrology is indeed a science whereas the use of an alternative term may bring to mind a different kind of activity. Another hypothesis that has intuitive plausibility, but which has not previously been tested, is that many people mistake astrology for astronomy—a simple semantic confusion. The extent to which this might occur is also likely to vary across countries according to the similarity of the two terms in different languages.

THE STARS DOWN TO EARTH

The explanations just outlined for belief, or apparent belief, in the scientificness of astrology are, in one form or another, based on deficits of understanding and information. There may be other reasons why some people more than others place faith in astrological predictions. One of the most interesting social psychological viewpoints on this question is found in the work of Theodor Adorno. In 1952–3, Adorno carried out a study of Caroll Righter's *Los Angeles Times* astrology column. The fruits of this did not appear in English until published in *Telos* in 1974 as 'The stars down to earth' (Adorno, 1974). In the study, referred to by the author as a 'content analysis,' Adorno analyses, somewhat haphazardly and selectively, the advice given to readers in the column over a period of several months. He identifies many of the aspects of astrological readings that other psychological research (e.g. Forer, 1949) confirmed were effective in making them convincing: the Barnum effect, the tendency to personalise general statements, and so forth. He is witheringly critical of astrology, dubbing it, with the rest of occultism, a 'metaphysic of dunces' and suggesting that 'a climate of semi-erudition is the fertile breeding ground for astrology' (Adorno, 1994 p.44). The claim is that it resembles other 'irrational creeds' like racism by offering a shortcut to (erroneous) knowledge that actually requires no intellectual effort or capacity (Dutton, 1995).

What is more interesting for the present study, though, is the connection drawn between astrology (and other forms of popular occultism) with authoritarianism, fascism, and modern capitalism. Adorno sees astrology as emphasising conformity and deference to higher authority of some kind. Nederman and Goulding sum this up concisely as 'Take things as they are, since you are fated for them anyway' (Nederman & Goulding, 1981). Adorno posits an 'astrological ideology' that he claims 'resembles, in all its major characteristics, the mentality of the 'high scorers' of the Authoritarian Personality' (Adorno, 1994). The work on 'Authoritarian Personality' by Adorno and colleagues has been much criticised since its appearance in 1950 (Adorno, Frenkel-Brunswik, Levinson, & Sanford, 1950; Kirscht & Dillehay, 1967) with particular criticism being directed towards the test items in the 'F-Scale' (Hyman & Sheatsley, 1954). Nevertheless, it is possible to deduce a reasonably clear empirical hypothesis from 'stars down to earth.' Those who value conformity and obedience, and tend towards uncritical acceptance of in-group moral authority, will be more likely to give credence to the claims of astrology.

Adorno also discusses the relationship of organised religion, or religious belief with astrological belief. He suggests that part of astrology's appeal is that it formalises the notion of some higher authority at work controlling life events yet does not come with the explicitly restrictive structure of formal religious adherence, churchgoing, and so on. This is part of what, for Adorno, makes astrological belief and capitalist individualism such well-suited bedfellows. That is to say that religious belief and astrological belief are both consistent with the same authoritarian trait of personality. If this is true, one might expect beliefs about astrology and about religion or God to be related.

HYPOTHESES AND QUESTIONS

The foregoing discussion leads to the derivation of the following hypotheses:

- H1 The suffix 'ology' means that people should tend to rate 'astrology' as more scientific than 'horoscopes'.

- H2a Because of potential confusion or elision of meaning between 'astronomy' and 'astrology,' we should expect there to be a positive correlation between how people rate the scientificness of these two subjects.

- H2b Assuming H2a to be correct, we should not expect to see the same positive correlation between ratings of horoscopes and astronomy because the potential for semantic confusion is much less.

- H3 Citizens who are more knowledgeable about science should be less likely to rate astrology as scientific.

- H4 Following Adorno's thesis, we should expect that people who score higher on a measure of authoritarianism will be more likely to rate astrology as being scientific.

In addition to the evaluation of these empirical expectations, there are two more general questions that are addressed in the analysis that follows:

- Q1 How is astrology viewed by Europeans in relation to other scientific and non-scientific subjects?

- Q2 How much of the variability in beliefs about astrology across Europe is related to country of citizenship?

- Q3 Are there Europeans who are both knowledgeable and credulous of astrology, and if so, what are consequences of this for engagement with science?

DATA AND MEASURES

The data for this study come from the *Special Eurobarometer 224* and *225* surveys, *Europeans, Science and Technology* and *Social Values, Science and Technology* (European Commission, 2005a; 2005b). Both these survey modules were fielded as part of the same face-to-face interview with citizens in 25 EU member states during the fall of 2004. Approximately 1,000 respondents were interviewed in each country, using a multistage probability design. (For more details on the survey methodology, see European Commission, 2005a).

The key dependent variable, belief in the scientificness of astrology, was measured by asking respondents how scientific they consider each of ten subjects to be, on a scale from 1 to 5 where 1 indicates "not at all scientific" and 5 indicates "very scientific". As part of the list of ten subjects a randomised half of the sample was asked about "astrology" and the other half about "horoscopes." The other nine subjects were physics, medicine, astronomy, economics, history, homeopathy, psychology, biology, and mathematics. (The exact question wordings and response alternatives for the English questionnaire for all of the measures employed can be found in European Commission, 2005a; 2005b).

Authoritarian values are measured with a single indicator. Respondents are shown a list of qualities that children might be encouraged to learn. One of these qualities is 'obedience.' Responses to this item are on a 4-point scale ranging from 'not at all important' to 'very important.' Also related to Adorno's work on authoritarianism is religious belief. To capture this, we use an item that asks whether a respondent believes in 'God,' a 'spirit' or neither. From this categorical variable, two dummy variables have been derived, indicating belief in God or belief in a spirit (both versus no belief). We also use a dummy variable indicating whether or not the respondent is Catholic. The majority of Europeans are Christians of some type, so distinguishing between Catholics and all others is a reasonable way of simply controlling for religious denomination in the analysis, in the absence of any particular hypotheses about denomination-based differences in beliefs about astrology.

Knowledge about science is measured in a number of ways in the survey, and in the analysis we use three separate indicators. Two of these tap into respondents' understanding of scientific process and method. An understanding of method is arguably central to being able to distinguish between scientific and pseudoscientific claims. People were asked 'what does it mean to study something scientifically?' The verbatim responses were coded into one of several mutually exclusive categories, based on what was said (in other words, not pre-coded). Typically this question has been a very good predictor of attitudes and beliefs about science with the mention of hypothesis testing and experimentation as the critical component. Here we use this as an indicator of greater scientific understanding along with another indicator based on the mentioning of 'measurement' in response to the question. The third indicator is a summated scale of correct responses to a series of 13 true/false quiz items that tap textbook type knowledge about scientific facts. For the purposes of this analysis, 'don't know' responses are coded with a zero, the same way as incorrect true/false answers. The scale has reasonable internal consistency with a Cronbach's Alpha coefficient of 0.72.

A range of other background characteristics were measured in the survey and used in the analysis. Respondent age was coded in bands: 16–24, 25–39, 40–54, 55+. Occupational status was measured with a dummy variable contrasting white-collar and management occupations with all others. This is a necessarily crude indicator as it is based on standardised Eurobarometer occupational coding that needs to be comparable across European states. However, as it is only being used as a control variable in the analyses that follow, it is not critical to obtain a more fine-grained estimate of the effects of occupational status on beliefs about astrology. Education is measured, again quite bluntly, with a variable that indicates whether or not the respondent left full-time education after the age of 20. This broadly distinguishes the graduate population, and it is this distinction that has in previous research been shown to be the most diagnostic of differences in attitudes and beliefs about science and technology (e.g. Miller, Pardo, & Niwa, 1997). Finally, the type of area in which the respondent lives is captured with a variable that indicates residence in a large town versus other types of areas. Typically one might expect urban populations to have different cultural and political orientations to rural and provincial populations, net of education and occupational differences; hence this variable is used as a control.

SPLIT BALLOT EXPERIMENT

Figure 17.1 shows the response distributions for the two experimental conditions. In one condition, respondents were asked how scientific they thought astrology was and in the other how scientific did they think horoscopes were. Quite clearly, many more Europeans think astrology scientific than horoscopes. Fifty-seven per cent think that horoscopes are "not at all scientific," and only 24 per cent believe the same about astrology. About one quarter of the sample believe astrology to be "very scientific," and only 7 per cent think that horoscopes are "very scientific." The difference in distributions is highly significant (Chi^2 3400 (4df), p<.001. n=23,473). Astrology is clearly viewed as greatly more scientifically credible than horoscopes, and this is in line with the expectations set out in H1. Another point to note is that there is a great deal more heterogeneity in beliefs about astrology than about horoscopes, with both 'very scientific' and 'not at all scientific' each attracting one quarter of all respondents.

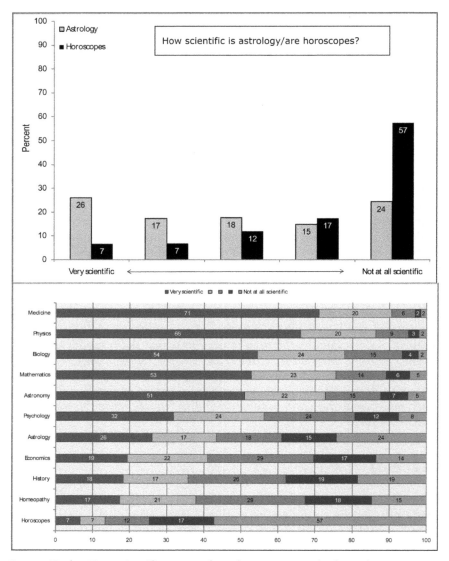

Figure 17.1b How scientific are astrology, horoscopes, and other subjects?

THE RELATIONSHIP BETWEEN ASTROLOGY, HOROSCOPES, AND OTHER SUBJECTS

The lower panel of Figure 17.1 shows the distribution of European beliefs about the scientificness of all 11 subjects included in the questionnaire (including both horoscopes and astrology). The chart is ranked in descending order according to the percentage of people thinking each subject is 'very scientific.' As can be seen, the list includes a range of more or less scientific subjects, including both horoscopes and astrology, along with homeopathy, as examples of pseudoscience.

Medicine is viewed as the most scientific subject, followed closely by physics and then biology. Horoscopes are least likely to be thought "very scientific," followed by homeopathy, history, and economics (17, 18, and 19 per cent respectively). In most respects this is not a surprising result. The natural sciences are at the top with social and behavioural subjects lower down. What is somewhat surprising, although in line with other surveys (European Commission, 2001b), is that astrology is considered to be more scientific than economics, and only just less so than is psychology. Figure 17.1 suggests some ambivalence about the scientific status of astrology if not homeopathy and horoscopes. A further exploration of the basis on which Europeans make judgments about the scientificness of these subjects was carried out using factor analysis. Two separate analyses, one for each split half of the sample, were performed using maximum likelihood estimation and oblique rotation. Figure 17.1 shows the factor loadings for the 3-factor model that resulted, based on an examination of the scree plot and substantive interpretability. An obliquely rotated solution was preferred because we assume that there are individual differences in the propensity to agree or disagree that any subject is scientific, as well as a tendency to discriminate between groups of similar subjects. The assumption is therefore that the factors will be positively correlated, which is what the oblique rotation achieves.

We have labelled the three factors 'hard science', 'soft science' and, for want of a better term, 'new age science'. The highest factor loading for 'hard science' is physics at 0.88, and the 'soft science' includes economics and history, anchored by economics, at 0.71. Homeopathy, astrology, and psychology form the 'new age science' group, although psychology has a weak cross-loading with 'soft science'.

Figure 17.2 plots these loadings in the first two dimensions, with the third dimension, 'new age science,' indicated by the size of the bubbles. The reason for the 'new age' label is that the three subjects that load on this factor are the kind of subjects that people expect to see in the self-help, pop-psychology or new age therapy sections of bookstores. The representation of psychology for the European public is perhaps less about, for instance, cognitive neuroscience and more about self-help for overcoming depression.

In Figure 17.2, the rotated solution for the sample that was asked about horoscopes is presented. The main difference is that only two factors are needed to describe the cognitive structure organising responses to the ten subjects.

There is no separate 'new age science' dimension, with horoscopes, homeopathy, and psychology joining history and economics to form a 'soft science' group. The 'hard science' subjects remain the same, with physics again anchoring the factor with the highest loading, at 0.78.

310 Nick Allum and Paul Stoneman

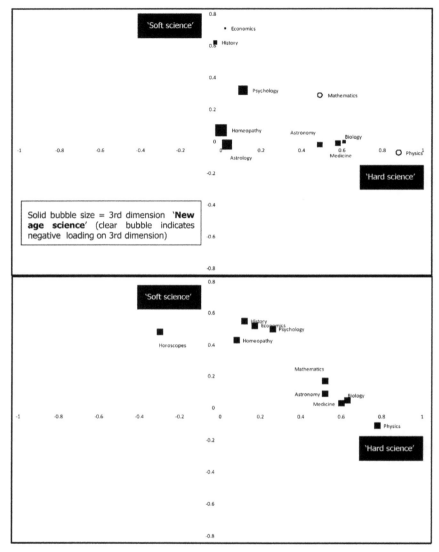

Figure 17.2 Factor loading plots for horoscopes and astrology conditions.

The factor loadings are plotted in Figure 17.2, where it is possible to clearly see the two groupings. Note also that horoscopes is the only 'soft science' subject that loads negatively on the hard science dimension. This was not the case for astrology, demonstrating further evidence that the latter perceived as being more like a science subject.

The results seen thus far indicate fairly unequivocally that while horoscopes and astrology are to all intents and purposes the same thing, at least insofar as the casual engagement of the average citizen is concerned, the two terms have rather different connotations.

Astrology probably 'sounds' more scientific than horoscopes, it is consequently evaluated as being more scientific and is viewed in a more similar way to psychology and homeopathy than are horoscopes. Having elaborated a description of how astrology is perceived, the following section turns to the question of what might underlie the variation in these perceptions amongst citizens and across European states.

VARIATION IN BELIEFS BETWEEN CITIZENS

The final part of the analysis is a multivariate investigation of social and psychological factors that might influence individuals' propensity to believe that astrology is a scientific subject. To do this we use a variant of an OLS multiple regression model.

This analysis is primarily to examine individual level factors associated with beliefs about astrology. However, there is also likely to be heterogeneity of beliefs amongst the 25 European countries, even after taking into account individual characteristics. Modelling this situation calls for country to be included as either a fixed or random effect. The first approach essentially gives each country its own dummy variable and regards each as a unique entity, so to speak. The second approach treats the countries in the data set as a random sample of potential countries that could have been included and estimates a single mean and variance for a continuous random variable that captures the heterogeneity across countries (Raudenbusch & Bryk, 2001). The model parameters can then be used to derive an estimate of each country's location in the distribution of this variable.

If the individual level effects are uncorrelated with this country random effect, the random effects estimator is unbiased and preferred over the fixed effects one, because it is more efficient. If this assumption is not met, the fixed effects approach is appropriate. The result of a Hausman test (Hausman, 1978), which tests for the presence of this correlation, was highly non-significant (Chi^2 5.86, 17df, p=0.99), so the final model presented here uses the random effects estimator.

Table 17.1 presents the estimates for the model predicting beliefs about astrology. Higher scores on the dependent variable indicate stronger belief that astrology is scientific. The included predictors account for 17 per cent of the variance in beliefs. Turning first to hypothesis H2a, the expectation was that there should be some correlation between beliefs about astronomy and beliefs about astrology because of semantic confusion. This is indeed the case here. The coefficient for astronomy is .35 with a very small standard error. Net of all other modelled influences, the more likely it is that citizens believe astronomy is scientific, the more likely they are to think that astrology is also scientific. It appears that the two subjects are not always well differentiated in European public imagination. In order to

Table 17.1 Random effects regression estimates (individual characteristics)

(n=11622)	B	SE	Z	p
Age (reference category 15-24)				
25–39	-.12	.04	-2.66	.01
40–54	-.14	.04	-3.34	<.01
55+	-.22	.04	-5.22	<.01
Female	.06	.03	2.18	.01
Higher education	-.19	.03	-5.75	<.01
Professional or management occupation	-.08	.04	-1.89	.06
Large town or city dweller	-.10	.03	-3.04	<.01
Religious belief				
Catholic	.13	.04	3.70	<.01
God	.08	.04	2.02	.04
Spirit	.12	.04	3.21	<.01
Right wing political orientation	.01	.004	2.26	.02
Science knowledge				
Quiz score	-.09	.01	-17.42	<.01
Mentions hypothesis testing	-.25	.06	-4.55	<.01
Mentions measurement	-.21	.05	-4.52	<.01
Authoritarian	.22	.02	10.36	<.01
Astronomy scientific	.32	.01	26.80	<.01
All scientific	.24	.02	12.41	<.01
(Intercept)	1.18	.14	8.67	<.01

counter potential objections to this conclusion, we have included an additional variable ('all scientific') that is calculated as the mean of respondents' belief scores on all the other subjects except for horoscopes and homeopathy. Without this control, it could be argued that the correlation between astronomy and astrology is due to individual differences in the propensity to express the view that *anything* is scientific. The inclusion of this variable, though, does not eliminate the positive coefficient for astronomy.

H2b, if supported, would corroborate further this interpretation. We would not expect to see the same positive relationship between astronomy and horoscopes, precisely because they do not sound similar. To test this, we fitted the same model using the other split half of the sample and designating horoscopes as the dependent variable. The pattern of results is very similar to that in Table 17.3 but, crucially, the coefficient for astronomy is very small, at .05 (SE=.01; Z=4.23; p<.01), compared to the estimate in the astrology model.

H3 concerns the relationship of scientific knowledge or literacy with perceptions of astrology. The expectation is that those who are better endowed with civic scientific literacy will be better able to distinguish science from pseudoscience. Therefore, we expect positive coefficients for the knowledge measures. Looking at Table 17.3, one can see that this is indeed the case. The coefficients for all three knowledge variables are negative and statistically significant, meaning that the more knowledgeable a person is, the less scientific they believe that astrology is. To get a sense of the magnitude of the effects, if one compared a citizen who scored at the mean on the quiz and mentioned neither hypothesis testing nor measurement in their open-ended answer, their expected rating of astrology would be one point higher on the 5-point scale (in the 'more scientific' direction) than a European mentioning both measurement and hypothesis testing and who scored at the maximum on the quiz, with all other variables held constant. Particularly interesting is the fact that science knowledge has an effect even after controlling for education.

The fourth hypothesis addressed in the analysis considers the relationship of authoritarian personality type with beliefs about astrology. The coefficient for the authoritarianism question is positive, at .22, and highly significant. For each one point increase on the 'importance of obedience' question the model predicts just under a one quarter point increase in the scientificness of astrology rating. So, controlling for all the other covariates, the predicted difference in belief about astrology between Europeans who do not think obedience is at all important to teach to children and those who think it very important is just slightly more than one scale point. Here, then, is empirical support for Adorno's linking of authoritarianism and openness to pseudoscience. Note again that this relationship is robust to the full range of other controls in the model, in particular age, education, and conservative or right-wing political orientation.

People who report believing in God or in a 'spirit of some kind' are more likely to think astrology is scientific. Catholics are also more likely to express this view. These results are again in line with what one might expect from Adorno's account of the appeal of astrology to those who have a propensity to defer to higher authority of different kinds, including religion. We have no clear explanation for why Catholics should be more likely to be credulous of astrology, but it should be borne in mind that this finding of course controls for country as well as for individual characteristics. This means that religious denomination here is not simply a proxy for divergent beliefs of citizens from predominantly Catholic and non-Catholic European states.

The coefficients for the other socio-demographic variables are worth noting. Women are slightly more likely than men to think astrology is scientific, as are people who self-identify as being politically on the right.

Those who live in large urban areas, who are better educated and in high-status occupations are all less likely to accord astrology scientific credibility. A rather interesting finding is that older people tend to be less credulous of astrology. It is the youngest age group, 15–24, that regards astrology as most scientific. This invites speculation as to whether it is a life cycle or a cohort effect we are seeing here. Do people become more sceptical as they age? Or are younger generations in general less sceptical than their parents? This is something for further research to establish.

VARIATION IN BELIEFS BETWEEN COUNTRIES

Table 17.2 shows the variance components estimates for both individual and country levels. This can be thought of as the proportion of unexplained residual variation between individuals and between countries. The first two columns of the table show these variance estimates and the corresponding percentages for a model where only the intercept is fitted. In other words, it simply partitions total variation into within-group and between-group (individual and country). As little as 13 per cent of the variation in beliefs about astrology is systematically related to country; the remainder is due to inter-individual variation. When the full model, with all the independent variables is estimated, the proportion due to country drops by around one third, to 8 per cent. This indicates that some of the apparently systematic country variation is due to compositional differences in populations on the individual characteristics entered into the model. In general terms, it would appear that the social-psychological factors that influence beliefs about astrology and science across Europe are broadly common to citizens from all countries.

Nevertheless, this is not to say that there is no systematic difference between countries at all. One way of exploring this is to obtain estimates of the unobserved country level random effect variable and to compare estimates across countries. These estimates are also known as 'empirical Bayes estimates' (Raudenbusch & Bryk, 2001). These can be thought of as unexplained country level residuals, expressed in standard deviations from the mean (zero), after taking into account all the individual level variables in

Table 17.2 Variance components for intercept-only and full models

	Intercept-only model		Full model	
	Variance (SE)	% of total variance	Variance (SE)	% of total variance
Individual level residual	2.08 (.03)	87	1.80	92
Country residual	.27 (.08)	13	.14	8

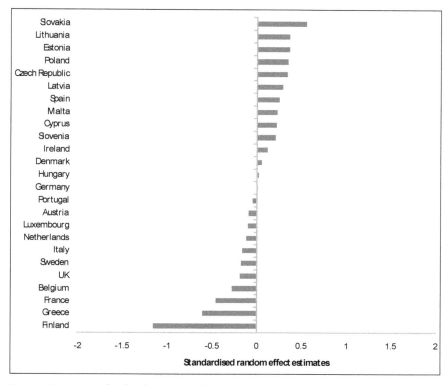

Figure 17.3 Standardised random effect estimates for 25 European countries.

the model. There is a surprisingly clear pattern to the variation. Controlling for the individual characteristics of citizens, all of the former Eastern bloc countries are more accepting of astrology as being scientific than the model would predict, whereas the majority of western European states have more sceptical citizens than would be expected given the individual attributes of their citizens. Figure 17.3 shows standardised random effects estimates for each of 25 European member states.

ASTROLOGY *AND* SCIENCE?

From the analysis presented so far, there are clues that belief in astrology, or perhaps para-science more generally, is not in mutual opposition to belief in science and scientific method. The 'hard' and 'soft' science factors identified are not orthogonal, and there are significant numbers of scientifically well-informed Europeans who nevertheless believe that astrology is also scientific. What can one make of this? One possibility is that astrology and science occupy different realms of thinking and practice within individual citizens and that an overarching adherence to the scientific worldview no matter what the context is a fiction at odds with the way people actually think and behave. This idea is embodied in the concept of cognitive polyphasia (Jovchelovitch, 2007; Jovchelovitch, 2008; Moscovici, 2008), which points to a plurality of knowledge systems coexisting, but not necessarily conflicting, within a single individual. This is an interesting way to describe what appears to be a widespread phenomenon, whereby strictly scientific knowledge systems do not appear to displace others, such as that represented by astrology.

Building on our previous analyses, it would be illustrative first to see the degree to which polyphasia exists across countries and how this relates to the pattern of cultural variation shown in the preceding section. Second, it would be interesting to see whether one or another knowledge system is most likely to govern people's behaviour in terms of their engagement with science. Which belief 'wins out,' and under what conditions? Is it the belief in the scientificness of maths and the natural sciences or the scientificness of the 'new age' sciences?

We generated a variable operationalising cognitive polyphasia by using the battery of items that measure how scientific people think various subjects are (ranging from 1 = not scientific at all to 5 = very scientific). Of interest here is the trade-off (or not) between believing that the 'hard sciences' are scientific versus believing that the 'new age' sciences of homeopathy and astrology are scientific. For some, there will be a clear trade-off; for others neither group of subjects will be seen as scientific or both will be deemed to have scientific merit. Given that very few people think that most or all of the hard sciences are 'unscientific' (only 1.5% of the working sample), and even fewer believe that the 'new age' sciences are uniquely scientific or that both are not scientific, the interesting groups to compare here are those who simply believe in the hard sciences and not the new age sciences (65% of the working sample) versus those that believe in both the hard and new age sciences (35%). As such, the reference group for the cognitive polyphasia variable is those who score 4 or 5 on the scientificness scale for the hard sciences AND score low on the new age scientificness scale (<4). The 'treatment' group, coded as one, are those who score 4 or 5 on the scientificness scale for the hard sciences but also score 4 or 5 on the new age scientificness scale.

COUNTRY DIFFERENCES

In Figure 17.4 we show the percentage of respondents who can be classified as 'polyphasic' in each country. Levels of cognitive polyphasia are reasonably high even in the West European countries where 'new age' sciences are less popular (with the exception of Finland). At least one-fifth of these

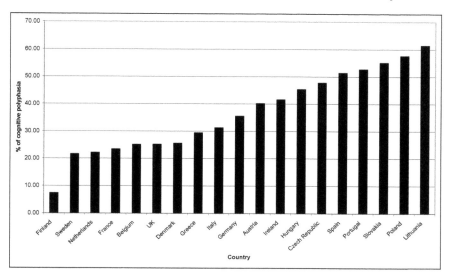

Figure 17.4 Levels of cognitive polyphasia in Europe.
N = 7982. *Note*: data from Luxembourg, Malta, Estonia, Cyprus, Slovenia, and Latvia were not included due to low numbers of observations (<100)

populations appear to believe in the scientific value of the hard sciences as well as astrology and homeopathy.

But what more can we say about such individuals? Do they genuinely appreciate the scientific value of the natural sciences, or does their concomitant belief in the new age sciences trump their scientific worldview? One way to find out is to contrast the effects on scientific engagement variables of believing in the scientific value of the natural sciences versus believing in both the natural and new age sciences. By focusing on behavioural variables, which are theoretically crucial for the notion of cognitive polyphasia, some evidence can be obtained to see which of the two valued 'scientific' beliefs displays the most dissonance with actual behaviour. This yields the following hypothesis:

If the scientific value of the new age sciences is dominant over natural sciences, then a negative correlation should exist between the polyphasia variable and science engagement variables.

THE DEPENDENT VARIABLES

As a general rule, behavioural variables provide researchers with a way to test how strong individual beliefs are. For example, if an individual says they value the protection of the environment then all things being equal we would expect the same individual to be more likely than others to engage in 'green' activities such as recycling waste on a regular basis. Similarly, if an individual values the natural sciences, then we would also expect them to engage in activities that would satisfy this interest, such as visiting a science museum or talking to friends about scientific issues. Fortunately, the Eurobarometer survey asked a series of questions relating to people's active engagement with science and technology. Two variables are derived for the subsequent analyses: an 'informal' science engagement variable derived from the frequency of reading scientific articles and talking to friends about science and a 'formal' science engagement variable derived from the number of visits to science museums, exhibitions, and libraries.

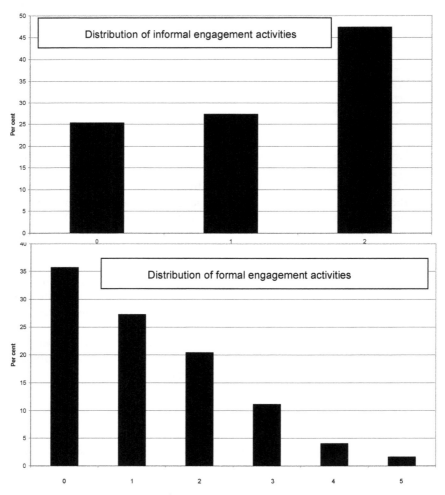

Figure 17.5 Engagement with science.

Beliefs about Astrology across Europe 319

As can be seen, the majority (64%) of the selected sample engage in at least one formal engagement activity, and a slightly higher figure is found for the informal engagement variable of reading and/or talking about science and technology (75%). As the concern here is to what extent cognitive polyphasia affects engagement with science per se (as opposed to the level of engagement), both of the above variables are recoded into binary outcomes, where 0 = no engagement mentioned, and 1 = at least one engagement activity mentioned. By running a regression analysis and controlling for other potential causes of engagement, the nature of the cognitive polyphasia towards various 'scientific' areas of study can be evaluated. In these models we also, as in the previous analyses, control for country by specifying it as a random variable.

Table 17.3 shows unstandardised coefficients, standard errors, and z-values for two models, one each for formal and informal engagement. The key parameter is the effect of cognitive polyphasia on engagement. In both cases the coefficient is negative. For informal engagement, the estimate is not statistically different from zero, whereas for formal engagement, the coefficient is -.18 and significant at p<.05. This provides some clue that when people employ dual knowledge frames, although this does

Table 17.3 Formal and informal engagement in science (unstandardised coefficients)

	Informal Engagement			Formal Engagement		
	B	SE	z	B	SE	z
Age 25-39	-0.31	1.16	-0.27	0.79	0.95	0.83
Age 40-54	-0.31	1.16	-0.27	0.12	0.95	0.13
Age 55+	-0.52	1.16	-0.45	-0.15	0.95	-0.16
Female	-0.48**	0.08	-6.00	0.40**	0.07	5.71
Higher education	0.58**	0.09	6.44	0.66**	0.08	8.25
Professional or management occupation	0.64**	0.16	4.00	0.90**	0.14	6.43
Large town or city dweller	0.18	0.09	2.00	0.21*	0.08	2.63
God	0.02	0.11	0.18	0.09	0.10	0.90
Spirit	0.26**	0.12	2.17	0.42*	0.11	3.82
Right wing political orientation	-0.03*	0.01	-3.00	-0.02	0.01	-2.00
Science knowledge quiz	0.19**	0.02	9.50	0.16**	0.02	8.00
Test hypotheses	0.32	0.20	1.60	0.43**	0.18	2.39
Measurement	-0.02	0.16	-0.13	-0.05	0.14	-0.36
Cognitive Polyphasia	-0.10	0.08	-1.25	-0.18*	0.08	-2.25
(Intercept)	0.07	1.18	0.06	-1.64	0.98	-1.67

not, by definition, prevent them being 'conventionally' engaged with science and technology, it does nevertheless serve to reduce it. We can tentatively conclude that cognitive polyphasia is present to some degree for some people, but that beliefs in the scientificness of astrology are not entirely comfortable bedfellows with strong beliefs in genuine scientific knowledge systems.

CONCLUSION

Europeans have a range of beliefs about astrology and its status as a scientific or quasi-scientific subject. The evidence from the experiment and from the observational data shows that there is considerable blurring over what the term means. There is a clear confusion between astronomy and astrology. There are also widely differing opinions on horoscopes as compared to astrology, even though one might consider the two as functionally equivalent to most intents and purposes. Astrology is regarded as more scientific than horoscopes. Perhaps previous research on science literacy that has gauged the credulousness of citizens about astrology has led to rather more pessimistic conclusions than are really warranted.

However, conventional science literacy clearly makes a difference. Net of a range of other potential confounding influences, the better one's understanding of scientific terms and concepts, as well as factual knowledge of science, the better one is able to distinguish science from pseudoscience. This immunising effect of scientific knowledge is perhaps not a surprising finding, but one that underlines the utility of these survey indicators in distinguishing between modes of citizen reasoning about science and confirms the importance of scientific literacy in helping Europeans make informed judgments about the validity of pseudoscientific claims. Although this study has not focused on health and consumer choices, it is quite possible that the model presented here holds across a range of pseudoscientific domains in which citizens are confronted with the need to make choices.

The result that authoritarian-type values are associated with greater credulity towards astrology is fascinating and in line with Adorno's prediction. Whether it is because of a general propensity to defer to any kind of authority, or whether it is bound up with anti-rational culture in the way that Adorno hypothesised is open to question. Further research could profitably be directed towards a greater elaboration of the mechanisms underlying the observations made in the present study. Perhaps linked to this finding are the systematic country variations. Former Eastern bloc states seem to be more accepting of astrology and more likely to consider it scientific. This holds true even when controlling for on science literacy, religion, education, political orientation, and values. Citizens of these countries have a recent history of state and civil society being organised along authoritarian lines, and this cultural norm may be reflected in the readier acceptance of astrology in Eastern Europe, over and above that due to variation in individual personality traits.

This investigation has explored some of the correlates and putative causes of different beliefs about astrology and its relationship with science. As a final note, it should probably be borne in mind that for most people, reading a horoscope is a leisure activity and not one of central importance either. Nevertheless, by understanding how Europeans differ in their perceptions of astrology and horoscopes, it is possible to gain some insight into the bases of how citizens evaluate scientific and pseudoscientific claims more generally.

REFERENCES

Adorno, T. (1974). The stars down to earth. *Telos,* 6(Spring).
Adorno, T. (1994). The stars down to earth: The *Los Angeles Times* astrology column. In S. Crook (Ed.), *The Stars Down to Earth and Other Essays on the Irrational in Culture.* London: Routledge.
Adorno, T., Frenkel-Brunswik, E., Levinson, D. J., & Sanford, R. N. (1950). *The Authoritarian Personality.* New York: Harper & Row.
Allum, N., Sturgis, P., Tabourazi, D., & Brunton-Smith, I. (2008). Science knowledge and attitudes across cultures: A meta-analysis. *Public Understanding of Science,* 17(1), 35–54.
Blackmore, S., & Seebold, M. (2001). The effect of horoscopes on women's relationships. *Correlation,* 19(2), 17–32.
Carlson, S. (1985). A double-blind test of astrology. *Nature,* 318, 398–399.
Dutton, D. (1995). Theodor Adorno on astrology. *Philosophy and Literature,* 19, 424–430.
European Commission. (2001a, December). *Europeans, Science and Technology.* Available: http://europa.eu.int/comm/public_opinion/archives/eb/ebs_154_en.pdf.
European Commission. (2001b). *Europeans, Science and Technology (55.2)— SPSS dataset.* Available: www.data-archive.ac.uk.
European Commission. (2005a). *Special Eurobarometer 224: Europeans, Science & Technology.* Brussels.
European Commission. (2005b). *Special Eurobarometer 225: Social Values, Science & Technology.* Brussels.
Eysenck, H. J., & Nias, D. K. B. (1982). *Astrology: Science or Superstition?* London: Maurice Temple Smith.
Forer, B. R. (1949). The fallacy of personal validation: A classroom demonstration of gullibility. *Journal of Abnormal Psychology,* 44, 118–121.
Hausman, J. A. (1978). Specification tests in econometrics. *Econometrica,* 46(6), 1251–1271.
Hyman, H., & Sheatsley, P. (1954). The authoritarian personality—a methodological critique. In R. Christie & M. Jahoda (Eds.), *Studies in the Scope and Method of the "Authoritarian Personality".* Illinois: Glencoe.
INRA. (1993, December). *Europeans, Science and Technology: Public Understanding and Attitudes.* Available: http://europa.eu.int/comm/public_opinion/archives/eb/ebs_154_en.pdf.
Jovchelovitch, S. (2007). *Knowledge in Context: Representations, Community and Culture.* London: Routledge.
Jovchelovitch, S. (2008). Rehabilitating common sense: Knowledge, representations and everyday life. *Journal for the Theory of Social Behaviour,* 38 (4), 431–448.

Kirscht, J. P., & Dillehay, R. (1967). *Dimensions of Authoritarianism: A Review of Research and Theory.* Lexington: University of Kentucky Press.

Miller, J. D. (2004). Public understanding of, and attitudes toward, scientific research: What we know and what we need to know. *Public Understanding of Science, 13*(3), 273–294. Available: http://pus.sagepub.com/cgi/content/abstract/13/3/273

Miller, J. D., Pardo, R., & Niwa, F. (1997). *Public Perceptions of Science and Technology: A Comparative Study of the European Union, the United States, Japan and Canada.* Bilbao: Fundacion BBV.

Moscovici, S. (2008). *Psychoanalysis: Its Image and Its Public.* Cambridge: Polity Press.

National Science Board. (2000). *Science and Engineering Indicators—2000.* Arlington, VA: National Science Foundation.

National Science Board. (2006). *Science and Engineering Indicators—2006.* Arlington, VA: National Science Foundation.

National Science Board. (2008). *Science and Engineering Indicators—2008.* Arlington, VA: National Science Foundation.

Nederman, C., & Goulding, J. (1981). Popular occultism & critical social theory: Exploring some themes in Adorno's critique of astrology & the occult. *Sociological Analysis: A Journal in the Sociology of Religion, 4* (Winter), 325–332.

Raudenbusch, S., & Bryk, A. (2001). *Hierarchical Linear Models: Applications and Data Analysis Methods.* (2nd ed.). Thousand Oaks, CA: Sage.

Schuman, H., & Presser, S. (1996). *Questions and Answers in Attitude Surveys: Experiments on Question Form, Wording, and Context.* Thousand Oaks, CA: Sage.

18 The Human-Animal Boundary in Switzerland
A Cross-Cultural Perspective
Fabienne Crettaz von Roten

The empirical research conducted on science-society relations over the past three decades has pointed out various explanatory factors such as scientific knowledge (Evans and Durant, 1995; Miller and Pardo, 2000), trust (Priest et al., 2003; Crettaz von Roten et al., 2003), and values related to nature (Gaskell et al., 2005; Peters et al., 2007). In their review of the field, Bauer et al. (2007: 88) suggest the interpretation of these results as "indicators of the cultural climate", therefore reconsidering the unit of analysis. As Switzerland lies at the crossroads of three cultures (French-, German-, and Italian-speaking culture), the purpose of this chapter is to map out cultural differences in perceptions of animals and nature among the three linguistic regions to improve our understanding of the relationship between science and society.

In the public perception of science, views of nature are sense-making 'tools', because science and nature are related in various ways. First, science and technological developments have supplemented and sometimes supplanted 'natural' ways, for example in food and reproduction, resulting in accusations of 'playing God' and of performing "an act of human hubris, exceeding natural and moral boundaries" (Peters et al., 2007: 198). Some scientific discoveries have raised the issue of human-animal boundaries in new terms: for example, the representation of the human species as a special case, set apart from the rest of the natural realm, is questioned with the transfer of a gene from one species to another. Too, some scientific developments have had severe environmental impacts, and the population is now concerned about unforeseen consequences of interfering with nature, with particular reluctance to condone the increasing (bio)technological domination of nature (Wagner et al., 2001; Michael, 2001). Finally, animals play a crucial role in science (medicine, pharmacology, and recently biotechnology), because they were and are models for human disease, sources of organs or cells, and are used in animal experimentation, xeno-transplantation, and cloning. Part of the debate on the use of animals in experimental research focuses on the human-animal boundary and on speciesism[1]: is experimenting on nonhuman primates vital to science? Has experimenting on dogs led to medical breakthroughs? Use of animals in science has raised many concerns and protests among European and American populations (Einsiedel, 2005; Crettaz von Roten, 2008).

Previous studies have established that the conception of nature is socially constructed and varies according to cultural and historical factors (Michael, 2000; Strauss, 2005). Bourg (1997) described the emergence of the nature/culture cleavage in modernity, which gave humans the right to control and subject nature, and how technology produced a triple setback of nature (around us, between us and within us). Franklin and White (2001) reported the changing pattern of relations with animals as becoming increasingly zoocentric and sentimentalized in post- or late modernity. According to Franklin (1999), the human-animal boundary has been dismantled via a growing misanthropy, ontological insecurity, and the notion of risk-reflexivity. Philosophers, historians, and social scientists have widely studied conceptions of nature, including, among others, whether nature is considered spiritual or sacred or not, and including the type of relationship with human society: separation, with opposition between nature and humans; connection, with dominion of nature by humans; or integration, with harmony between nature and humans (Frank, 1997; Strauss, 2005).

Cultural models of Public Understanding of Science (PUS) have been identified by analyzing public perceptions in different national or regional contexts. Bauer et al. (1994) introduced the postindustrialization model of PUS, which interprets the changing pattern of relationships between interest, knowledge, and attitudes toward science by transition from industrial to postindustrial society. Based on the difference of dynamics of industrial development across European regions, Allum, Boy, and Bauer (2002) reconsidered the postindustrialism hypothesis at the level of regions. In comparing the reception of genetically modified foods in Europe and the US, Gaskell et al. (1999: 386) concluded that "[D]ifferent histories of media coverage and regulation go together with different patterns of public perceptions, and these in turn reflect deeper cultural sensitivities".

Switzerland lies in the heart of Europe, at the crossroads of three cultures (French-, German-, and Italian-speaking). In this paper, we propose a 'linguistic definition' of culture by which people speaking the same language over a contiguous region are members of the same culture, sharing a common set of values and representations, and people speaking a different language belong to another culture.[2] Swiss multiculturalism has been widely studied. On the political side, Kriesi et al. (1996) showed differences in voting patterns on issues related to the country's foreign policy, to openness toward foreigners, to the reinforcement of the federal State, to restriction of personal freedoms, etc. Among the humanities, a study on social integration has revealed that the German-speaking Swiss tend to be more civically engaged (through voluntary work) than their neighbors of the Italian- and French-speaking regions (Freitag and Stadelmann-Steffen, 2009). In fact regions tend to orient themselves to the bordering country with which they share a language. Steckeisen et al. (2002) explored two academic cultures of professors in higher education institutions of the French-speaking and German-speaking regions.

Analysis by Peters et al. (2007) showed that appreciation of nature explains part of the well-known difference in attitudes toward food biotechnology in the US and Germany. In their conclusion, the authors raised some questions (ibid.: 216): "Do societies differ in their ability to absorb innovations? What role does culture play in this process?" Starting with these questions, we first hypothesize cultural differences in perceptions of science, animals, and nature in Switzerland: more precisely, that French-speaking people tend to be more positive toward science and German-speaking to be more protective toward nature and animals. In fact, these cultural differences could be linked: French-speaking culture may be more apt to assimilate technical innovations than German-speaking culture because of a greater tolerance of human intervention into nature. Even if this may be true, we hypothesize, second, a cultural difference in approval of the use of animals in science once attitudes toward science are included in the model.

This paper first outlines public perceptions of science, animals, and nature in the three linguistic regions of Switzerland, and then focuses on the different perceptions of the use of animals in science (i.e. animal cloning for medical research) by testing explanatory factors related to our hypotheses.

MAPPING SWISS CULTURAL CLIMATES

This chapter analyzes an existing data set, the Eurobarometer Science and Technology 2005 63.1, which was conducted face-to-face on 1,000 people, aged fifteen and over, from the three linguistic regions of Switzerland. If at national level, the error rate is 3.1%, it reaches 3.65% in the German-speaking part (n = 719), 6.59% in the French-speaking part (n = 221), and 12.65% in the Italian-speaking part (n = 60).

The analyses focus on two sets of variable: attitudes (toward science, technology, nature, etc.) and sociodemographics (gender, age, education, and linguistic region).

Attitudes toward science were measured by thirteen five-point Likert-type items (from 1 "strongly disagree" to 5 "strongly agree", DK recoded as "neither agree, nor disagree"), such as "Science and technology make our lives healthier, easier and more comfortable" or "The public is sufficiently involved in decisions about science and technology". Attitudes toward governance of science were investigated by two additional questions: who, according to the public, should make decisions about science (either experts or average citizens) and what criteria should guide such decisions (either the analysis of the risks and benefits involved or moral and ethical considerations).

Attitude toward technology, called technological optimism, was measured by the number of new technologies among fifteen (solar energy, computer and information technology, etc.) which will have, according to the respondent, a positive effect on our way of life in the next twenty years.

Attitudes toward nature were measured by five four-point Likert-type items (from 1 "strongly disagree" to 4 "strongly agree", DK recoded as "neither agree, nor disagree"), such as "We have a duty to protect nature, even if this means limiting human progress" or "We have a duty to protect the rights of animals whatever the cost". Attitude toward animal experimentation was measured by a five-point Likert-type item (from 1 "strongly disagree" to 5 "strongly agree", DK recoded as "neither agree, nor disagree"): "Scientists should be allowed to experiment on animals like dogs and monkeys, if this can help resolve human health problems". Attitudes toward animal cloning was measured by a four-point item (from 1 for "never agree", 2 for "agree only in exceptional circumstances", 3 for "agree only if it is highly regulated and controlled", to 4 for "agree in all circumstances", DK recoded as missing): "Tell me to what extent, if at all, you approve cloning animals such as monkeys or pigs for research into human diseases".

Finally, in addition to gender (coded 1 for men and 2 for women), the following three sociodemographics variables were used: age (in years); education (highest level achieved, transformed into two dummies: one differentiating the compulsory level [coded 1] from the other levels [coded 0] and one differentiating the upper secondary level [coded 1] from the other levels [coded 0]); and linguistic region (transformed into two dummies: one differentiating the French-speaking region [coded 1] from the other regions [coded 0] and one differentiating the German-speaking region [coded 1] from the other regions [coded 0]).

To test cultural differences, we performed analyses of variance for quantitative variables and Chi-squared tests for categorical variables. To explain the attitudes toward the use of animals in science (i.e. animal cloning for medical research), we performed a multiple regression model.

Attitudes Toward Science

Attitudes toward science were on the whole positive in Switzerland in 2005: 87.8% agree that "Scientific and technological progress will help to cure illnesses such as AIDS, cancer", 82.2% believe that " Science and technology make our lives healthier, easier and more comfortable", 68.3% accept the idea that "Even if it brings no immediate benefits, scientific research which adds to knowledge should be supported by Government", and a relative majority (43%) approve the the statement that "The benefits of science are greater than any harmful effects it may have" (Crettaz von Roten, 2006). However, we observe few visions of omnipotent science (a mere 7.1% believe that "Science and technology can sort out any problem" and only 36.9% think that "One day science will be able to give a complete picture of how nature and the universe work"), which is moderate technological optimism compared to European countries (the Swiss average is 9.6 against an average of 11.1 in EU25). Attitudes toward scientists are ambivalent (64.2% agree that "Because of their knowledge, scientists have a power that makes them dangerous", and a relative majority [43.4%] think that "Scientists are responsible for the misuse of their discoveries by other people") and about the involvement of the public (34.7% agree that "The public is sufficiently involved in decisions about science and technology", but 47.5% disagree).

Besides this complex structure of attitudes toward science in Switzerland, a second thorough examination of the data shows significant cultural differences in some specific attitudes (Table 18.1). In the first set of items (1 to 4) measuring general attitudes toward science, there is generally no cultural difference, except for the item related to health, where the mean score for the Italian-speaking people is higher than the one for French- and German-speaking people, indicating that Italian-speaking people are more convinced that such developments will help cure illnesses.

Table 18.1 Linguistic Regions Difference in Agreement with Science-related Items (Mean, Standard Deviation in Parenthesis, Anova F Statistic and Significance)

Variable	German-sp.	French-sp.	Italian-sp.	F	Sig
1. Science and technology make our lives healthier, easier and more comfortable.	4.10 (0.79)	3.89 (1.04)	3.94 (0.93)	2.51	0.083
2. Scientific and technological progress will help to cure illnesses such as AIDS, cancer, etc.	4.21 (0.85)	4.24 (0.83)	3.69 (1.25)	3.83	0.022
3. Even if it brings no immediate benefits, scientific research which adds to knowledge should be supported by the Government.	3.64 (1.11)	3.85 (1.09)	4.01 (1.00)	2.31	0.100
4. The benefits of science are greater than any harmful effects it may have.	3.30 (1.01)	3.46 (0.92)	3.26 (0.99)	1.04	0.354
5. Science and technology can sort out any problem.	1.70 (0.91)	2.05 (1.26)	1.58 (1.11)	5.30	0.005
6. One day science will be able to give a complete picture of how nature and the universe work.	2.70 (1.33)	3.13 (1.20)	3.05 (1.17)	4.81	0.009
7. There should be no limit to what science is allowed to investigate on.	2.26 (1.27)	2.73 (1.38)	2.44 (1.22)	5.34	0.005
8. Technological optimism	9.45 (2.84)	10.16 (2.95)	9.06 (3.35)	5.66	0.004
9. Because of their knowledge, scientists have a power that makes them dangerous.	3.75 (1.04)	3.26 (1.25)	3.75 (1.30)	8.51	0.000
10. Scientists are responsible for the misuse of their discoveries by other people.	3.11 (1.29)	2.81 (1.31)	2.38 (1.34)	4.86	0.008
11. The public is sufficiently involved in decisions about science and technology.	2.87 (1.16)	2.49 (1.28)	2.89 (1.25)	4.19	0.016

Table 18.2 Linguistic Regions Difference in Type of Governance (% by Criteria, Chi-Square-test Statistic and Significance)

Variable	German-sp. (n= 747)	French-sp. (n=210)	Italian-sp. (n=43)	T (df 4)	Sig
Advice of experts—general public's views	54.2–30.4	52.1–28.0	55.8–18.6	6.33	0.176
Risks and benefits criteria—moral and ethical criteria	21.0–65.9	31.4–45.2	39.5–37.2	39.99	0.000

In the mean scores of the second set of items (5–8) measuring scientific and technician utopianism, we observe that French-speaking people perceive science as being more omnipotent on average: they are more likely to believe that science can sort out any problem, or will give a complete picture of how nature and the universe work, and therefore they are more inclined to think that this ambitious scientific project should not be limited. Finally, French-speaking people are more optimistic about technology.

Moreover, mean scores of the German-speaking people on items 9–10 indicate less positive attitudes toward scientists: they are more convinced that scientists have a power that makes them dangerous, and they more frequently hold them responsible for the misuse of their discoveries by other people than the other respondents.

Finally, French-speaking people are less convinced that the public is sufficiently involved in decisions about science and technology (item 11): they wish for more democratic government through more public involvement. To deepen this last point, differences related to the governance of science were investigated (Table 18.2).

The cultural differences are not on who decides (no difference, as significance reaches 0.176) but on what criterion (significance of 0.000). In the German-speaking region, the percentage of "moral and ethical criteria" is higher (65.9%) than in other regions (respectively 45.2% and 37.2%), and the answers are asserted more definitely (the percentage of "Don't know" is smaller [13.1%] than in other regions [respectively 23.4% and 23.3%]).

Attitudes Toward Nature and Animals

On the whole, Swiss people are very protective toward nature and animals: 93.1% believe that it is a duty to protect nature, even if the price to pay is to limit progress, 73.7% disagree with the exploitation of nature for the sake of human well-being, 78.9% believe that it is a duty to protect animal rights whatever the cost. On the question of the necessity to exploit nature—for the sake of humankind—the results are much more mixed: 48.8% approve the exploitation of nature, 46.3% disapprove of it. This ambivalence is also observed on human effects on nature: 52.6% believe that nature will survive human actions, but 46.3% reject the idea.

This general picture of support for nature conceals dissimilarities, in particular among linguistic regions: attitudes toward nature show significant

Table 18.3 Linguistic Regions Difference in Agreement with Attitudes toward Nature and Animals (Mean, Standard Deviation in Parenthesis, Anova F Statistics and Significance)

Variable	German-sp. (n= 747)	French-sp. (n=210)	Italian-sp. (n=43)	F	Sig
1. We have a duty to protect nature, even if this means limiting human progress.	4.54 (0.72)	4.28 (0.92)	4.42 (0.93)	8.96	0.000
2. We have a right to exploit nature for the sake of human well-being.	1.66 (1.04)	3.24 (1.35)	3.83 (1.38)	214.44	0.000
3. Nature will be able to survive human actions.	3.39 (1.49)	2.66 (1.42)	2.69 (1.44)	22.53	0.000
4. Exploiting nature may be unavoidable if humankind is to progress.	2.96 (1.39)	3.04 (1.38)	2.29 (1.43)	5.37	0.005
5. We have a duty to protect the rights of animals whatever the cost.	4.04 (1.10)	3.87 (1.21)	3.89 (1.36)	2.00	0.136

cultural differences, except where the protection of animals is concerned (Table 18.3). The German-speaking people are more protective and more optimistic toward nature: on average, they are more willing to protect nature, they agree less with its exploitation, and they are more likely to think that nature will survive human actions.

An important similarity between the regions lies in the fact that French- as well as German-speaking people are more realistic or fatalistic regarding the exploitation of nature than the Italian-speaking people: in the French region, we exploit nature and we have the right to do it, whereas in the German region, we exploit nature, but we don't have the right to do it.

Finally, attitudes toward animals may be studied in the context of animal experimentation. The relative majority of Swiss people are against experimentation on dogs and monkeys for human health research: 49.6% refuse it whereas only 34.7% approve it (Crettaz von Roten, 2008): there is no significant cultural difference in the way research on animals is perceived. Between 2001 and 2005, the percentage of people against animal research increased in each linguistic region, especially in the French-speaking one, where it rose by nearly half (from 32.4% to 48.5% in 2005).

Attitudes Toward Animal Cloning for Medical Research

Biotechnologies have provided other uses of animals in science with a wide range of applications—animal pharming (Pardo, 2009), animal cloning for research or for food production purposes, etc.—and have fundamentally challenged the human-animal boundary (Hobson-West, 2007). The Swiss attitudes toward cloning animals such as monkeys or pigs for medical research are ambivalent: 49.6% disapprove of animal cloning, 22.5% approve of it only in exceptional circumstances, 23.1% agree with it only if it is highly regulated and controlled, and 3.5% support it in all circumstances. Overall, French-speaking people approve significantly more animal cloning than the German and Italian speakers.[3]

How does one explain the variation in attitudes toward animal cloning? Following the existing literature and our hypothesis, we tested a model with three explanatory factor families: sociodemographics (among which figures linguistic region), factors related to science, and factors related to nature and animals. The regression model, which explains 29% of the variation in the approval of animal cloning, is multicausal and includes all explanatory factor families (Table 18.4). In this table, a positive Beta coefficient indicates that an increase of the factor in question implies a broader acceptance

Table 18.4 Modelization of Approval of Animal Cloning for Medical Research

	Coefficient Beta	Sig
Education		
Compulsory level	-0.105	0.021
Upper secondary level	-0.021	n.s.
Gender (1 = Men, 2 = Women)	-0.050	n.s.
Age	-0.050	n.s.
Linguistic region		
French-speaking region	0.244	0.005
German-speaking region	0.157	n.s.
Science and technology make our lives healthier, easier and more comfortable.	0.123	0.004
Even if it brings no immediate benefits, scientific research which adds to knowledge should be supported by Government.	0.084	0.049
Technological optimism	0.138	0.001
Decisions about science and technology should be based primarily on an analysis of the risks and benefits involved (1) on the moral and ethical issues involved (2).	-0.118	0.004
Decisions about science and technology should be based primarily on an analysis of the advice of experts about risks and benefits involved (1) on the general public's view about risks and benefits (2).	-0.097	0.019
The public is sufficiently involved in decisions about science and technology.	0.083	0.038
We have a right to exploit nature for the sake of human well-being.	0.103	0.043
Scientists should be allowed to experiment on animals like dogs and monkeys, if this can help resolve human health problems.	0.204	0.000
N=485 ; adequacy test $F_{14, 471}$ = 13.39, sig = 0.000	R^2 = 0.29	

of animal cloning, controlling for the other independent factors, whereas a negative Beta coefficient indicates that an increase of the factor in question implies a more restrictive frame for animal cloning.

People who are more educated and French-speaking people are more likely to express wide support of animal cloning. The broad approval of animal cloning is also linked with positive attitudes toward science and technology, with a governance of science valuing scientific evidence over moral and ethical considerations, and valuing the advice of experts over the general public's view, with less concern about participation in the decision-making process, and with a higher degree of acceptance of the exploitation of nature and of animal experimentation. Gender and age are not significant once these ten variables are incorporated into the model.

DISCUSSION AND CONCLUSIONS

In the last twenty years, animals and nature have increasingly attracted public attention, in particular the blurring of the human-animal boundary, pandemic flu related to animals (avian flu, porcine flu), and risk related to wild and dangerous animals. Therefore, social scientists have engaged in the study of how the public perceives animals and nature. In parallel, parliaments and public agencies have devised a regulatory framework that copes with public attitudes and moral concerns related to science but that also tries not to constrain too much scientific and technological developments (for example, in genetic research on the modification of plants and animals). What role does culture play in this context?

This contribution has established the basic outlines of the Swiss attitudes regarding science, animals, and nature, which can be characterized as positive toward science, and protective toward animals and nature. Moreover, the findings of this study have shown differences in the way science, nature, and the use of animals in science are each perceived in the three linguistic regions of Switzerland, which supports our hypotheses.

Among the attitudes toward science, cultural differences are not observed on general attitudes, but on scientific and technician utopianism, on scientists and governance of science. The French-speaking people are more utopian, positive toward scientists, and participative, whereas German-speaking people lay more emphasis on the "moral and ethical criteria" of science governance. These combined results draw a picture of a German-speaking region that is more critical toward science and scientists, and therefore wishes for a holistic governance of science, taking into consideration both ethical and moral criteria. These results are in line with the postindustrialization model of PUS: less developed countries are more optimistic about science; more developed countries are more critical toward science (Bauer et al., 1994). The Swiss linguistic regions may be characterized according to an economical dimension: historically the German-speaking region is more developed economically with most big pharmaceutical industries and a banking center, whereas the French- and Italian-speaking regions have higher unemployment rates.[4] Therefore the French- and Italian-speaking regions count on scientific innovations with some success, for example many biotechnological industries are now established in the French-speaking region.

There are strong cultural differences on attitudes toward nature and animals: the German-speaking region takes a more protective and optimistic attitude toward nature, whereas the French-speaking one adopts more positive attitudes toward animal cloning for medical research. Our results are in line with the Diekmann et al. (2009) study, which establishes a lower percentage of attribution of the same moral rights to animals and humans and a lower perception of risk related to biotechnology applied in medicine and research among French-speaking people. And these attitudes lie within the general framework of the environmental protection attitudes of German-speaking people documented for decades: a stronger environmental concern and a higher level of environmental knowledge and of environment-friendly behavior (Diekmann and Franzel, 1997; Diekmann et al. 2009).[5] Burton-Jeangros et al.'s (2009) study on animals' representation in the media found an increase between 1978 and 2008, and showed individual cultural differences: French-speaking people more often bring up the human-animal boundary; Italian- and French-speaking people emphasize the question of the management of human-animal territory.

Finally, approval of animal cloning in medical research is explained by perceptions of science, nature, and animals, and by education and linguistic regions, in line with the literature (Einsiedel, 2000; Pardo, 2009). Therefore belonging to a linguistic area (distinguishing the French-speaking region from others) is significant once attitudes toward science, technology, nature, and animals are included in the model. This result proves the relevance of analyses on linguistic regions in Switzerland and, more generally, raises the question of the level of analysis in PUS studies (country, regions, etc.). More PUS studies with a cross-cultural perspective should be undertaken, but if possible, avoiding the methodological problems found in this study.

In fact, some non-significant cultural differences may rest on methodology. First, due to split-ballot variables,[6] the sample size for the analysis of attitudes toward science items is 500 respondents, which is insufficient to analyze cross-cultural effects (cultural effects are small or medium but not large, therefore the sample size needs to be bigger [Cohen, 1992]).[7] Next, indicators of attitudes toward nature and animals suffer from low reliability and tricky wording. The inter-item reliability of the five items related to animals and nature is poor (Cronbach's Alpha = 0.31). Further, among different questions measuring the acceptance of animal experimentation (Hagelin et al., 2003), the EB 2005 used an item related to species that are most debated: monkeys, at the boundary between humans and animals, and dogs, man's best friend. It is well-known that the type of animal and the nature of goals (medical, cosmetics, etc) determine whether attitudes fall into acceptance or rejection. Future studies on the use of animals should use a set of indicators mixing species and research purposes.

From the above elements of discussion, mapping the relation between science, nature, and culture in Switzerland is very interesting. The political system of direct democracy allows the public to launch a campaign to collect signatures for a so-called initiative on any subject. Nature, the protection of animals, animal experimentation, GMOs, etc. were the object of such initiatives. Analyses of the results yielded by votes frequently show linguistic divergence (Kriesi, 1996). This raises the crucial question of the possibility to elaborate scientific and environmental policies, which can be understood, interpreted, and accepted in the same way in the three linguistic regions. Therefore, we need to conduct more research to improve our understanding of the complex structure of attitudes toward science, nature, and culture.

NOTES

1. That is, arbitrary discrimination on the basis of species memberships. Previous studies have illustrated the fact that the use of dogs and monkeys results in lower support for experimentation than other species such as small rodents (Hagelin et al., 2003).
2. The Swiss Federal Statistical Office defines linguistic region according to the majority language of the population of the respective districts. The results of the 2000 Swiss Census indicate that 63.7% of the population primarily speak German, 20.4% French, 6.5% Italian, and 0.5% Romansch (the latter group is recoded in this study with the German-speaking group).
3. The French-speaking mean is 2.12, the German-speaking mean 1.72 and the Italian-speaking mean 1.64; the Anova test is significant: $F = 7.97$ and significance <0.001.
4. In December 2005, the unemployment rate was 5.7% in the Italian-speaking region; it was between 4.2% and 7.3% in the French-speaking canton; but it was between 1.2% and 3.9% in the German-speaking cantons (SECO, 2005).
5. The finer points of the human-animal boundary seem to reveal various attitudes toward the environment (Burton-Jeangros et al., 2009).
6. The split-ballot technique consists in the application to each half of a given sample of one or more questions allowing more items to be covered. In our case, attitudes toward science and toward some scientific developments were realized in split-ballot.
7. For example, this may explain why there is no significant cultural difference in the way research on animals is perceived in our study, while there is a significant difference in Diekmann et al. (2009).

REFERENCES

Allum, N., Boy, D., & Bauer, M. (2002). European Regions and the Knowledge Deficit Model. In M. Bauer & G. Gaskell (Eds.), *Biotechnology: The Making of a Global Controversy* (pp. 224–243). Cambridge: Cambridge University Press.

Bauer, M., Allum, N.,& Miller, S. (2007). What can we learn from 25 years of PUS survey research? Liberating and epanding the agenda. *Public Understanding of Science*, 16(1), 79–95.
Bauer, M., Durant, J., & Evans, G. (1994). European Public Perceptions of Science. *International Journal of Public Opinion Research*, 6(2), 163–186.
Burton-Jeangros, C. et al. (2009). Les representations des animaux dans les medias suisses d'information, 1978–2008. Rapport final, Université de Genève.
Bourg, D. (1997). *Nature et technique. Essai sur l'idée de progrès.* Paris: Hatier.
Cohen, J. (1992). *Statistical Power Analysis for the Behavioural Sciences*, 2nd ed. Mahwah: Lawrence Erlbaum.
Crettaz von Roten, F. (2006). Les Suisses et les sciences en 2005 : structure et facteurs explicatifs des attitudes. In. J.-Ph. Leresche, M. Benninghoff, F. Crettaz von Roten, & M. Merz (Eds.), *La fabrique des sciences. Des institutions aux pratiques* (pp. 283–304). Lausanne: PPUR.
Crettaz von Roten, F. (2008). Mapping Perception of Animal Experimentation: Trend and Explanatory Factors. *Social Science Quarterly*, 89(2), 537–549.
Crettaz von Roten, F., Hof, P., & Leresche, J.-Ph. (2003). Le public suisse face à la science et aux institutions: des relations à explorer. *Revue Suisse de Science Politique*, 9(3), 97–111.
Diekmann, A. & Franzel, A. (1997). *Les Suisses et l'environnement*. Berne: OFS.
Diekmann, A., Meyer, R., Mühlemann, C., & Diem, A. (2009). Schweizer Umweltsurvey 2007—Analysen und Ergebnisse. Bericht für BFS und BAFU.
Einsiedel, E. (2000). Cloning and its Discontents—a Canadian Perspective. *Nature Biotechnology*, 18, 943–944.
Einsiedel, E. (2005). Public Perceptions of Transgenic Animals. *Rev. Sci. Tech. Off. Int. Epiz.*, 24(1), 149–157.
Evans, G. & Durant, J. (1995). The Relationship Between Knowledge and Attitudes in the Public Understanding of Science in Britain. *Public Understanding of Science*, 4, 57–74.
Frank, D. (1997). Science, Nature, and the Globalization of the Environment, 1870–1990. *Social Forces*, 76(2), 409–437.
Franklin, A. (1999). *Animals and Modern Cultures. A Sociology of Human-animal Relations in Modernity.* London: Sage.
Franklin, A. & White, R. (2001). Animals and Modernity: Changing Human-Animal Relations, 1949–98. *Journal of Sociology*, 37(3), 219–238.
Freitag, M. & Stadelmann-Steffen, I. (2009). Les mondes du bénévolat—l'engagement bénévole en Suisse et dans le contexte transnational. In Ch. Suter et al. (Eds.), *Rapport social 2008. La Suisse mesurée et comparée* (pp. 170–190). Zürich: Seismo.
Gaskell, G., Bauer, M., Durant, J., & Allum, N. (1999). Worlds Apart: The Reception of GM Foods in the United States and Europe. *Science*, 285, 384–387.
Gaskell, G., Eyck, T., Jackson, J., & Veltri, G. (2005). Imagining Nanotechnologies: Cultural Support for Technological Innovation in Europe and the United States. *Public Understanding of Science*, 14(1), 81–90.
Hagelin, J., Carlsson, H., & Hau, J. (2003). An Overview of Surveys on How People View Animal Experimentation: Some Factors That May Influence the Outcome. *Public Understanding of Science*, 12(1), 67–81.
Hobson-West, P. (2007). Beast and Boundaries: An Introduction to Animals in Sociology, Science, and Society. *Qualitative Sociology Review*, III(1), Retrieved April 2010 (http://www.qualitativesociologyreview.org /ENG/archive_eng.php).
Kriesi, H., et al. (1996). *Le clivage linguistique: problèmes de compréhension entre les communautés linguistiques en Suisse.* Berne: OFS.
Michael, M. (2000). *Reconnecting Culture, Technology and Nature.* London: Routledge.

Michael, M. (2001). Technoscientific Bespoking: Animals, Publics and the New Genetics. *New Genetics and Society*, 20(3), 205–224.

Miller, J. & Pardo, R. (2000). Civic Scientific Literacy and Attitude to Science and Technology: A Comparative Analysis of the European Union, the United States, Japan, and Canada. In M. Dierkes & C. Grote (Eds.), *Between Understanding and Trust: The Public, Science and Technology* (pp. 265–285). Reading: Harwood Academic Publishers.

Pardo, R. (2009). Public Views and Attitudes to Pharming. In E. Rehbinder et al. (Eds.), *Pharming. Promises and Risks of Biopharmaceuticals Derived from Genetically Modified Plants and Animals* (pp. 121–178). Berlin: Springer.

Peters, H., et al. (2007). Culture and Technological Innovation: Impact of Institutional Trust and Appreciation of Nature on Attitudes toward Food Biotechnology in the USA and Germany. *International Journal of Public Opinion Research*, 19(2), 191–220.

Priest, S.H., Bonfadelli H., & Rusanen M. (2003). The 'Trust Gap' Hypothesis: Predicting Support for Biotechnology across National Cultures as a Function of Trust in Actors. *Risk Analysis*, 23(4), 751–766.

SECO (2005). La situation sur le marché du traail en décembre 2005. Bern, Directorate of Labour, State Secretariat for Economics Affairs.

Strauss, L. (2005). Concepts of Nature. In S. P. Restivo (Ed.), *Science, Technology and Society. An Encyclopedia* (pp. 351–357). Oxford: Oxford University Press.

Streckeisen, U., Diem, M., & Franzen, A. (2002). La problématique des échanges scientifiques entre professeurs d'universités dans le contexte des relations Suisse alémanique—Suisse romande. Neuchâtel: Office federal de la statistique.

TNS Opinion & Social (2005). Special Eurobarometer 224/Wave 63.1. Europeans, Science and Techonology. Survey requested by Directorate General Research and coordinated by the Directorate General Press and Communication (European Commission), Brussels.

Wagner, W. et al. (2001). Nature in Disorder: the Troubled Public of Biotechnology. In G. Gaskell & M. Bauer (Eds.), *Biotechnology 1996–2000. The Years of Controversy* (pp. 80–95). London: NMSI Trading, Science Museum.

19 Religious Belief and Attitudes about Science in the United States

Scott Keeter, Gregory Smith and David Masci

Religion and science have long had an uneasy coexistence, at least in the popular imagination. This perception has been fueled in part by a number of famous episodes in history that have pitted scientists, like Galileo and Darwin, against the prevailing religious establishments of their time.

The United States has made its own contributions to this perception. In 1925, for instance, a shy science teacher from Dayton, Tennessee, named John Thomas Scopes made headlines and history when he was tried and convicted for teaching evolution to his high school students. To this day, the "Scopes Monkey Trial" remains a potent symbol of what many see as an inherent conflict between faith and science (Larson, 1997). But to what extent does this conflict really exist? And is there clear evidence for it in American public opinion today?

The United States is perhaps the most religious of the advanced industrial democracies. At the same time, American scientists are recognized as leaders in many areas of scientific research and application. This combination of widespread religious commitment and leadership in science and technology makes the potential for conflict in the United States very great. Indeed, the potential impact of a serious conflict on the ability of science to function could be significant, given the high level of religious commitment among the US public and much of its political leadership.

The analysis that follows will argue that although Americans respect science and scientists, they are not always willing to accept scientific findings that squarely contradict their religious beliefs. At the same time, such conflicts are not common in the United States today (Numbers, 2009). True, more than eighty years after the Scopes trial, the subject of evolution offers a very concrete (and ongoing) example of such a conflict. Faith also appears to play a less pronounced, but still powerful, role in shaping views of the nature of homosexuality. But on other questions, religion is a less powerful predictor of views of science; for instance, compared with views of evolution and homosexuality, views of global warming are not highly correlated with religious variables.

Furthermore, there are no additional obvious examples of issues where scientists and people of faith disagree on critical facts. There are areas that one day could become the source of a factual dispute between scientists and some religious Americans. For instance, some scientists publicly claim that the most recent research on the human brain shows that it and it alone is the seat of consciousness and personhood and that this evidence disproves the existence of a soul or spirit. If this idea were to become widely accepted and publicized, it could, and indeed probably would, prove to be another area of conflict between religion and science. But currently, the debate over "the death of the soul" is not stirring significant opposition from religious people and groups, primarily because there is no scientific consensus on the issue (as there is with evolution) and no real constituency for increasing public knowledge of this debate (as there is with questions concerning the nature of homosexuality).

Finally, there are religious differences of opinion about the ethics of some kinds of scientific research and its application. In the field of bioethics, public opinion is divided about cloning, embryonic stem cell research, end-of-life issues, and genetic testing. But here, the disputes do not concern questions of fact, as they do in the debates over evolution and homosexuality. Instead, the bioethics debates involve purely moral and ethical questions, which have been well documented and so will not be discussed at great length in this paper.

EVOLUTION

In the last century, the most persistent and sharpest clash between religion and science in the United States has centered on the issue of evolution. Public opinion polling finds that there has been virtually no change over the past several decades in the percentage of the US public—approximately 40% to 50%—who reject the very idea of natural evolution, largely on the grounds that it conflicts with biblical accounts of creation.

The Pew Research Center developed a sequence of questions to gauge opinion about evolution. Because of concerns that some respondents may feel a need to support a creationist account in order to express their belief in God, in its original formulation this sequence first asked half of the respondents about the belief in God or a higher power, and then asked the believers (95% of those asked) if God (or the higher power) was responsible for the creation of life on earth. The other group did not receive these introductory questions about belief. All respondents were then asked a two-part question about the development of human and other life. A comparison of the treatment and control groups found no difference in the percentage who believe in evolution; being first asked about belief in God neither increased nor reduced the percentage of those who believe in creationism. Thus subsequent Pew Research Center surveys have omitted the belief lead-in question.

The two-part evolution question also attempts to gauge support for the idea that a supreme being guided evolution with the goal of creating life in its present form. This idea, similar to the notion of "intelligent design," was appealing to many of those who said they believe that life has evolved. The sequence was worded as follows:

> "Some people think that humans and other living things [have evolved over time/have existed in their present form since the beginning of time]. Which of these comes closest to your view? And do you think that [Humans and other living things have evolved due to natural processes such as natural selection/a supreme being guided the evolution of living things for the purpose of making humans as they are today]?" (Items in brackets were rotated randomly.)

Overall, 31% of respondents in a May 2009 Pew Research Center poll rejected evolution, choosing the option that humans and other living things have existed in their present form since the creation (see Table 19.1). About six-in-ten (61%) believe that evolution has occurred, but many think it was guided by a supreme being or higher power (22%). Just 32% believe in evolution through natural selection (Pew Research Center 2009).

The strongest opposition to the idea of evolution comes from evangelical Christians, most of whom accept the Bible as literally true and see a direct conflict between the biblical creation account and scientific accounts of evolution. Most white evangelicals (57%) and nearly half of black evangelicals (46%) say that life did not evolve. Just 34% of white evangelicals and 45% of black evangelicals believe in evolution, and fewer still think evolution occurred through natural selection.

Among other religious groups, majorities believe in evolution: this includes 65% of Catholics, 69% of white mainline Protestants, and 85% of the unaffiliated. But mainline Protestants and Catholics who believe in evolution are themselves divided over the question of whether evolution occurred through natural selection or was guided by a supreme being for the purpose of creating human life in its present form. Overall, 38% of mainline Protestants believe in natural selection, and 25% believe a supreme being guided the process. Among Catholics, 33% subscribe to the idea of natural selection, and 25% think evolution was divinely guided. Only among the unaffiliated does a majority accept natural selection: 60% of respondents with no religious affiliation believe that life evolved through natural selection.

Part of the reason for the extent of creationist influence in the United States is the stronger degree of certainty of belief about the origins and development of human life possessed by those who reject the theory of evolution compared with those who believe in evolution (Pew Research Center 2005). More than six-in-ten (63%) of those who accept the creationist account say they are "very certain" of their views, while those who believe evolution occurred are far less certain (32% very certain). Those who accept evolution by natural selection are the least certain (28%). Relatedly, 69% of biblical literalists are very certain about their view on the development of life, compared with about one-third among those who do not take the Bible literally.

Table 19.1 Opinions on Evolution and Climate Change, by Religious Affiliation

	Total	White Evangelical	White Mainline	Catholic	Unaffiliated
Humans and other living things have . . .	%	%	%	%	%
Existed in present form only	31	57	23	27	11
Evolved over time	61	34	69	65	85
Guided by supreme being	22	20	25	25	15
Through natural selection	32	9	38	33	60
Don't know how evolved	7	4	6	7	10
Don't know	8	7	8	8	4
	100	100	100	100	100
Do scientists agree about evolution?					
Yes	60	43	64	63	79
No	28	45	22	27	15
Don't know	11	12	14	10	5
	100	100	100	100	100
Earth getting warmer?					
Yes	85	75	87	87	89
Due to human actions	49	38	54	52	53
Due to natural causes	36	37	33	35	36
No, not warming	11	19	8	10	7
Don't know	4	6	4	3	4
	100	100	100	100	100
How serious a problem is global warming?					
Serious (net)	73	61	74	73	79
Very serious	47	35	43	51	49
Somewhat serious	26	26	31	22	30
Not too serious	11	15	11	14	9
Not a problem	13	21	14	12	9
Don't know	2	2	1	1	3
	100	100	100	100	100
Do scientists agree about global warming?					
Yes	56	48	58	59	59
No	35	41	32	35	34
Don't know	9	11	9	6	6
	100	100	100	100	100
Sample size	2001	417	356	477	321

Source: PEW RESEARCH CENTER May 2009.

The rejection of evolution is not entirely a result of a lack of awareness of the scientific consensus on the subject. For instance, 43% of white evangelicals say that scientists are in agreement about the theory of evolution, although only 34% of evangelicals accept the theory themselves (Table 19.1). Nor is the rejection of evolution a result of political or ideological beliefs. Republicans and conservatives are more apt than Democrats or liberals to deny that evolution occurs, but this correlation is mostly a result of the large number of evangelicals with creationist views in the Republican Party and among conservatives.

Indeed, a binary logistic regression model shows that even after controlling for education, political partisanship, and ideology, awareness of the scientific consensus on the topic of evolution, and a host of other demographic variables, religion remains a highly significant predictor of views of evolution (see appendix for details on the regression model). With all other religious variables held at 0 and all other variables held at their mean value, for instance, the model predicts that the likelihood that committed evangelical Protestants (those who go to church regularly) will believe that humans have evolved over time is a mere .35, making this group .53 points less likely to believe in evolution compared to the unaffiliated (Table 19.2). Other religious groups are

Table 19.2 Predicted Probability of Holding Selected Beliefs, by Religious Tradition, Party Affiliation and Perceptions of Scientific Consensus

	Evolution[1]	Man-made Global Warming[1]	Homosexuality is Immutable[2]	Homosexuality is Innate Trait[2]
Committed Evangelical	0.35	0.42	0.25	0.18
Other Evangelical	0.63	0.48	0.53	0.32
Committed Mainline	0.64	0.62	0.62	0.53
Other Mainline	0.81	0.53	0.76	0.61
Committed Catholic	0.71	0.45	0.63	0.49
Other Catholic	0.79	0.54	0.77	0.62
Unaffiliated	0.88	0.51	0.71	0.61
Republican	0.73	0.34	0.57	0.36
Democrat	0.70	0.63	0.64	0.52
Independent	0.70	0.48	0.53	0.41
Scientists do not agree on issue	0.60	0.30	NA	NA
Scientists agree on issue	0.77	0.67	NA	NA
Sample size for regressions	1,850	1,917	865	852

Entries are predicted probabilities of holding a particular belief, controlling for gender, race, ethnicity, age, education, income, party affiliation, ideology, and religious affiliation.

Source: PEW RESEARCH CENTER, [1]May 2009; [2]July 2006.

also significantly less likely to believe in evolution relative to the unaffiliated, although by lesser degrees. The model confirms that awareness of the scientific consensus on evolution is also closely associated with the belief that evolution has occurred, but merely knowing about the scientific consensus is not sufficient to induce acceptance of the theory of evolution. Political partisanship is a comparatively weak predictor of views about evolution.

HOMOSEXUALITY

Unlike evolution, there is no scientific consensus on the causes of homosexuality. Consequently, there is little agreement about whether or not homosexuality is a personal preference. Still, the share of Americans, particularly well educated Americans, who see homosexuality as something people are born with is growing. At the same time, conservative and religious groups continue to strongly contest this view, even in the face of scientific research showing that homosexuality may be associated with certain genetic features or exposure to hormones during gestation. Furthermore, they object to the idea that homosexuals cannot change their sexual orientation or be "cured." (Pew Research Center, 2006).

Opinions about the nature of homosexuality have changed somewhat since the 1980s (Table 19.3). In a 2006 Pew Research Center poll, 36% said that homosexuality is innate (up from 20% in 1985, and from 30% in 2003). Compared with 2003, more people in 2006 said that homosexuality cannot be changed (from 42% to 49%). But the majority of the public still rejects the idea that homosexuality is something that people are born with, and see it instead as either a product of the way people are brought up (13%) or as "just the way that some people prefer to live" (38%).

Although the number of Americans who see homosexuality as something people are born with has increased only modestly since 2003, this view is now much more widely held among certain groups in the population than it was three years ago. Among college graduates, for instance, there has been a double-digit increase since 2003 in the view that homosexuality is innate (from 39% to 51%). Similar increases are seen among liberals (46% to 57%), mainline Protestants (37% to 52%), and among those who seldom or never attend church (from 36% to 52%).

In contrast to these groups, majorities of white evangelicals (51%) and black Protestants (52%) continue to view homosexuality as a choice. White evangelicals, in particular, have changed very little in their views on this question over the past three years.

Though most Americans reject the notion that homosexuality is an innate trait, a plurality of the public (49%) views sexual orientation as a

Table 19.3 Beliefs about Homosexuality, by Education, Ideology, Religious Affiliation, and Attendance

	Homosexuality is...				Homosexuality...		
	Something people are born with	Linked to upbringing	Just how some prefer to live	Don't know	Can be changed	Cannot be changed	Don't know
	%	%	%	%	%	%	%
Total, July 2006	36	13	38	13=100	39	49	12=100
October, 2003	30	14	42	14=100	42	42	16=100
December, 1985	20	22	42	16=100	NA	NA	NA
College graduate	51	9	28	12=100	31	58	11=100
Some college	39	15	32	14=100	40	52	8=100
High school or less	26	14	46	14=100	43	42	15=100
Conservative	21	20	46	13=100	52	36	12=100
Moderate	38	9	37	16=100	37	49	14=100
Liberal	57	7	27	9=100	21	71	8=100
Total Protestant	29	15	41	15=100	45	42	13=100
White evangelical	17	15	51	17=100	56	29	15=100
White mainline	52	13	22	13=100	22	67	11=100
Black Protestant	20	19	52	9=100	60	30	10=100
Total Catholic	44	10	33	13=100	31	56	13=100
White, non-Hispanic	48	10	29	13=100	26	61	13=100
Unaffiliated	48	7	29	16=100	27	59	14=100
Religious Attendance							
Weekly or yearly	25	17	44	14=100	54	34	12=100
Monthly or less	36	8	43	13=100	34	52	14=100
Seldom or never	52	13	23	12=100	22	68	10=100

Source: PEW RESEARCH CENTER July 2006.

characteristic that cannot be changed, a seven percentage-point increase since 2003.

Views of whether homosexuality can be changed have both a political and a religious component. A small majority of conservatives (52%) says homosexuality can be changed, whereas the overwhelming majority of liberals (71%) disagrees. Similarly, substantial majorities of white evangelicals (56%) and black Protestants (60%) say that homosexuality can be changed, whereas majorities of white mainline Protestants (67%), Catholics (56%), and the unaffiliated (59%) say homosexuality cannot be changed.

As with views of evolution, multivariate analysis reveals that religion variables are strong predictors of views of the nature of homosexuality even after controlling for education, partisanship, and a number of demographic variables (Table 19.2). Most notably, evangelicals are much less likely than other groups to accept either the notion that homosexuality is an innate characteristic or that it is a permanent, unchangeable trait. But it also appears that religion is a less important predictor of views of homosexuality than it is of views of evolution. Unlike with evolution, there are only small differences in the views of homosexuality among non-evangelicals; whereas highly committed Catholics and both highly committed and less-committed mainline Protestants were significantly less likely than the unaffiliated to believe in evolution, none of these groups differs significantly from the unaffiliated in their views of the nature of homosexuality. And political partisanship is a somewhat stronger predictor of views of homosexuality than of views of evolution; Democrats are significantly more likely than independents and Republicans to say that homosexuality is innate and unchangeable. (The survey did not ask about perceptions of the degree of scientific consensus regarding the nature of homosexuality.)

GLOBAL WARMING

Religious differences are somewhat smaller on another controversial issue, global warming. An overwhelming majority of those polled in May 2009 by Pew Research (85%) believe that the earth is getting warmer; just 11% say the earth is not getting warmer (Table 19.1). Sizable majorities of each of the largest religious groups agree: 75% of evangelical Protestants; 87% of Catholics and mainline Protestants; and 89% of those unaffiliated with a religion (Pew Research Center, 2009).

Most of those who believe that the earth is getting hotter also believe that human activity, such as the burning of fossil fuels, is responsible: based on the total sample, 49% say this, and 36% say it is mostly a result of natural patterns in the earth's environment. But there are somewhat larger differences across religious groups on this question: 52% of Catholics, 53% of the religiously unaffiliated, and 54% of white mainline Protestants believe the earth is getting hotter and think this is because of human activity, whereas fewer evangelicals think this (38%).

Although majorities of all religious groups (from 61% among white evangelicals to 79% among the unaffiliated) view global warming as a serious problem, considerably fewer regard it as a *very* serious problem: just 47% overall and as few as 35% among evangelicals say this (Table 19.1). Global warming also ranks near the bottom of policy priorities for the public when asked about a long list of issues.

There also are differences among religious groups in the perception that there is a scientific consensus on global warming, but even among evangelicals, almost half (48%) thinks there is a scientific consensus, a finding that may help explain why some evangelical leaders have begun to pay more attention to environmental issues (Table 19.1). Overall, 56% of the public says that scientists agree that global warming is occurring and that it is caused by human activity; 35% think there is no scientific consensus. Among other religious groups, more see scientific agreement: 58% among mainline Protestants and 59% among both Catholics and the religiously unaffiliated.

Multivariate analysis of the correlates of views of global warming reveals that after controlling for demographic and political variables, and especially for awareness of the scientific consensus with respect to global warming, religion is not closely related to views of global warming (Table 19.2). In particular, the magnitude of the difference between evangelicals and the unaffiliated in views of global warming is much smaller than that seen on views of evolution and homosexuality. The predicted probability of believing in man-made global warming is .09 higher among the unaffiliated than among highly committed evangelical Protestants; by comparison the predicted probability of agreeing with the scientific mainstream on the issue of evolution is 53 points higher among the unaffiliated than highly committed evangelicals, 43 points higher on the question of whether homosexuality is innate, and 46 points higher on the question of whether homosexuality is changeable.[2]

At the same time, perception of a scientific consensus on global warming is a relatively stronger predictor of views on this issue than on views of evolution. Whereas the predicted probability of believing that evolution occurred is 17 points higher among those aware of the scientific consensus on evolution than those unaware of it, the probability of believing that man-made global warming is occurring is 47 points higher than among those who believe there is a scientific consensus on global warming. Political variables are also strong predictors of views of global warming, with Republicans and political conservatives much less likely to believe that human-caused global warming is occurring.

Religious Belief and Attitudes about Science in the United States 345

The apparent low relevance of religion in shaping views on global warming is corroborated in respondent self-reports about the importance of various factors with respect to another environmental issue: whether environmental regulations do more harm than good. Asked to compare religious beliefs as an influence with five other possible factors, fewer than one-in-ten cite their religious beliefs as the biggest influence on their views on environmental regulation.

DO RELIGION AND SCIENCE CLASH IN THE UNITED STATES?

Religion is clearly at the heart of some of the scientific disagreements we have reviewed. And many in the public see a conflict between religion and science. Indeed, a 2009 Pew Research Center survey found that 55% of the public believes that science and religion are often in conflict (Pew Research Center, 2009). Notably, however, it is the more secular and less traditionally religious—not the very religious—who are most likely to hold this view.

It is true that scientists themselves tend not to be religious. According to the Pew Research's 2009 survey of US members of the American Association for the Advancement of Science, a large and heterogeneous scientific society, just 33% said they believed in God. An additional 18% say they believe in a higher power or universal spirit. Thus nearly half of US scientists are neither spiritual, nor conventionally religious. Only 3% share the religious affiliation of Francis S. Collins, an evangelical Christian, who was appointed by President Obama to direct the National Institutes of Health. By contrast, about one-in-four among the US public is an evangelical, and overwhelming majorities of Americans believe in God. Thus the potential for conflict over belief is present.

Yet the better test of a public clash between religion and science may be to ask people if they believe that science contradicts their own religious beliefs. According to the Pew Research 2009 survey, 36% of respondents say that science sometimes conflicts with their own religious beliefs. Not surprisingly, people for whom religion is highly salient are more likely to see such a conflict. But even among evangelicals—the most traditionally religious group in the United States, many of whom accept a literal interpretation of the Bible—only about half say science sometimes conflicts with their religious beliefs.

But a more precise gauge of the consequences for conflict between religion and science is whether or not science is held in high regard by people who do not accept some scientific perspectives, or believe that science clashes with their own beliefs. In fact, large majorities of traditionally religious Americans nevertheless hold very positive views of science and scientists. Even people who accept a strict creationist view regarding the origins of life are mostly favorable toward science. And those who say that science sometimes conflicts with their religious views are just as likely to say science has a mostly positive impact on society as those who see no conflict with their religion.

More generally, it should be noted that the public image of science in the United States is very positive, and there has been little change in this perception over the past three decades. According to the National Science Foundation, public attitudes about science are more favorable in the United States than in Europe, Russia, or Japan, despite great differences across these cultures in the level of religiosity (National Science Foundation, 2008).

Of course, some in the public do have reservations about science. A small majority of 56% agreed in a 2006 NSF survey that "scientific research doesn't pay enough attention to the moral values in society." And half (50%) agreed in 2001 that "we depend too much on science and not enough on faith." Religiously committed individuals are more likely than others to hold these attitudes. A similar number (51%) said that "scientific research has created as many problems for society as it has solutions," although questions asked by the Pew Research Center in 2009 about the relative benefits and harm of research on nuclear energy, human genetics, and space exploration contradict this finding. For each of these types of research, the percentage saying that they have done more good than harm was at least twice as large as the percentage who said "more harm than good."

DISCUSSION

Our review of three important issues on the public policy agenda in the United States suggests that although there is a potential for broad religiously based conflict over science, the scope of this conflict is quite limited. Only on one issue does a significant portion of the public deny a strong scientific consensus for religious reasons: evolution. The significance of this disagreement should not be understated, but it is decidedly unrepresentative of the broader set of scientific controversies and issues.

As already noted, it is difficult to find many other major policy issues on which there are strong religious objections to scientific research. Religious concerns do arise in connection with a number of areas of life sciences research, such as the effort to develop medical therapies from embryonic stem cells. But these are not rooted in disputes about the truth of scientific research, and can be found across the spectrum of religious sentiment. Devout Catholics join with evangelical Protestants in opposing stem cell research, though they may diverge on many other issues. And members of other religious traditions very different from these, as well as some secular people, may object to medical research conducted on animals.

Second, there are important differences among the three cases described here. On evolution, there is a clear and strong objection to the scientific consensus among people who accept a literal interpretation of the Bible. Significant numbers within other religious groups believe that evolution occurred, but was divinely guided. The latter sentiment may or may not pose a problem for scientific enterprise, in that much of it does not explicitly dispute the scientific fact that evolution has occurred.

In contrast, beliefs about global warming appear to be only tangentially related to religious beliefs. On the issue of homosexuality, where a clear scientific consensus has yet to form, and where significant cultural traditions may continue to exert an influence on individual attitudes, religious beliefs are strongly related to opinions but even the nonreligious are conflicted.

It is also worth remembering that religion has no special claim to beliefs that run counter to strongly corroborated factual evidence. Psychological research finds that people sometimes believe what they want to believe when what they want to believe is compelling to them—for whatever reason, whether religious or not. In this respect, the examples of religious faith trumping scientific truth do not constitute a special case of the rejection of strong evidence of truth. Indeed, polls also routinely show that large numbers of Americans believe in quasi-religious phenomena, such as astrology and ESP, in spite of the fact that scientists routinely contend that there is no evidence to support the validity of any of these claims. One also can point to numerous examples in highly charged situations outside the metaphysical or supernatural realm where erroneous beliefs persisted in the face of strong evidence to the contrary: the case of WMDs in Iraq, the guilt or innocence of Duke University lacrosse players vs. their accuser, President Obama's birthplace, even the question of whether O.J. Simpson was guilty of murder in the eyes of blacks and whites in the public.

CONCLUSION

The late paleontologist Stephen Jay Gould spoke of the disciplines of science and religion as separate, operating in different realms to answer different questions. He described this complementary relationship as "non-overlapping magisteria" or "NOMA." Gould wrote about religion and science and hoped that the concept of NOMA could provide a way to prevent what he saw as needless conflict:

> I believe, with all my heart, in a respectful, even loving concordat between our magisteria—the NOMA solution. NOMA represents a principled position on moral and intellectual grounds, not a mere diplomatic stance. NOMA also cuts both ways. If religion can no longer dictate the nature of factual conclusions properly under the magisterium of science, then scientists cannot claim higher insight into moral truth from any superior knowledge of the world's empirical constitution. This mutual humility has important practical consequences in a world of such diverse passions. (Gould, 1997: 21)

Despite the well-known counterexamples, including Galileo's forced recantation in 1633 of his views, battles over the teaching of evolution in the public schools, and the recent publication of books such as Richard Dawkins *The God Delusion* and Christopher Hitchens' *God is Not Great*, religion and science have lately tended to avoid each other's spheres. As *New York Times* science writer William J. Broad wrote, "For its part, organized religion has for centuries found means of accommodation, approval and even support for science to leaven its sometime resistance. The early Roman Catholic Church adapted cathedrals across Europe to serve as solar observatories. Muslims of the Middle Ages pioneered the forerunners of optics and algebra" (Broad, 2006). The survey evidence described here suggests that the public instinctively understands the idea of nonoverlapping magisteria, even if it would not express it in such eloquent terms.

However the public comes to its views, the public's knowledge and perception of science should be of interest to policymakers, because surveys show that public acceptance or rejection of scientific findings are closely related to the public's policy preferences. Those who reject evolution are more likely to favor teaching creationism along with evolution in the public schools compared with those who accept evolution. Those who believe that homosexuality is an innate characteristic are much more likely than others to favor gay marriage, civil unions, and allowing gay and lesbian couples to adopt children. And, not surprisingly, those who believe that global warming is occurring as a result of human activity are substantially more likely than others to view global warming as a problem that requires immediate government attention and action.

APPENDIX

To investigate the relationship between religion and views of science, we estimated four binary logistic regression models, the results of which are reported in Table 19.4.

Table 19.4 Details of Logistic Regression Predicting Beliefs about Evolution, Homosexuality, and Climate Change

	Believe in evolution[1]	Believe homosexuality is innate trait[2]	Believe homosexuality is immutable[2]	Believe in man-made global warming[1]
Demographic Predictors				
Male	.391***	-0.594***	-0.791***	-.284***
Black	.090	-0.413	-0.220	-.290
Other nonwhite	-.600**	0.245	-0.073	.451*
Hispanic	-.319	-0.506	-1.038***	.404**
Age	.001	0.017***	-0.003	-.003
Education	.269***	0.262***	0.074	.139***
Income	-.027	0.140***	0.088**	-.018
Political Predictors				
Republican	.024	-0.209	0.167	-.595***
Democrat	.144	0.408**	0.431**	.618***
Conservatism	-.326***	-0.588***	-0.535***	-.335***
Scientific Agreement	.819***	--	--	1.566***
Religious Predictors				
Evangelical, low commitment	-1.481***	-1.174***	-0.805**	-.093
Evangelical, high commitment	-2.624***	-1.917***	-2.015***	-.316*
Mainline, low commitment	-.580**	0.013	0.240	.107
Mainline, high commitment	-1.434***	-0.307	-0.417	.465*
Catholic, low commitment	-.663	0.048	0.276	.148
Catholic, high commitment	-1.100***	-0.466	-0.407	-.247
Other religious affiliation	-1.064***	-0.866**	-0.521	.255
Constant	1.084	-0.442	1.869	-.344
Nagelkerke R-squared	.333	0.360	0.322	.340
Sample size	1,850	852	865	1,917

*p<.1, **p<.05, ***p<.01.
Source: PEW RESEARCH CENTER, 1May 2009, N=2,001; 2 July 2006, N=996.

The dependent variables are specified as follows:

Believe in evolution: coded 1 for those who believe that humans have evolved over time, and 0 for those who believe that humans and other living things have existed in their present form since the beginning of time

Believe homosexuality is innate trait: coded 1 for those who believe homosexuality is something that people are born with, and 0 for those who say homosexuality is something that develops because of the way people are brought up or just the way that some people prefer to live

Believe homosexuality is unchangeable: coded 1 for those who believe that homosexuality cannot be changed, and 0 for those who believe that homosexuality is a trait that can be changed

Believe in man-made global warming: coded 1 for those who believe that the earth is warming and that this is caused by human activity and 0 for those who do not believe the earth is warming or who believe that warming is caused by natural processes

The independent variables are specified as follows:

Male is a dummy variable coded 1 for males and 0 for females. *Black* is a dummy variable measuring race, coded 1 for blacks and 0 for all others. *Other nonwhite* is also a dummy variable measuring race, coded 1 for nonwhites who are not black, and 0 for all others.[3] *Hispanic* is a dummy variable coded 1 for Hispanics and 0 for all others. *Age* is a continuous variable measuring respondents' age in years, ranging from 18 to 97. *Education* is a seven-category variable ranging from 0 for those with less than an eighth-grade education to 6 for those with an advanced degree. *Income* is a nine-category variable ranging from 0 for those with an annual income of less than $10,000 to 8 for those with an annual income of $150,000 or more.

In the evolution model, *scientific agreement* measures agreement with the statement that "there is agreement among scientists that humans evolved over time," coded 1 for those who agree with the statement and 0 for all others. In the global warming model, *scientific agreement* measures agreement with the statement that "scientists agree that the earth is getting warmer because of human activity," coded 1 for those who agree with the statement and 0 for all others.

A series of dummy variables combine measures of religious identity (e.g., evangelical Protestant, mainline Protestant, and Catholic) with a measure of religious commitment. In the evolution and global warming models, those who attend church weekly are considered to be highly committed, and those who attend less frequently constitute the less religiously committed. In the homosexuality models, the religious commitment variable is a combination of measures of church attendance and religious salience. Those who attend church frequently and say religion is important in their lives are considered to be highly committed, and those who do not attend church frequently or say religion is not very important constitute the less religiously committed.

All predicted probabilities in the text were generated by allowing the variable in question to vary from its minimum value to its maximum value, with all other variables held at their mean value.

NOTES

1. Scott Keeter is Director of Survey Research for the Pew Research Center. Gregory Smith and David Masci are Senior Researchers at the Pew Forum on Religion & Public Life.
2. The only religious group that is significantly less likely than the unaffiliated to believe in man-made global warming is high commitment evangelicals, but in absolute terms they are very similar to the other religious groups. Highly committed mainline Protestants do stand out for their high level of belief that the earth is warming due to human causes, but the size of this group is quite small—accounting for roughly 6% of the US population and only 119 of the 1,917 cases in the data upon which this model is based. It is therefore not possible to say with a high degree of confidence that mainline Protestants are substantially more likely than other groups to accept scientific claims about global warming.
3. In the evolution and global warming models, the categories "black" and "other nonwhite" exclude Latinos.

REFERENCES

All data analyzed for this chapter are available for download from the Pew Research Center website at http://people-press.org/dataarchive/.

Broad, William J. 2006. "The Oracle Suggests a Truce Between Science and Religion." New York Times. February 26, 2006. Accessed July 23, 2010, at http://www.nytimes.com/2006/02/28/science/28essa.html.

Dawkins, Richard. 2006. *The God Delusion*. New York: Houghton Mifflin.

Gould, Stephen Jay. 1997. "Nonoverlapping Magisteria." *Natural History* 106 (March): 16–22.

Hitchens, Christopher. 2007. *God is Not Great: How Religion Poisons Everything*. New York: Warner.

Larson, Edward J. 1997. *Summer for the Gods: The Scopes Trial and America's Continuing Debate Over Science and Religion*. Cambridge, Mass.: Harvard University Press.

National Science Foundation. 2008. "Science and Engineering Indicators." Accessed July 23, 2010, at http://www.nsf.gov/statistics/seind08/c7/c7h.htm.

Numbers, Ronald L. 2009. *Galileo Goes To Jail and Other Myths about Science and Religion*. Cambridge, Mass.: Harvard University Press.

Pew Research Center. 2005. "Religion a Strength and Weakness for Both Parties." Report released August 30, 2005. Accessed July 23, 2010 at http://people-press.org/report/254/religion-a-strength-and-weakness-for-both-parties.

Pew Research Center. 2006. "Pragmatic Americans Liberal and Conservative on Social Issues." Report released August 3, 2006. Accessed July 23, 2010, at http://people-press.org/report/283/pragmatic-americans-liberal-and-conservative-on-social-issues.

Pew Research Center. 2009. "Public Praises Science; Scientists Fault Public, Media." Report released July 9, 2009. Accessed July 23, 2010 at http://people-press.org/report/528/.

20 Worldviews, Frames, Trust and Perceptions of Stem Cells across Europe

Rafael Pardo

SCIENTIFIC LITERACY AND CULTURAL APPROPRIATION OF SCIENCE

Few areas of science make it to the forefront of media and public perceptions and, once there, find themselves the object of sustained attention. At most times in most societies, science tends to merit only episodic attention and a fuzzy or generic positive appreciation, more latent than explicit, which is only brought to light by surveys designed to capture public attitudes to science. Against this background of long periods of silence or inattentiveness and a kind of mute approval on the part of society, there emerges from time to time a science or technology area with a high level of salience and the potential for controversy. The picture on those infrequent occasions is one in which general attitudes toward science with a clearly positive sign, or valence, commingle with critical perceptions (ranging from reservations to active opposition by way of ambivalence) toward new scientific or technological developments.

The life sciences, biotechnology in particular, have stood as the prime exponent of a high-visibility, controversial area since the last decades of the twentieth century. And stem cell research exhibits these traits more than most biotechnology subsets, having captured the attention of the media, a number of influential interest groups and the general public even before it has left the laboratory. This subset of biotechnology research is characterized by a high level of expectations regarding its goals (finding effective treatments for some of the major diseases prevalent in advanced societies) and an equally intense level of dispute surrounding the means involved, i.e. the creation and use of early human embryos, one of the most potent symbols of life and of the limits on scientific intervention to shape or alter natural processes.

If the "scientific literacy" variable can directly account for only a small portion of general views and predispositions toward science at large (Sturgis, 2004; Pardo and Calvo, 2002, 2004, 2006a), we could expect its independent influence to be weaker or even nonexistent with regard to the most controversial areas of science; that is, those that involve and touch on a number of central dimensions of the culture of a given time. In these cases, the core explanatory role that canonic models in the PUoS field assign to familiarity with the cognitive content of science is occupied by variables of an essentially cultural, symbolic and psychosocial nature. Among them are worldviews concerning some of the major dimensions of how we see and act in the world, trust in key actors and, also, highly specific frames that are semantically close to the target of the attitudes.

Although it is more than a century since religious-moral creeds withdrew from the conceptual space of most areas of scientific research, in the present decade religion and science are still fighting it out over the conceptual modeling of human evolution from antecedent animal species versus creationist or "intelligent design" accounts and over the image of the human embryo and the ethical standards to govern its creation and use. In the case that concerns us here, the scientific community has not confined itself to launching frequent messages about the therapeutic advances potentially to be gleaned from embryonic stem cell research, but has also promoted the image of a few-day-old human embryo as a biological entity ("a mere cluster of cells") sharing none of the key characteristics of more developed life stages, and therefore not deserving the same level of protection. At the same time, religious organizations, particularly the Catholic Church, have endowed the few-day-old embryo with a moral status comparable to that of a human being, flatly rejecting not only its production but the use of embryos left over from IVF treatments (Waters and Cole-Turner, 2003).

The "scientific literacy" variable in the domain of biology can be understood as a proxy for a public perception of the embryo and its utilization analogous to the one held by the scientific community, i.e. without attribution to the embryo of properties usually absent and unnecessary in the context of scientific research. Conversely, the "religiosity" variable (religious belief and its activation through recurrent rites and practices) will tend to remove the embryo from the minimalist, neutral framework of science and insert it in a framework steeped in cultural symbolism and moral judgment.

The insistence on referring mainly, if not solely, to the "scientific literacy" variable to explain public attitudes to science has its roots in an enlightenment assumption (social resistance to novelties and progress as a function of ignorance), insufficiently analyzed (what are the exact mechanisms connecting an abstract cognitive dimension like scientific knowledge and an evaluative one, such as attitudes to science and its effects?) and weakly supported by the empirical evidence. At the other extreme, the argument that there is no link between these two planes, which has gained much currency in recent years, rests on evidence that is either shallow or has been statistically misinterpreted (see Miller, 2001; Bauer, Allum and Miller, 2007; Pardo and Calvo, 2008). Both views cling to an analytical axis that obstructs the field of vision and theoretical development of the PUoS field—which is, after three decades of intense research, still struggling to find itself conceptually. In order to provide more solid foundations for this field of study, we need to concentrate on building "theories of the middle range" (Merton, 1968). Such middle-range theories must be integrated by variables reflecting dimensions of the more general *culture* of the society, where science is embedded and operates, along with the psychosocial constructs (fear, stigma, expectations, trust,) documented in other segments of the social sciences literature as shaping attitudes toward a broad spectrum of formal objects. Some of these variables will be of a general nature, such as worldviews and values, with a long semantic or conceptual distance to the attitudinal object, and others highly specific, comprising frames that are close or semantically akin.

In this chapter we explore a limited range of those variables potentially useful for building an explanatory model of attitudes to embryonic stem cell research integrated by four types of variables: 1) "worldviews" (religiosity and two facets of views on science), 2) a specific ethical frame connected both with religion and science (the moral status of the embryo), 3) a socio-psychological construct (trust in the scientific community) and 4) biological literacy. We test the explanatory role of those variables on data from a multicountry study on attitudes to embryo research and biotechnology.[1]

CAN THE PUBLIC'S ATTITUDES TOWARD COMPLEX ISSUES BE A FORMAL OBJECT OF STUDY?

Despite the unusual degree of attention received from the media, churches, certain interest groups and policy-makers, stem cell research remains a highly complex issue that is hard to grasp for the majority of individuals lacking scientific training in biology. Accordingly some commentators have seized on the earliest polls on stem cells as an example of the superficial nature of public opinion studies (Clymer 2001) and as incapable of offering more than the weakest evidence on a question beyond the most basic understanding of the public. Political scientist George F. Bishop has argued that what US stem cell polls really reflect is the image of a "phantom public" (Bishop 2005, 39ss). The reasons for these damning judgments are essentially the issue's remoteness and complexity, its relative novelty and pollsters' reliance on simple measurements.

Yet at the same time, embryo research to obtain stem cells arouses high expectations of medical benefits for many individuals and touches on core aspects of the "moral landscape" of society. Few people have no view at all on matters that directly or indirectly involve or evoke the image of a human embryo, even though they know little or nothing about the specifics of its creation and use in the laboratory.

For the great majority of the population, issues and dilemmas of an evaluative nature neither present themselves as abstract or isolated matters nor are evaluated by reference to a single criterion or principle (embryo rights, medical benefits). At a social level, only single-issue groups or, at an individual level, people with strong ideological views take a position based on a unique evaluative angle and this only in cases of high salience issues involving their core values. For most other individuals, ethical questions arise in specific contexts composed or integrated by several overlapping domains, in which multiple and diverse values and ethical principles may apply, giving rise to a certain level of inconsistency or, at least, to a loosely coupled array of criteria for making up their mind and adopting a particular stance.

The encapsulation of attitudes toward embryo research in the strict domain of science will happen only with professional scientists and perhaps a very small segment of individuals, attuned to science by extensive training or cognitive and evaluative proximity to it (knowledge professionals, science amateurs or enthusiasts). Most other individuals will perform, in reaching a judgment, an informal cost-benefit analysis of the issue at stake and connect or embed that specific analysis in a more general evaluative framework, i.e. with relevant aspects of the general cultural domain of which science is just a part, even if a very significant one. This is the realm of general schemas or worldviews and specific frames (such as the views on the embryo) semantically connected to the target of the attitude.

A SCALE OF ATTITUDES TOWARD EMBRYONIC STEM CELL RESEARCH

In most debates on the acceptability of using human embryos as a source of human stem cells, the status and rights of the embryo are not balanced against the advancement of pure knowledge, but against the finding of effective cures for devastating diseases affecting many individuals.

A robust measurement of the dependent variable was constructed by means of a 10-item summated scale that covers a wide spectrum of positions or attitudes—representative of those expressed in public debates—regarding the consequences (hoped-for benefits, undesired effects) and requirements (destruction of embryos, creating life for scientific experimentation purposes) of research with human embryos.[2]

This scale exhibits strong reliability for the whole set of nine countries, $\alpha=0.88$, and for each nation individually, with a range from $\alpha=0.86$ for the UK to $\alpha=.0.90$ for Germany. This aggregate measure marks a significant advance over the few existing studies on stem cell research, which rely on a single direct question about how far the interviewee agrees or disagrees with the use of embryos for obtaining stem cells (Nisbet, 2005; Liu and Priest, 2009).

FROM SCIENTIFIC LITERACY TO WORLDVIEWS AND FRAMES

Attitudes toward the use and creation of embryos belong to a domain crosscut by evaluative vectors, specific and general, embedded in the culture of a given society and period. This being so, we can hardly expect a primarily cognitive variable—biological literacy—to compete head-on as a direct explanatory factor of attitudes with primarily axiological variables, both cultural (frames, worldviews) and psychosocial (trust), although it may be associated with some of these variables and, as such, possess an indirect explanatory power.

The moral status of the embryo frame. Cognitive science has taken the concept of *frame* or *schema* as a central construct representing the way individuals organize their information about the world. A frame represents an object as a group of properties, and when the image of a particular object is recalled from memory, a cluster of its most characteristic attributes are immediately retrieved (see Minsky 1975; Ringland 1988). The identification of the relevant frame has proved to be a powerful resource for explaining attitudes to certain objects. Such frames may lie more or less close to the attitude target in semantic terms, and may include cognitive and evaluative elements.

In the case that concerns us, the variable of most semantic proximity to the dependent variable is, without doubt, a bioethical frame known in the literature as the "moral status of the embryo:" a construct referring to where the embryo is placed in relation to other moral agents, and these agents' relations and obligations regarding the embryo (Warren, 2000).

The questionnaire used in our study offered four options on the status of the embryo that appear under different guises in current debates among better informed population segments and stakeholders: "a human embryo that is a few days old . . ." [. . .] 1) "is a mere cluster of cells, and it makes no sense to discuss its moral condition"; 2) "has a moral condition halfway between that of a cluster of cells and that of a human being"; 3) "is closer in its moral condition to a human being than to a mere cluster of cells"; 4) "has the same moral condition as a human being."

Since this variable is central to our analysis of attitudes to embryo research, before a multivariate model is introduced, it is worth exploring its linkage with variables relating to the two main cultural forces at work in its modeling, namely religion and scientific advances in biology and medicine.

Religion and the moral status of the embryo frame. Clearly, the main divide runs between individuals who believe in one of the major religions and those who declare themselves "nonbelievers" (see Table 20.1). If we focus on the two extreme positions for the conceptual frame that concerns us, we can observe a common pattern through all the European countries included in our study (the exception being the UK, where the differences are weak): religiously affiliated individuals (it matters less if they are Roman Catholic or Lutheran Protestant) tend to believe that "a few-days-old human embryo has the same moral condition as a human being." In general, the nonbelievers maintain a position closer to scientific facts and images of the embryo, and do not share believers' assumptions about the existence of a precise moment when life begins.

Table 20.1 Views on the Moral Status of the Embryo by Religious Beliefs.

		Cluster of Cells	Halfway Cluster Cells— Human Being	Closer to a Human Being	Same Moral Status as a Human Being	DK	Difference "Same Moral"— "Cluster Cells"
Denmark	Protestants	36.3	23.0	12.5	22.7	5.5	-13.6
	Nonbelievers	45.4	24.9	8.8	14.7	6.1	-30.7
Italy	Catholics	17.7	15.6	14.5	40.2	11.9	22.5
	Nonbelievers	36.9	19.8	7.3	23.0	13.0	-13.9
Spain	Catholics	16.8	18.6	11.8	33.5	19.3	16.7
	Nonbelievers	35.9	18.9	14.2	20.1	10.9	-15.8
France	Catholics	14.6	20.7	18.8	35.6	10.3	21.0
	Nonbelievers	34.3	24.6	15.4	19.8	5.9	-14.5
Poland	Catholics	13.4	14.2	9.9	39.5	22.9	26.1
	Nonbelievers	30.2	33.1	8.6	22.1	6.0	-8.1
UK	Anglican	28.0	14.9	16.2	25.3	15.5	-2.7
	Evangelical	20.6	23.5	13.3	22.7	19.9	2.1
	Nonbelievers	28.6	16.5	9.2	20.4	25.3	-8.2
Netherlands	Catholics	19.3	29.0	19.0	24.3	8.4	5.0
	Protestants	9.4	25.4	21.6	35.1	8.5	25.7
	Nonbelievers	26.3	36.7	11.8	16.5	8.8	-9.8
Germany	Catholics	9.6	21.6	25.2	34.9	8.7	25.3
	Protestants	6.2	20.4	23.7	32.0	17.7	25.8
	Nonbelievers	9.9	25.1	24.7	24.9	15.3	15.0
Austria	Catholics	4.9	21.1	23.4	33.8	16.8	28.9
	Nonbelievers	9.0	24.7	22.3	28.6	15.4	19.6

Replacing the "religion" variable with "religiosity" (combining the belief in a religion with the observance of the corresponding rites and practices of worship, that is, with the different levels of activation of the religious perspective) gives rise to stronger differences in views of the "status of the embryo."[3] Believers who pray (or engage in religious practices) on a daily basis are, by a large margin, those likeliest to see the embryo as having the same status as a human being, with 53.9% opting for this category. At the other extreme, nonbelievers and nonpracticing believers come first in the cluster of cells category (30%) then in the category halfway cluster cells-human being (28%; see Table 20.2). This variable, with more discriminatory power than religious beliefs, will be the one used in the explanatory model.

Table 20.2   Beliefs about the Moral Status of the Embryo by Religiosity

	Cluster of Cells	Halfway Cluster Cells— Human Being	Closer to a Human Being	Same Moral Status as a Human Being
Never prays or practices+ Nonbelievers	30.0	28.2	17.0	24.7
Believes and practices from time to time	21.6	26.5	19.6	32.4
Believes and weekly practice	15.0	22.5	19.5	43.0
Believes and daily practice	14.0	15.9	16.3	53.9

χ^2 = 728.3, df =9, Signif. = 0.000

Biological literacy and the moral status of the embryo frame. The other main force beside religious-moral beliefs shaping the image of the human embryo and its status is the advance of biology. Accordingly it is of interest to study the possible association between views of the embryo (more biological or, alternatively, more based on religious-moral criteria) and the subset of scientific understanding that we might call biological literacy.

In order to gauge the public's familiarity with basic elements of biology and, more specifically, biotechnology (as well as related knowledge on environmental matters), a battery of 22 items was constructed.[4] Dichotomizing responses into correct/incorrect we get a summated scale with a Cronbach's α= 0.81 for the nine countries together and similar values in each individually (with reliability highest in Spain, α=0.83, and lowest in Italy, α=0.73). The scores obtained in all nine countries are then grouped into the following three segments: high biological literacy (16 to 22 points), making up 23.3% of the total sample; medium literacy (9 to 15 points), making up 54.2%; and low literacy (0 to 8 points), making up the remaining 22.4%.

In tune with expectations, there is a significant association between level of literacy and view of the embryo's status, such that the propensity to see a few-days-old embryo as "a cluster of cells" increases with a person's scientific knowledge (Table 20.3).

As well as this general pattern, there are other more specific profiles that bear mention. First, the differences between high and low literacy groups are attenuated in the cases of Germany and Austria, which stand apart from the rest both in their vision of the embryo and its use in research, due perhaps to the so-called "ripple effects" (see Kasperson et al., 2000) of the brutal eugenics program applied by the Nazi regime. Second, responses in the Netherlands are concentrated in the "halfway" category, with a strong differentiation between high and low biological literacy groups (43.5% and 12.5% respectively). Finally, Italy and Poland show a specific distribution,

360 Rafael Pardo

Table 20.3 Beliefs about the Moral Status of the Embryo by Biological Literacy (%)

		Cluster of Cells	Halfway Cluster Cells—Human Being	Closer to a Human Being	Same Moral Status as a Human Being	DK	Difference Same Moral—"Cluster Cells"
Denmark	Low biological liter.	22.4	12.2	7.3	27.8	30.3	5.4
	High biological liter	42.2	28.0	10.5	17.8	1.5	-24.4
UK	Low biological liter.	13.9	10.8	9.1	28.3	37.8	14.4
	High biological liter.	38.8	16.9	15.4	18.6	10.2	-20.2
Spain	Low biological liter.	15.3	13.8	10.9	32.3	27.7	17.0
	High biological liter.	31.8	21.2	8.9	32.4	5.6	0.6
France	Low biological liter.	14.4	15.1	7.3	33.0	30.2	18.6
	High biological liter.	30.8	27.9	19.1	19.7	2.6	-11.1
Italy	Low biological liter.	19.7	13.0	10.6	31.3	25.4	11.6
	High biological liter.	22.5	15.4	18.3	38.8	4.9	16.3
Netherlands	Low biological liter.	17.3	12.5	11.5	26.3	32.4	9.0
	High biological liter.	22.0	43.5	17.8	14.2	2.4	-7.8
Poland	Low biological liter.	10.2	9.5	7.5	36.5	36.3	26.3
	High biological liter.	18.0	26.9	14.9	35.1	5.1	17.1
Germany	Low biological liter.	7.3	7.7	24.9	33.8	27.0	26.5
	High biological liter.	9.7	26.0	25.4	28.3	10.6	18.6
Austria	Low biological liter.	5.6	14.7	20.3	33.8	25.6	28.2
	High biological liter.	7.8	32.2	28.3	26.1	5.7	18.3

with the majority in both countries opting for a view of the embryo similar to that of the Catholic Church, evidencing the strong influence of its doctrine even among population sectors with greater biological knowledge.

WORLDVIEW ON SCIENTIFIC PROGRESS AND ATTITUDES TO STEM CELL RESEARCH

General attitudes and worldviews do have a significant function as "orienting dispositions" and are relevant for explaining evaluative positions toward specific objects (Slovic, 2000; Pardo, Midden and Miller, 2002; Sjøberg, 2004). In principle, frames and salient beliefs closer in content or meaning to the formal object x as target of attitudes will, all else being equal, have a greater influence or discriminatory power than more distant and abstract worldviews.

The following are among the worldviews identified in the literature: fatalism, hierarchy, individualism, egalitarianism and technological enthusiasm (Slovic, 2000). What Slovic has termed "technological enthusiasm" is in fact the worldview expressed in Holton's elegant wording as "the rightful place of science in our culture," "the way in which science—as a body of knowledge, as a source of technical applications, as a generator of models for thinking and acting [. . .] is viewed and used" (Holton, 1996)—a worldview which, in turn, provides an evaluative angle to assess specific science subsets or technological developments.

In the case of attitudes toward the use of embryos for research, two facets of the worldview about science at large and its social role are particularly relevant. The use of the plural form "facets" is deliberate, to express the idea that views about science may have at least two axiological axes: a perception of the beneficial effects ("promise of science") and a perception of undesired effects ("reservations") (Miller and Pardo, 2000; Pardo and Calvo, 2006a, 2008).

"Reservations about science" is represented by seven items, making up a summated scale of strong reliability regarding both the nine countries together (Cronbach's $\alpha=0.81$) and each individually (from Germany's $\alpha=0.75$ up to the UK and Austria's $\alpha=0.85$).[5] In the analysis presented here, we use the "reservations" indicator, plus another indicator that captures a positive view of science, but from a more precise angle departing somewhat from the conventional "promise of science" indicator. This indicator, which can be called "expectations about techno-scientific developments," measures perceptions of the future consequences of recent scientific and technological advances and it was constructed using a 10-item battery.[6] The reliability of this scale is good for most countries and the consolidated sample (Cronbach's $\alpha=.77$), although in two countries it comes down considerably (Spain, $\alpha=85$, Poland, .81, UK, .80, Austria, .80, Italy, .79, Denmark, .72, Netherlands, .71, France, .69 and Germany, .66).

The hypothesis is that individuals with a generally or consistently optimistic view of techno-scientific progress (high scores on expectations and low scores on reservations) would also have a more favorable view of a particular, even if controversial, new scientific area such as stem cell research.

TRUST AND PERCEPTIONS OF EMBRYONIC STEM CELL RESEARCH

The multidisciplinary literature on social capital has shown the key role played by trust in interpersonal relations, the functioning of society's institutional fabric and even economic growth. From a sociological standpoint, trust comes into its own as a core component of social conduct in situations of high complexity, as is generally the case with those involving science and technology, where it operates as a cognitive shortcut or cue for decision-making by reducing information needs. Its basic function is the reduction of cognitive complexity (Earle and Cvetkovich, 1999). Hence, strong trust in the scientific community may substitute for or compensate a modest level of scientific literacy when evaluating emergent research areas.

Trust ($Trust_1$, Vertrauen) can be understood as an expectation of "fair play" by individuals, social groups and organizations that recognize and respect the interests of others. To capture this canonical dimension of trust, interviewees were asked, "Could you tell me how much trust you have in what each of the following groups of professionals do or say?" using a scale from 0 to 10, where 0 means that you do not trust them at all, and 10 means that you trust them a great deal."

Another facet of trust ($Trust_2$, Zutrauen), refers to the belief in the competence or capacity of a professional person or group to carry out their function or tasks in accordance with the high performance standards demanded by a postindustrial society that is fundamentally knowledge driven (US Department of Energy, 1993; Earle and Cvetkovich, 1995). A direct measure of this second facet of trust was obtained through a specific item occurring in a battery measuring views on scientists with the format of a 7-point semantic differential scale.[7] In addition, a proxy for this second dimension of trust is provided by a question about the perception of the contribution to the improvement of living conditions and human progress made by different professional groups, among them the scientific community.[8]

Our hypothesis is the existence of a positive association between trust "in what scientists do or say" and attitudes toward embryo creation for biomedical research. Less obvious is the assumption that the confidence dimension in the case that concerns us—public trust in the ability of scientific researchers—stands clearly on the side of more favorable attitudes to research with embryos. Precisely some of the deepest fears about scientists in general and biomedical researchers in particular, from those reflected in literature and film to social representations of the deviant conduct of scientists in popular culture, concern the mix of competence (the ability to do certain things) and the hubris of venturing into domains that should be left to God's will or the laws of nature (depending on whether this nonintervention principle is interpreted religiously or secularly) (Haynes, 1994; Turney, 1998).

AN EXPLANATORY MODEL OF ATTITUDES TOWARD EMBRYONIC STEM CELL RESEARCH

To estimate the relative explanatory power of each independent variable, we conducted a multiple regression analysis according to the enter method. The y-variable is a summated scale of attitudes to embryonic stem cell research, and the set of explanatory variables, $x_1, x_2, \ldots, x_q$ is made up of eight variables: "Status of the embryo," "Reservations about science," "Expectations about techno-scientific developments," "Trust in the scientific community," "Proxy for Confidence in the scientific community," "Religiosity," "Biotechnological literacy" and "Gender" (the last variable, under the assumption that since the degree of personal and biological "involvement" with the embryo varies in its intensity, and even nature, from women to men, we might expect to find women holding a view of the embryo that is not purely biological and more opposed to the use of embryos for purposes other than their own development and well-being). All variables meet the condition of noncolinearity.[9]

The multiple correlation of the scale of attitudes toward embryo research with the eight explanatory variables is $R=.62$ in the total 9-country sample, running from .68 in the Netherlands to .52 in Italy: a value that can be considered medium for an explanatory model in the perceptions of science framework (see Table 20.4). Seven of these variables are statistically significant at $p<0.01$ for all nine nations together (the only nonsignificant coefficient corresponding to "Biological literacy"), but the "Confidence" and "Gender" coefficients are very low, evidencing a poor predictive power for the dependent variable in the presence of the remaining variables.

If a ninth variable is added to the multiple regression equation, chosen from among the socio-demographic characteristics "Age," "Education," "Subjective social class" and, also, "Political/ideological orientation," the multiple correlation undergoes no variation, nor is the coefficient of this ninth variable in any case significant (except "Age" in the case of the nine nations together, though not individually with the exception of France).

A favorable attitude toward embryo experimentation depends directly in greatest measure on viewing the embryo as a "cluster of cells," to a lesser extent on having "no reservations about science and technology," maintaining "high expectations about techno-scientific developments" and "trusting scientists" and more weakly on "not being a believer" or, if so, "not following the prayer practices" of the religion in question and, more weakly still, on having a "high appreciation of scientists' contribution to improving society" (proxy for Confidence). Attitudes of rejection, at the opposite extreme, flow from a view of the embryo as having the "same moral condition as a human being," having "reservations about science," "low expectations about techno-scientific developments," "distrust in scientists," "assiduously practicing a given religion" and having a "low appreciation of the role of scientists."

As seen, biological literacy and, more so, degree of religiosity are associated with the view held of the moral condition of the embryo. Science and religion are cultural constructs supplying some of the dominant worldviews in most societies. The frame regarding a few-days-old embryo, be it based not on observable attributes but aprioristic belief, in the case of religious-moral creeds, or adopting a minimalist, parsimonious view, in the case of science, pervades society and exerts its sway on public attitudes toward research with embryos to obtain stem cells. It may be that, like

Table 20.4 Explanatory Model of Attitudes to Biomedical Embryo Research

	R	Status	Reserv.	Expectations	Trust	Religiosity	Confidence	Gender	Biological literacy
						β coefficients			
All 9 countries	.62	.39(*)	-.18(*)	.17(*)	.14(*)	-.12(*)	.03(*)	-.03(*)	.002
						β coefficients			
UK	.61	.42(*)	-.18(*)	.04	.12(*)	-.11(*)	.11(*)	-.05	-.003
						β coefficients			
Italy	.52	.32(*)	-.13(*)	.17(*)	.15(*)	-.12(*)	.01	-.04	-.003
						β coefficients			
France	.64	.39(*)	-.14(*)	.20(*)	.19(*)	-.15(*)	.03	.02	-.017
						β coefficients			
Germany	.61	.32(*)	-.16(*)	.26(*)	.14(*)	-.10(*)	.05	-.01	-.021

Denmark	R	Status	Reserv.	Expectations	Trust	Religiosity	Confidence	Gender	Biological literacy
				β coefficients					
	.63	.34(*)	-.23(*)	.21(*)	.14(*)	-.05	.03	-.04	-.032

Netherlands	R	Status	Reserv.	Expectations	Trust	Religiosity	Confidence	Gender	Biological literacy
				β coefficients					
	.68	.42(*)	-.17(*)	.17(*)	.18(*)	-.12(*)	.01	.01	.004

Poland	R	Status	Reserv.	Expectations	Trust	Religiosity	Confidence	Gender	Biological literacy
				β coefficients					
	.57	.39(*)	-.21(*)	.15(*)	.07	-.09	-.02	-.11(*)	-.070

Austria	R	Status	Reserv.	Expectations	Trust	Religiosity	Confidence	Gender	Biological literacy
				β coefficients					
	.63	.45(*)	-.17(*)	.10(*)	.12(*)	-.10(*)	.02	-.05	.038

Spain	R	Status	Reserv.	Expectations	Trust	Religiosity	Confidence	Gender	Biological literacy
				β coefficients					
	.60	.30(*)	-.26(*)	.09(*)	.20(*)	-.13(*)	.05	-.01	0.032

Coefficients signaled * are "statistically significant at the level p<0.01" other issues and objects before it, the early embryo will shed this cultural wrapping of a religious cast, and come to be viewed with just the properties biology assigns it in each growth stage. But the change will not happen automatically. It will require "other voices" now largely unheard to come more strongly to the fore, like those of patients and their associations. And the scientific community will need to do its bit by adopting a more sophisticated approach to communication with the public that takes account of people's mental models for making sense of complex issues and also projects the axiological and cultural component of its research, and not just the bare cognitive side.

CONCLUSIONS

In the last fifteen years, views and attitudes toward the genetic modification of plants and animals have occupied most of the attention of social scientists engaging in the study of public perceptions of science. And so too with parliaments and public agencies, who have faced the task of devising a regulatory framework that does not constrain one of the most dynamic and promising scientific and technological areas of this turn of the century, but nonetheless takes on board the public's fears and cultural concerns about the modification of nature's processes and interspecies barriers made possible by the advance of biotechnology.

The basic profile of the public's attitudes toward biotechnology in advanced societies is well known, and can be characterized by little or no familiarity with genetics, a rejection of, or, alternatively, a failure to perceive most of the alleged potential benefits to be derived from the ends of such research and a deep distrust of the means employed, the genetic engineering of the blueprint of plant and animal life. In general, the so-called "red" biotechnologies, of a biomedical nature, are favorably perceived or, at least, do not meet with significant reservations, whereas "green" biotechnologies, focusing on the genetic modification of plants for agriculture and the production of foods (not for pharmaceuticals), are critically perceived (Gaskell et al., 1997; Priest, 2001; Pardo, Midden and Miller, 2002; Bauer and Gaskell, 2002; Sturgis, Cooper, Fife-Schaw and Shepherd, 2004; Sturgis, Cooper and Fife-Schaw, 2005; Gaskell and Bauer, 2006; Pardo and Calvo, 2006b; Pardo et al. 2009). The more or less active resistance of the first half of the 1990s has given way at the turn of the century to a moderate opposition or even a positive evaluation of some applications, and, at any rate, to a more flexible perspective that discriminates according to the goal of research and specifics of the means utilized.

In this context, research with embryonic stem cells has been viewed by the public and the regulator as a separate case with a profile not automatically comparable to that of other red biotechnological developments. In a short time, stem cells have made the leap from the laboratory to the mass media, attaining an unusually high degree of salience. Biomedical science policies, generally outside the arena of public debate, have broken out of the usual circle of experts and regulators and caught the attention of broad sectors of the public and a large spectrum of organizations. The scientific community working in this area has seen their research activity subjected to restrictions of differing severity and reacted accordingly. The cartography of public perceptions of embryo research is therefore of special interest, though to date social science analyses directly addressing this issue have been very few in number, particularly multicountry studies with an empirical base (Pardo, 2003; Nisbet, Brossard, and Kroepsch, 2003; Nisbet, 2005; Pardo and Calvo, 2008; Liu and Priest, 2009; Einsiedel, Premji, Geransar et al., 2009; Caulfield, Zarzeczny, McCormick et al., 2009a, 2009b).

The human embryo is a powerful symbol of life and the very idea of its use and, more so, its creation for biomedical research purposes is embedded in a complex space, occupied by varied and deep-rooted evaluative structures and social imagery, shaped by the culture of a particular period, particularly religious-moral creeds, in collision with the self-contained vision of contemporary biology and genetics. Other actors and organizations like the women's movement and patients' associations also play their part in shaping visions of the embryo, but in this case (in contrast to the earlier debate on abortion law), their influence has so far been slight. Unlike other areas of contemporary science, which are culturally appropriated in a similar fashion within each country and with no marked differences between advanced societies, research with embryonic stem cells attracts starkly contrasting views, which complicates its regulation in pluralist societies like those of twenty-first-century Europe. Furthermore, the specific historical experience of Germany with regard to eugenics has had a ripple effect appreciable nowadays as feelings of caution and restraint toward wholly different lines of genetic research.

Attitudes of favorability to research with human embryonic stem cells tend to depend not so much on the goals pursued, but on general worldviews associated with religion and science, and above all on a specific frame also linked with them and close to the attitudinal object, known in the bioethics literature as the "moral status of the embryo." Those who take a strictly biological view of the few-days-old embryo (a cluster of cells) tend to be favorably disposed to its use in biomedical research, whereas those who share the view of certain religions (the embryo as a person or, at least, as an entity endowed with full potential to become a person) are characterized by their opposition to the same. In other words, just knowing where an individual stands in relation to this bioethical frame allows a first prediction or classification of his or her attitude toward research with embryonic stem cells. A more refined prediction (with less error) can be obtained by incorporating in the explanatory model other theoretically and statistically significant influences.

Parallel to the embryo frame's leading influence, there is a direct association between people's religiosity and, also, general views on science, and their attitudes toward embryo research. However specific and controversial this area of research may be, it is nonetheless a subset of science, such that the worldview on the effects of science at large plays a significant part in forming opinions of acceptance or rejection. In particular, two facets of this worldview, reservations about science and expectations about current techno-scientific developments are directly associated with attitudes toward embryo research for obtaining stem cells. The variable trust in the scientific community also finds a place in the explanatory model for these attitudes. Biological literacy, though clearly influential in the image held of the embryo, does not exercise a direct or independent effect on attitudes to embryo research. This last result suggests that the failure reported at times in the literature to find an association between the public's scientific knowledge and their attitudes toward science, especially when evaluating science subsets marked by controversy, may be because analysts have confined themselves to the direct relationship between both variables, instead of exploring the relations potentially at work through the medium of frames or schemata like the status of the embryo. Introducing these specific constructs into models of public perceptions of science, together with some of the main worldviews operating in turn-of-the-century culture and a more robust measurement of the canonical variables of "interest in," "knowledge of" and "attitudes to" science, could contribute significantly to improving the field's theoretical development program and its connection with frontier research in other disciplines.

NOTES

1. The theoretical and methodological design of the First BBVA Foundation Study on Attitudes to Biotechnology in Europe and the United States was the work of the Social Studies Department at the BBVA Foundation under the direction of the author, who also wrote the embryo research module of the questionnaire with the assistance of an interdisciplinary group on embryo research organized by the Europäische Akademie zur Erforschung und Beurteilung von Folgen wissenschaftlich-technischer Entwicklungen in Bad Neuenahr-Ahrweiler (Germany) (see D. Solter, D. Beyleveld, M.B. Friele, J. Hołówka et al., *Embryo Research in Pluralistic Europe*. Berlin-Heidelberg: Springer). Félix Calvo cooperated with the author in the analysis of the data. The BBVA Foundation financed the study. The fieldwork was conducted between October 2002 and February 2003 by TNS. Sample size in each country: 1,500 individuals. The consolidated sample size is 13,500. The questionnaire was administered through face-to-face interviews with a random representative sample in each country. Sampling error for a 95.5% confidence level is, in the worst scenario ($p=q=0,5$), equal to $\pm 2.58\%$..
2. The exact wording of the question and the items are as follows: "Please rate your agreement or disagreement with the following statements about stem cells on a scale from 0 to 10, where 0 means you disagree completely, and

Worldviews, Frames, Trust and Perceptions of Stem Cells   369

10 means you agree completely. You may, of course, give any score between 0 and 10. 1) Research with human embryos that are a few days old is an unacceptable interference into the natural processes of life. 2) Restricting research with embryos that are a few days old means opposing progress and welfare. 3) It is profoundly wrong to damage the least developed forms of human life, such as embryos that are a few days old. 4) The medical benefits for many human beings that can perhaps be obtained in the future thanks to research with embryos that are a few days old are much more important than the embryos' rights. 5) We have already gone too far in the manipulation of life, and we should not continue along this path. 6) I completely trust that scientists will use research with embryos that are a few days old in a responsible fashion. 7) Allowing research with embryos that are a few days old in order to obtain stem cells for use in medicine will open the door to other morally reprehensible uses. 8) Research with stem cells from embryos that are a few days old should be supported as a means of finding efficient cures for diseases such as Parkinson's, Alzheimer's, or diabetes as soon as possible. 9) If research with stem cells obtained from embryos that are a few days old is permitted, we will end up creating monsters. 10) Research with embryos that are a few days old is morally acceptable if it leads to medical benefits for many people in the future.
3. The four religiosity categories we get from combining the variables "Believe in a religion" and "Practice of prayer" are 1 Nonbelievers + Nonpracticing believers, 2 Believers occasionally practicing prayer, 3 Believers who practice prayer weekly and 4 Believers who practice prayer daily.
4. For each of the items below, respondents were asked to define "the extent to which you believe they are true or false" according to the following scale: "absolutely true," "probably true," "probably false" and "absolutely false": 1) All plants and animals have DNA. 2) Humans have somewhat less than half of their DNA in common with chimpanzees. 3) All humans share exactly the same DNA. 4) The father's gene determines whether a newborn child will be a boy or a girl. 5) Human beings evolved from preceding animal species. 6) Ordinary tomatoes, the ones we normally eat, do not have genes, whereas genetically modified tomatoes do. 7) Today it is not possible to transfer genes from animals to plants. 8) Today it is not possible to transfer genes from humans to animals. 9) Genetically modified animals are always much larger than ordinary animals. 10) If someone eats a genetically modified fruit, there is a risk that the person's genes might be modified too. 11) During the first months of pregnancy, it is possible to apply a genetic test to determine whether or not the baby will have Down Syndrome. 12) Cloning is a form of reproduction in which offspring results from the union of egg and sperm. 13) The great majority of bacteria are harmful to humans. 14) It is possible to extract stem cells from human embryos without destroying the embryos. 15) A few months after an organ transplant, the recipient's organism adapts to the new organ without any problems. 16) The greenhouse effect causes the planet's temperature to rise. 17) Nuclear power plants destroy the ozone layer. 18) The use of aerosol sprays creates a hole in the ozone layer. 19) The greenhouse effect is caused by the use of carbon and gasoline. 20) All radioactivity is produced by humans. 21) All exposure to radioactivity causes death. 22) All pesticides and chemical products used in agriculture cause cancer in humans.
5. The seven items making up the "reservations about science" scale are: 1) Science and technology make our way of life change too fast. 2) Scientists should not interfere with or change the workings of nature. 3) People would be better off if they lived a simpler life. without so much science and technology. 4) Science and technology have created a world that is full of risks for people. 5) Technological

progress creates a completely artificial and inhuman way of life. 6). Science and technology destroy people's moral values. 7) Technological progress is one of the main reasons for the current high levels of unemployment.

6. The question used to capture expectations about science is drawn from one used in the Eurobarometer series on perceptions of biotechnology, and is worded: "I will read a list of new technologies to you. Please tell me, in each case, if you think this technology will improve our quality of life in the next 25 years, if it will have no effect, or if it will worsen the quality of life": 1) Solar energy, 2) Computers, 3) Biotechnology, 4) Telecommunications, 5) New materials, 6) Genetic engineering, 7) Space exploration, 8) Animal cloning, 9) Internet, 10) Nuclear energy.
7. The wording of the item on the competence of scientists, with the semantic differential format, is as follows: "People hold different opinions about scientists. On this card there is a list of opposing words which may be used to describe scientists. For each pair of words, I would like you to tell me which side come closest to your own opinion. You may use this card to indicate the position which best matches your opinion."

Competent   1 2 3 4 5 6 7   Incompetent

8. The wording of this proxy on the competence of scientists is as follows: "I will read you a list of groups of professionals. Please rate each group according to its contribution to the improvement of living conditions and human progress on a scale from 0 to 10, where 0 means that the group does not contribute at all to the improvement of people's lives, and 10 means it contributes a great deal."
9. No equation gave rise to colinearity or multicolinearity, with tolerance values for individual variables of $T_i > 0.1$ and variance inflation factors (VIF)<10.

REFERENCES

Bauer, M.W., N. Allum and S. Miller (2007) 'What can we learn from 25 years of PUS survey research? Liberating and expanding the agenda', *Public Understanding of Science*, 16: 79–95.

Bauer, M.W. and G. Gaskell (eds.) (2002) *Biotechnology. The Making of a Global Controversy*. Cambridge: Cambridge University Press.

Bishop, G.F. (2005) *The Illusion of Public Opinion. Fact and Artifact in American Public Opinion Polls*. Lanham-Boulder-New York: Rowman & Littlefield Publishers.

Caulfield, T. et al. (2009a) 'The stem cell research environment: A patchwork of patchworks', *Stem Cell Rev and Rep*, 5: 82–88.

Caulfield, T. et al. (2009b) 'International stem cell environments: A world of difference', *Nature Reports Stem Cells*. Published online: 16 April 2009 | doi:10.1038/stemcells.2009.61

Clymer, A. (2001) 'The Unbearable Lightness of Public Opinion Polls', *New York Times*, July 22, sec. 4, p. 3.

Earle, T.C. and G.T. Cvetkovich (1999) 'Social trust and culture in risk management', in G. Cvetkovich & R. E. Löfstedt (eds.), Social Trust and the Management of Risk. London: Earthscan, 9–21.

Earle, T.C. and G.T. Cvetkovich (1995) *Social Trust. Toward a Cosmopolitan Society*. New York: Praeger.

Einsiedel, E., S. Premji, R. Geransar, N.C. Orton, T. Thavaratnam and L.K. Bennett (2009) 'Diversity in public views toward stem cell sources and policies', *Stem Cell Rev and Rep*, 5: 102–107.
Gaskell, G. and M.W. Bauer (2006) *Genomics & Society. Legal, Ethical & Social Dimensions*. London and Sterling, VA: Earthscan.
Gaskell, G. et al. (1997) 'Europe ambivalent on biotechnology', *Nature*, 387: 845–847.
Haynes, R.D. (1994) *From Faust to Strangelove. Representations of the Scientists in Western Literature*. Baltimore and London: The Johns Hopkins University Press.
Holton, G. (1996) 'What Place for Science at the "End of the Modern Era"?' in *Einstein, History and Other Passions. The Rebellion against Science and the End of the Twentieth Century*. Reading, MA; Menlo Park, CA: Addison-Wesley Publishing Company, 3–39.
Kasperson, R.E., O. Renn, P. Slovic, H.S. Brown, J. Emel, R. Goble, J.X. Kasperson and S. Ratick (2000) 'The social amplification of risk: A conceptual framework', in P. Slovic, *The Perception of Risk*. London and Sterling, VA: Earthscan, 232–245.
Liu, H. and S. Priest (2009) 'Understanding public support for stem cell research: Media communication, interpersonal communication and trust in key actors', *Public Understanding of Science*, 18(6): 704–718.
Merton, R.K. (1968) 'On Sociological Theories of the Middle Range', in *Social Theory and Social Structure*. New York-London: The Free Press, 39–72.
Miller, J.D. and R. Pardo (2000) 'Civic scientific literacy and attitude to science and technology: A comparative analysis of the European Union, the United States, Japan, and Canada', in M. Dierkes and C. von Grote (eds.) (2000) *Between Understanding and Trust. The Public, Science and Technology*. Australia-Canada-France: Hardwood Academic Publishers, 81–129.
Miller, S. (2001) 'Public understanding of science at the crossroads', *Public Understanding of Science*, 10: 115–120.
Minsky, M. (1975) 'A Framework for Representing Knowledge', in P.H. Winston (ed.), *The Psychology of Computer Vision*. New York: McGraw-Hill, 211–277.
Nisbet, M. (2005) 'The competition for worldviews: Values, information and public supprot for stem cell research, *International Journal of Public Opinion Research*, 17(1): 90–112.
Nisbet, M.C., D. Brossard and A. Kroepsch (2003). 'Framing science: The stem cell controversy in an age of press/politics', *The Harvard International Journal of Press/Politics*, 8: 36–70.
Pardo, R. (2003) 'Attitudes toward Embryo Experimentation in Europe', in D. Solter, D. Beyleveld, M.B. Friele, J. Hołówka et al., *Embryo Research in Pluralistic Europe*. Berlin-Heidelberg: Springer, 157–203.
Pardo, R. and F. Calvo (2002) 'Attitudes toward science among the European public: A methodological analysis', *Public Understanding of Science*, 11: 155–195.
Pardo, R. and F. Calvo (2004) 'The cognitive dimension of public perceptions of science: methodological issues', *Public Understanding of Science*, 13: 203–227.
Pardo, R. and F. Calvo (2006a) 'Mapping perceptions of science in end-of-century Europe', *Science Communication*, 28(1): 3–46.
Pardo, R. and F. Calvo (2006b) 'Are Europeans really antagonistic to biotech?' *Nature Biotechnology*, 24(4): 393–395.
Pardo, R. and F. Calvo (2008) 'Attitudes Toward Embryo Research, Worldviews, and the Moral Status of the Embryo Frame' *Science Communication*, 30(1): 8–47.
Pardo, R. et al. (2009) 'The role of means and goals in technology acceptance. A differentiated landscape of public perceptions of pharming', *EMBO Reports*, 10(10): 1069–1075.
Pardo, R., C. Midden and J.D. Miller (2002) 'Attitudes toward biotechnology in the European Union', *Journal of Biotechnology*, 98: 9–24.

Priest, S.H. (2001) *A Grain of Truth. The Media, the Public, and Biotechnology*. Lanham-Boulder-New York-Oxford: Rowman & Littlefield Publishers.

Ringland, G. (1988) 'Structured Object Representation—Schemata and Frames', in G.A. Ringland and D.A. Duce (eds.), *Approaches to Knowledge Representation: An Introduction*. New York: John Wiley & Sons Inc., 81–100.

Sjøberg, L. (2004) 'Principles of risk perception applied to gene technology', *EMBO Reports*, 5: 47–51.

Slovic, P. (2000) 'Trust, Emotion, Sex, Politics and Science: Surveying the Risk-assessment Battlefield', in P. Slovic, *The Perception of Risk*. London and Sterling, VA: EarthScan, 390–412.

Sturgis, P. (2004) 'Science in society: Re-evaluating the deficit model of public attitudes', *Public Understanding of Science*, 13(1): 55–74.

Sturgis, P., H. Cooper and C. Fife-Schaw (2005) 'Attitudes to biotechnology: Estimating the opinions of a better-informed public', *New Genetics and Society*, 24(1): 33–58.

Sturgis, P., H. Cooper, C. Fife-Schaw and R. Shepherd (2004) 'Genomic Science: Emerging Public Opinion', in A. Park, J. Curtie, K. Thomson, C. Bromley and M. Phillips (eds.), *British Social Attitudes. The 21st Report*. London-Thousand Oaks-New Delhi: Sage Publications-NatCen.

Turney, J. (1998) *Frankenstein's Footsteps. Science, Genetics and Popular Culture*. New Haven and London: Yale University Press.

US Department of Energy (1993) *Earning Public Trust and Confidence: Requisites for Managing Radioactive Waste*. Washington, D.C.: US DOE.

Warren, M.A. (2000) *Moral Status: Obligations to Persons and Other Living Things*. Oxford-New York: Oxford University Press.

Waters, B. and R. Cole-Turner (eds.) (2003) *God and the Embryo. Religious Voices on Stem Cells and Cloning*. Washington, D.C.: Georgetown University Press.

Part V
Complementary Data Streams

21 Measuring PEP/IS, a New Model for Communicative Effectiveness of Science

Hak-Soo Kim

Since the mid-20th century, graduate programs have produced the bulk of our scientists. In turn, these scientists have produced a mass of scientific knowledge. This knowledge, for all its usefulness, has also encountered much neglect and misuse.

Scientific knowledge is the product of a great deal of R & D investment, and so, neglect and misuse of it is a tragic waste of our resources, intellectual and financial. We need scientific knowledge in order to survive in this uncertain and risky world. It can help us solve our many problems, inherited or emerging.

A traditional learning theory assumes that human beings interested in something about their surroundings would obtain knowledge of it and then would be favorably disposed toward using that knowledge (e.g., McGuire 1985). So, more interest followed by more knowledge that is followed by a positive attitude has been considered the key to improving public understanding of science. This learning paradigm has been applied to any and every subject matter in education. However, here we are concerned that science is being presented as a problem in literacy or public relations, rather than shown to be a vital resource for problem solving.

This linear effects continuum is often treated as identical to the effects of simple exposure. That is to say, if we enable someone to be exposed to something, s/he will have interest in, knowledge of, positive attitude toward, and involvement in it. Thus, for example, more classes in sciences (physics, chemistry, mathematics, etc.) have been expected to enhance youths' interest in, knowledge of, positive attitude toward, and academic or vocational selection of those fields (e.g., Miller 2004).

But, do you suspect that this turns out to be the case? Have youth come to love science more in the UK, US, or South Korea? This notion of PUS (public understanding of science) has pushed the exposure effects goal toward the general public, not just the young. Normatively, modern (science-rich) society is supposed to make use of scientific decision-making for public policy, and so, it is argued that public understanding of science will enable the public's democratic participation in policy decision-making. The UK and EU have been champions of the PUS campaign. Scientists were mobilized to become direct campaigners in the PUS movement to disseminate scientific knowledge (e.g., Edwards 2004). More scientific knowledge was made available to the public. However, do you think that the general public turns out to be more knowledgeable of and more favorable toward science?

The US *Science & Engineering Indicators* survey series seems to provide the best evidence of "failure" for this exposure effects goal. Indicators of knowledge and attitude are found to be consistent over time, showing no improvement. We need to question the theory and the methodology that the *Indicators* are based on. Now! Because the UK, the EU, South Korea, and other developed and developing countries have adopted the US *Indicators* survey tools (e.g., Bauer, Durant, and Evans 1994; Korea Science Foundation 2006; Einsiedel 1994) and produced massive data sets, only to show the continued failure of the exposure effects goal of science communication.

It goes without saying that researchers responsible for producing those indicators will argue for the success, not failure, of the exposure effects goal of science communication. As a matter of fact, they try hard to find plausible evidence of success and mobilize advanced statistical techniques and control variables (however slight the relationships) (e.g., Miller and Pardo 2000; Pardo and Calvo 2002; Sturgis and Allum 2004; Sjøberg 2007).

Questioning the traditional learning theory and the PUS perspective on science communication, this article offers a brief introduction to Kim's PEP/IS model (2007a) for communicative effectiveness about science. It shows a potential measurement technique for PEP/IS, based on his recent article (2007b).

PEP/IS

Kim's new model, PEP/IS (Public Engagement with a Problem or an Issue relative to Science), is based on two principles, one on behavior and the other on communication. The principle on behavior derives from Carter's theory of behavior (2003, 2010; Kim 2003). It posits that body and behavior are essentially independent but functionally interdependent, and behavior (as process) has its own unique structure, completely different from bodily structure. A simple processual structure of behavior is the four sequential relating modes of exposing, focusing attention, cognizing, and moving. For example, interest, knowledge, and attitude are each possible (perhaps wishful) products (effects) of the behavioral process, not part of it. That is why we might sometimes (but not often) be successful in gaining interest, knowledge, or attitude (change).

The communication principle is that communication basically functions as information exchange, not persuasion. We communicate whether or not we reach agreement (Carter 1965; Kim 1986). This informative function of communicating must stimulate exposing, focusing attention, and cognizing, and these in turn stimulate moving. For example, as "receivers" of the media, we can expose ourselves to, focus our attention on, and/or cognize something scientific therein. But, now, will we give science our support?

Unfortunately, it is not easy to complete the behavioral process, no matter how much communicating "senders" do. Only strong engagement tends to lead us to complete the behavioral process. Here comes the importance of the concept of the *problematic situation*. As we face a problem, we begin to engage with it and transform it into a problematic situation. Then, we attempt to complete the behavioral process, to solve the problem. Otherwise, we might end up only exposing ourselves to the problem.

Engaged behavior from the actor's point of view starts with this concept of a problematic situation. The problematic situation may, however, have already been presented as an issue. The difference between problem and issue is important. Issue presumes solutions to be already available and competing for our support, whereas problem leaves open the possibility of constructing a new solution (Carter, Stamm, and Heintz-Knowles 1992; Kim, Carter, and Stamm 1995). Thus, issue hastens decision-making, as in voting. That is why an issue may bring out engagement more strongly and sharpen partisan confrontation.

Even though decision-making may lead us away from innovative problem solving, a divisive issue could lead us to look back to the problem for a "third way," to which informative communication and constructive cognition can contribute.

When we are engaged with a problem, we attempt to solve it. When we are engaged with an issue, we attempt to resolve it. Problem solving and issue resolution seem to be necessary for survival. However, solving a problem or resolving an issue is not easy and demands (hard) cognizing beyond focused attention. Cognizing not only leads us to define what a problem or an issue might be, but also to the relevance of science to a problem or issue.

Scientific communication will be more successful when it talks about problems and issues, not when it talks about itself. We need to participate in the public's engagement. Science needs to understand the public if the public is to understand science. Hence: PEP/IS.

Engagement with a problem or an issue and subsequently science might also bring a different kind of behavioral outcome: an impression of science. Impression is an ideational product. Problems, like other things, may evoke impressions. And people may act on their impressions, not just on facts and/or values. (This mechanism seems to help explain why pseudosciences can appeal to the desperate.) An impression might also be a prior condition to seeking knowledge of and even entering a major or career in science.

Now, we need to test how closely the level of problem or issue engagement relates to the level of science engagement. The measure used for these variables might be considered as an index of PEP/IS per se already achieved via communicating.

MEASURING PEP/IS

Based on two previous national surveys (Kim, Lee, and Hong 2002; Kim, Park, Park, and Hong 2003), we sampled ten social problems (not including issues, this time) that the public considered serious. They are inflation, elderly citizen support, corruption, water pollution, rich-poor gap, energy shortage, global warming, unemployment, labor-management conflict, and war. We presented them to a group of 603 nonscientist adults and a group of 610 professional scientists, who were 30 years old or older. We sampled in the metropolitan area around Seoul using the area-stratified sampling method. Face-to-face interviews were conducted in October, 2005. The level of a nonscientist's or a scientist's engagement (i.e., exposing, focusing attention, or cognizing) with each problem was measured by checking one of the following four items: S/he has not heard of it (1); s/he has heard of it but not paid attention to it (2); s/he has much interest in it and pays attention to its information (3); s/he has much interest in it, seeks information on it, and makes efforts to solve it (4). The scale number reflects a respondent's level of engagement with a problem (P-engagement).

We assumed that the scientists, economically stable and highly intellectual, and thereby, more homogeneous, might be more concerned with societal or collective (less personal) problems than the nonscientists. Certainly, they would have more concern with science. Therefore, comparing the nonscientists' responses with the scientists' responses was supposed to demonstrate our PEP/IS model's validity more clearly.

We also asked the two groups of nonscientists and scientists how much science could contribute to solving each problem. Their engagement (i.e., final cognizing) with science, relative to problem solving, was measured by checking one of the following four items: science can never contribute to solving a designated problem (1); science can seldom contribute to solving it (2); science can pretty much contribute to solving it (3); science can very much contribute to solving it (4). The scale number reflects a respondent's level of engagement with science as a problem solver (SPS-engagement). This level of cognition relating to science might reflect a respondent's general belief in science, but was here supposed to be evoked by a specific problem and its relevance to a specific science and/or scientific content as a problem-solver. However, in this study, we did not inquire into the specific science or scientific content.

Table 21.1 shows the P- and SPS-engagement levels for the nonscientists and the scientists, and the differences (t-tests) between nonscientist and scientist engagement levels. First, the two groups together were found to engage more with problems of inflation, energy shortage, elderly citizen support, unemployment, and rich-poor gap than with other problems (see the Total *Mean* column). They engaged more with science relative to solving energy shortage, water pollution, and global warming than to other problems (see the Total *Mean* column).

Table 21.1 P-Engagement and SPS-Engagement for Nonscientists and Scientists by Problem

Engagement	Problem	Total Mean	Nonscientists			Scientists			t
			n	M	SD	N	M	SD	
Engagement with problem (P-Eng.)	Inflation	2.79	603	2.88	0.643	609	2.71	0.616	4.55**
	Elderly support	2.71	601	2.78	0.716	609	2.65	0.648	3.40**
	Corruption	2.60	603	2.58	0.696	609	2.62	0.655	-1.25
	Water pollution	2.65	601	2.51	0.666	610	2.80	0.680	-7.48**
	Rich-poor gap	2.70	602	2.73	0.697	608	2.68	0.605	1.20
	Energy shortage	2.73	603	2.47	0.687	610	2.98	0.659	-13.26**
	Global warming	2.61	600	2.34	0.711	609	2.88	0.672	-13.46**
	Unemployment	2.71	602	2.75	0.692	609	2.66	0.621	2.45*
	Labor-management conflict	2.36	599	2.33	0.708	607	2.40	0.605	-1.98*
	War	2.54	603	2.49	0.754	606	2.59	0.628	-2.46*
Engagement with science (SPS-Eng.)	Inflation	2.74	602	2.60	0.766	609	2.89	0.693	-6.92**
	Elderly citizen support	2.70	610	2.62	0.793	609	2.77	0.682	-3.59**
	Corruption	1.97	602	1.91	0.724	608	2.02	0.696	-2.84**
	Water pollution	3.33	600	3.29	0.694	609	3.37	0.617	-2.15*
	Rich-poor gap	2.17	600	2.07	0.762	608	2.26	0.705	-4.61**
	Energy shortage	3.44	602	3.32	0.692	610	3.55	0.580	-6.14**
	Global warming	3.21	598	3.12	0.788	609	3.30	0.645	-4.39**
	Unemployment	2.53	601	2.36	0.833	608	2.71	0.735	-7.62**
	Labor-management conflict	1.89	597	1.87	0.692	606	1.91	0.589	-0.94
	War	2.46	603	2.54	0.976	605	2.37	0.863	3.12**

*$p<0.05$ (two-tailed); **$p<0.01$ (two-tailed).

Second, we found that, at a statistically significant level (upper t column), the nonscientists engaged more with such problems as inflation, elderly citizen support, and unemployment than the scientists, whereas the scientists engaged more with problems of global warming, energy shortage, water pollution, corruption, war, and labor-management conflict than the nonscientists (see shaded areas). This indicates that the nonscientists engage more with economic problems, and the scientists engage more with collective problems, perhaps due to their more stable economic condition.

And, at a statistically significant level (lower *t* column), the scientists thought more than the nonscientists that science would contribute to solving all those problems except war and labor-management conflict. However, both the nonscientists and the scientists thought that science would contribute least to solving the problem of labor-management conflict. As expected, overall, the scientists showed significantly more engagement with science relative to problem solving than the nonscientists.

Next, Table 21.2 shows that P-engagement is positively correlated with SPS-engagement for both nonscientists and scientists. The correlations are statistically significant except for corruption in the nonscientists' point of view. However, scientists thought that science could contribute to solving the corruption problem.

This table indicates that the more engaged the nonscientists or scientists are with a problem, the more engaged they are likely to be with science relative to solving it. This applies most to the problem of global warming for the nonscientists, but most to the problem of elderly citizen support for the scientists. The latter case contrasts sharply with the nonscientists' view.

Finally, we transformed those PEP/IS indices (i.e., correlations in Table 21.2) into Z-scores and tested the difference between the nonscientists and the scientists (Garson 2006). Table 21.3 shows the results of the test. There is no significant difference for PEP/IS indices between the nonscientists and the scientists except for the problems of elderly citizen support and water pollution. The scientists showed significantly higher positive PEP/IS indices for those two problems than the nonscientists. That is, the more engaged the scientists are with problems of elderly citizen support and water pollution

Table 21.2 Correlations between P- and SPS-Engagements for Nonscientists and Scientists by Problem

Problem	Nonscientists		Scientists		Total	
	Pearson r	n	Pearson r	N	Pearson r	n
Global warming	0.253*	598	0.291*	608	0.294*	1206
Rich-poor gap	0.237*	600	0.193*	608	0.211*	1208
Unemploy-ment	0.209*	601	0.261*	608	0.211*	1209
Energy shortage	0.197*	602	0.215*	610	0.250*	1212
War	0.187*	603	0.185*	604	0.178*	1207
Elderly citizen support	0.174*	599	0.321*	608	0.226*	1207
Labor-management conflict	0.171*	597	0.135*	605	0.157*	1202
Inflation	0.143*	602	0.219*	608	0.148*	1210
Water pollution	0.125*	600	0.269*	609	0.202*	1209
Corruption	0.074	602	0.109*	608	0.093*	1210

*$p<0.01$ (two-tailed).

Table 21.3 Test of Difference between the Nonscientists' and Scientists' PEP/IS Index Correlations, by Problem

Problem	Nonscientists Z-score	n	Scientists Z-score	n	SE *	Difference Test Z-score**
Global warming	0.26	598	0.30	608	0.057737	-0.69
Rich-poor gap	0.24	600	0.19	608	0.07688	0.87
Unemployment	0.21	601	0.27	608	0.057664	-1.04
Energy shortage	0.20	602	0.22	610	0.057592	-0.35
War	0.19	603	0.19	604	0.057711	0.00
Elderly citizen support	0.17	599	0.33	608	0.057713	-2.77***
Labor-management conflict	0.17	597	0.14	605	0.057833	0.52
Inflation	0.14	602	0.22	608	0.05764	-1.39
Water pollution	0.13	600	0.28	609	0.057665	-2.60***
Corruption	0.10	602	0.11	608	0.05764	-0.17

*SE = SQRT[(1/(n_1-3)) + (1/(n_2-3)] (n_1, n_2 : the numbers of each group's samples).
**Difference test Z-score = (Nonscientists Z-score − Scientists Z-score)/SE.
***$p<0.01$ (two-tailed).

compared to the nonscientists, the more engaged they are likely to be with science relative to solving these problems.

When this difference test was applied across the totality of these problems, we found no significant difference (Z-score = -1.55) for the two PEP/IS indices between the nonscientists and the scientists. Further analysis also found no significant difference between males and females except for the problem of water pollution. The females showed a significantly higher P- and SPS-engagements correlation for water pollution than the males (Difference test Z-score = -2.28).

And, less educated and older nonscientists showed significantly higher P- and SPS-engagements correlations for energy shortage than higher educated (Difference test Z-score = 2.54) and younger ones (Difference test Z-scores = 3.60 between 50 or above and 30s; 3.08 between 50 or above and 40s). They seem to expect more from science's potential to solve the energy shortage problem. This test did not apply to the scientists, because their socio-demographic background is more homogeneous.

CONCLUSION AND DISCUSSION

Here we have argued that most of the established PUS and scientific literacy research, based on learning theory, takes the information provider's viewpoint and looks, pretty much in vain, for communication effects such as interest, knowledge, and/or attitude. In fact, it ignores the information consumer's full behavioral process, identifies message exposure with (gained) knowledge and/or attitude, and then tries hard to obtain some traces of those effects with advanced statistical techniques, never questioning its theoretical validity. This simple "exposure effects" measurement model has been denied by the research record.

Kim's new model, PEP/IS, is based on three key points: 1. Behavior as process consists of basic act components: exposing, focusing attention, and cognizing, then moving; 2. Completing those acts begins with facing a problem and is made possible by informative contributions from communicating; 3. Communicating functions primarily as information transmission or exchange, not persuasion (e.g., attitude change).

So, the new model is for communicative effectiveness from the information consumer's viewpoint, not the effects of communication from the information provider's viewpoint.

From the above, PEP/IS derives two sequential engagements: first, engagement with a problem, P-Engagement; second, engagement with science's potential for solving the problem, SPS-Engagement. Insofar as the public usually considers science to contribute to problem solving, P-Engagement must occur prior to SPS-Engagement. Both engagements are highly achievable by transmitting information about that problem and science's potential for solving it.

Science can help solve the problem of science communication by showing its relevance to solving those other problems that have captured the public's engagement. Of course, the higher levels of those two engagements must have resulted from past, present, and sporadic communicating. (This could be demonstrated in future research.) Thus, the close interrelationship between those two engagements can be considered to be an integrated index of PEP/IS (for a particular problem) and as evidencing some prior communicative effectiveness for, if not by, science.

This article illustrates a new measurement technique for PEP/IS, yielding two correlated measures of engagement. However, these indices do not show what sciences could do to help solve our problems. As we come to know how sciences could help, we would be able to confirm specifically what points of communication contributed to bringing up the level of PEP/IS for various problems. Another curiosity is what difference those sciences bring relative to solving a problem, compared with other functional equivalents such as government, education, civic movement, etc. These questions are to be asked in future research.

As mentioned earlier, this PEP/IS model expects the public, as an outcome of the behavioral process, to build impressions of science, that are diverse, spontaneous, and powerful, not limited to knowledge or attitude (see Kim 2007a). We believe that those impressions are likely to guide people into science, whether science learning or a scientific career. This PEP/

IS model also guides us toward inventing new communicative practices, beyond merely exposing scientific knowledge to the public. For example, science journalism does not need to limit its readership to a community of scientists, nor its reports to the scientific community's new discoveries or activities.

Another example in practice is that, through the recognition of the German headquarters of Bayer, a multinational corporation, Bayer Korea has recently adopted our PEP/IS concept as the foundation for a social contribution project for Korean youths and began to sponsor our TOPS Round ("Think Our Problem relative to Science" **Round** Table; www.topsround.com), which is held once a month for college students. This open interdisciplinary meeting discusses a collective or community problem and uses science to solve it. We invite a scientist to participate. We find that youth easily and closely engage with the community problem and, moreover, with science. We will see what happens, through future research and observations.

REFERENCES

Bauer, M., J. Durant, and G. Evans. 1994. European public perceptions of science. *International Journal of Public Opinion Research* 6 (2): 163 86.
Carter, R. F. 1965. Communication and affective relations. *Journalism Quarterly* 42 (2): 203–212.
Carter, R. F. 2003. Communication: A harder science. In *Communication, A Different Kind of Horserace: Essays honoring Richard F. Carter*, edited by B. Dervin and S. H. Chaffee, 369–376. Cresskill, NJ: Hampton Press.
Carter, R. F. 2010. art, Art, and communication. In *Audiences and the Arts: Communication perspectives,* edited by L. Foreman-Wernet and B. Dervin, 265–277. Cresskill, NJ: Hampton Press.
Carter, R. F., K. R. Stamm, and C. Heintz-Knowles. 1992. Agenda-setting and consequentiality. *Journalism Quarterly* 69 (4): 868–877.
Edwards, C. 2004. Evaluating European public awareness of science initiatives. *Science Communication* 25 (3): 260–271.
Einsiedel, E. F. 1994. Mental maps of science: Knowledge and attitudes among Canadian adults. *International Journal of Public Opinion Research* 6 (1): 35–44.
Garson, G. D. 2006. Correlation. Available: http://www2.chass.ncsu.edu/garson/pa765/correl.htm.
Kim, H.-S. 1986. Coorientation and communication. In *Progress in Communication Sciences VII*, edited by B. Dervin and M. J. Voigt, 31–54. Norwood, NJ: Ablex.
Kim, H.-S. 2003. A theoretical explication of collective life: coorienting and communicating. In *Communication, A Different Kind of Horserace: Essays honoring Richard F. Carter*, edited by B. Dervin and S. H. Chaffee, 117–134. Cresskill, NJ: Hampton Press.
Kim, H.-S. 2007a. PEP/IS: A new model for communicative effectiveness of science. *Science Communication* 28 (3): 287 13.
Kim, H.-S. 2007b. Science communication via engagement with a collective problem: Exploring a new science communication model, PEP/IS. *Korean Journal of Journalism & Communication Studies* 51(4): 181–198.

Kim, H.-S., R. F. Carter, and K. R. Stamm. 1995. Developing a standard model of measuring the public understanding of science and technology. *Journal of Science and Technology Policy* 7 (2): 51–78.

Kim, H.-S., J.-H. Lee, and H.-H. Hong. 2002. Korean public understanding of science and technology: A national survey through a new conceptualization. *Journal of Technology Innovation* 10 (1): 124 47.

Kim, H.-S., J.-S. Park, S.-C. Park, and H.-H. Hong. 2003. *Developing a new measurement model for the youths' understanding of science and technology: A national survey*. Report on a policy research project to the Ministry of Science and Technology, the Korean government, Seoul, Korea.

Korea Science Foundation. 2006. *The Korean public understanding of science survey results*. Report submitted by Korea Gallup.

McGuire, W. J. 1985. Attitudes and attitude change. In *Handbook of Social Psychology*, edited by G. Lindzey and E. Aronson, 233–346. New York: Random House.

Miller, J. D. 2004. Public understanding of, and attitudes toward, scientific research: What we know and what we need to know. *Public Understanding of Science* 13 (3): 273–294.

Miller, J. D., and R. Pardo. 2000. Civic scientific literacy and attitude to science and technology: A comparative analysis of the European Union, the United States, Japan, and Canada. In *Between Understanding and Trust: The public, science and technology*, edited by M. Dierkes and C. v. Grote, 81–129. Amsterdam: Harwood Academic Publishers.

Pardo, R., and F. Calvo. 2002. Attitudes toward science among the European public: a methodological analysis. *Public Understanding of Science* 11: 155–195.

Sjøberg, S. 2007. *ROSE, the relevance of science education*. Available: http://www.ils.uio.no/english/rose/key-documents/questionnaire.html.

Sturgis, P. S., and N. Allum. 2004. Science in society: Re-evaluating the deficit model of public attitudes. *Public Understanding of Science* 13: 55–74.

22 Suggesting Cultural Indicators Derived from Exhibitory Science

Bernard Schiele

THE MUSEUM COMMUNICATIONS CONTRACT

The starting idea[1] is simple: characterize, as succinctly as possible, how knowledge is disseminated in the research field, then compare it with the school method–called "formal dissemination"–and finally compare it with the method instituted by the science exhibition–called "non-formal" dissemination. Why make such a comparison? On one hand, to discern the respective properties of these three types of dissemination, and, on the other, to spark a reflection on the modalities of non-formal appropriation.

Two contexts preside over the transformation of scientific discourse into exhibition discourse: the museum, whose mandate is to reconstruct the logical and historic coherence of science and technology, and the exhibition itself, which gives a form to this mandate. The exhibition effort draws upon the work of the museum institution. The shaping of content by the designers, like the strategies of information appropriation adopted by the visitors, takes place in these two institutional contexts: that of a museum with the institutional and organizational framework that it offers, and that of the concrete exhibition with its own constraints. So when we look at the exhibition as an object of study, we are confronted by two perspectives, each with its own coherence.

First, the science exhibition, analyzed as media, is apprehended according to its internal organization, that is, to what characterizes it as media and immediately distinguishes it from other media forms (radio, multimedia, newspaper, etc.). It is assumed that it combines and creates the conditions of a specific communications situation, different from that of television, for example. It is also assumed that the content and their shape (exhibition layout) are dependent on production conditions specific to the museum field. Consequently, the meaning that visitors attribute to the exhibition stems from a communicational contract determined by the exhibition's constitutive rules and the context in which these rules play out. We refer to the exhibition's symbolic operativity, that is, "the social function stemming from characteristics of the exhibition layout" (Davallon, 1986:

270) to designate both the mediation it enables and the resulting effort by visitors to give it meaning.

The term communications contract refers to "a reality both 'external' and 'internal' to language. It is an 'external' reality insofar as it depends on physical and psycho-social constraints in relation with the actional purpose contained in a given situation (. . .); it is 'internal' insofar as it involves achieving the exchange through speech behaviors, which are conditioned in large part by the situation" (Charaudeau, 1991: 15). The "inter-understanding" between the partners in a communications relationship results from abiding by "contract conditions." In the case of an exhibition, it is the manner in which an authority—the designers—sets out ideas and objects and associates them with audio-visual and text-visual information in a way that interests and captivates target visitors.

Secondly, the exhibition is analyzed from the point of view of the uses it represents to the visitors. This analysis, focused on visitor behavior, seeks to identify its dimensions. It also seeks to discern the communications effects induced by the exhibition, that is, the information that visitors retain and the interpretations they make during the course of their visit. These communications effects result from the linkage of visit strategies to the content and forms of the exhibition device. The understanding of the exhibition's pedagogical role—and in the case concerning us, the relationship to knowledge initiated by the mediation device—occurs at the meeting point of these two dimensions, that is, at the point of linkage of the modalities of mediation and those of appropriation.

FROM THE LABORATORY TO THE EXHIBITION

It is obvious to everyone that a science exhibition cannot be confused with a laboratory, where scientific knowledge is developed, nor with a classroom, where this knowledge is formally disseminated, although it claims to contribute in its own way to raising science literacy. The exhibition reworks scientific knowledge: it reformulates the scientific discourse (source-discourse) into exhibition discourse (target-discourse). But this is not a simple translation: communications constraints inherent in any attempt at non-formal dissemination of science modulate the reformulation. Furthermore, because tis an integral part of the domain of mass media, the media offer it a framework for selection and handling of scientific themes and content. This is why we define the exhibition as a particular modality of reformulation of scientific discourse, ruled over by the mass media field.[2]

The primary objective of the scientific field is to produce new knowledge [See Category 1—Table 22.1]. This is the purpose of laboratories, research centres, the Harvards, Stanfords, Princetons, etc. The purpose of the school, taken in the broad sense (from elementary to university), is the

Suggesting Cultural Indicators Derived from Exhibitory Science

formal assimilation and mastery of knowledge by the learners[3] and that of media, the public dissemination of science and technology information. Each field (scientific, school, media) is characterized by rules, specific constraints, and relative autonomy. The concept of field, developed by Bourdieu (1997), designates

> the universe in which are inserted the agents and institutions that produce, re-produce or disseminate art, literature or science. This universe is a social work like the others, but one which obeys more or less specific social laws. The concept of field is there to designate this relatively autonomous space, this microcosm endowed with its own laws. Although, like the macrocosm, it is subject to social laws, they are not the same. While it never completely escapes the constraints of the macrocosm, it enjoys a partial autonomy, more or less pronounced. And one of the big questions that will be asked about scientific fields (or sub-fields) will be precisely the degree of autonomy that they enjoy. One difference that is relatively simple, but not always easy to measure or quantify between the different scientific fields which are called disciplines, will in effect be their degree of autonomy. (Bourdieu, 1997: 14–15)

In this science field, researchers and specialists [2] produce messages intended for other scientists [4]. In the school and media fields, the messages are addressed to students and the public, respectively, i.e., to non-specialists. This is why these messages, called target-discourses, have different characteristics than primary discourse (source-discourses) [3]. For example, the presentation of research results from researcher to researcher follows a canonic presentation plan (publication or talk): *introduction, method and material, results, discussion, and conclusion*. Furthermore, the authors remain prudent: the method and results obtained are justified carefully, and generalization, if any, remains confined to the field of investigation.

Note also that it is imperative for the scientific text to use the language of a field of specialty [5] (Bally, 1951; Guilbert, 1973; Jacobi and Schiele, 1990; Kocourek, 1991). The vocabulary used obeys the rule of correspondence: specialized terms are monosemic (or mono-referential) [6] and hence different from the polysemic terms of everyday conversation. These scientific terms have another property lacking in everyday language. In everyday language, the "meaning" of words depends on co-text (the series of words around the pivotal term) and context (the communication situation, type of discourse, relationship of the communicators, etc.). The reader or listener attributes a "meaning" to the words in keeping with this whole ensemble. This is not the case in scientific language: the terms are autonomous (or tend to autonomy depending on the degree of formality of the field). Their meaning does not hinge on co-text or context (Jacobi, 1999). This is why learning a discipline always involves learning its vocabulary. Acquiring scientific knowledge means acquiring the coherent handling of terms and formalisms, the set of which constitutes the discourse of a science (Roqueplo, 1974). Consequently, the technical term is an effective and economical tool for the researcher. Its meaning is immediately full and complete for another specialist in the field. It is called jargon only by those who are not specialists!

But knowledge demands more than mastery of the language. For one thing, the language must submit to the test of experience, because the correct manipulation of terms and concepts, as essential as this is, is not enough. "(The) fundamental problem of scientific knowledge lies neither in the development of a language nor in the variation of experiences, (...) but in the collaboration of a linguistic expression and a manipulation" (Granger, 1967: 32). This is because reality reveals itself to understanding only when manipulated by experience [7] and where the concepts used are defined operatively. An operative definition "includes the description of a regular procedure for identifying, measuring and more generally attaining and identifying the defined concept" (Ullmo, 1969: 24). He states that "science seeks its objects, it constructs them, it develops them; it does not find them 'ready-made' or given in perception or immediate experience. The world of science is a construction; the methods of this construction constitute the first step of science, and it is not the least difficult step" (Ullmo, 1969: 23). In other words: "all scientific definition is an experience" (Bachelard, 1963: 65).

On the other hand, the properties of reality, revealed by the experiment, must emerge from repeatable relationships. It is through these relationships that "scientific beings who will constitute the objects" (Ullmo, 1969: 27) are attained and designated. This is why the scientific exchange requires disclosure of the conditions of the experiment itself: i.e., of the process leading to the same relationships. Having developed the capacity to inhabit an experimental protocol, the researchers read it for what it is: an instruction manual. "From the moment when they effectively practice a discipline, they have acquired, in the domain which is theirs, this 'truth structure' that allows them to experimentally interpret any experience and narrative of experience. It is precisely for this that scientific theory and discourse engage for their part on the tangible reality that experiment manipulates" (Roqueplo, 1974: 91). Now, for outsiders to the scientific field, the whole difficulty lies in the "incommunicability of the practice that underpins strictly scientific discourse" (Roqueplo, 1974: 95). In other words, the understood experience, seen or read by a layman, remains a narrative for him. Schoolwork precisely consists of knitting together discourse and practice in way that they reciprocally involve and apprehend each other.[4]

Table 22.1 Scientific, Didactic, and Media Fields

CATEGORIES	SCIENTIFIC FIELD - LABORATORY -	DIDACTIC FIELD -SCHOOL-	MEDIA FIELD -MUSEUM –EXHIBITION-
[1] Goal	Produce knowledge	Teach-learn	Disseminate-popularize
[2] Producer	Researcher, specialist	Teacher, educator	Mediator
	The laboratory (and by extension all places of knowledge production)	The school: (the education world in the broad sense)	The museum and comparable institutions
	Statement of the source- discourse	Transposition of the source-discourse into target-discourse (pedagogic)	Reformulation of the source-discourse into target-discourse (media).
[3] Receptor	Researcher, specialist	Students	General public (specialist, amateur, neophyte, novice)
[4] Message (type of discourse)	Primary discourse: communications aimed at other researchers (formal production of statements circulated in the scientific field)	Didactic-driven discourse: communications (texts, science manuals) geared to learners	Non-formal education-driven discourse: communications (mass media) aimed at the general public or segments of it
	Structural anteriority of the knowledge content	Structural subordination of the object to be taught	Structural subordination of the object to be popularized
	Object of learning	Object of teaching	Object of popularization
[5] Method of dissemination	Formal dissemination of knowledge by means of a specialty language (terminologies)	Formal dissemination of knowledge by means of didactic transposition	Non-formal dissemination of knowledge by means of media reformulation
[6] Vocabulary	Correspondence (monosemy)	Correspondence-polysemy	Polysemy
[7] Practice	Linkage of theoretical practice to empirical practice	Narrative joined to its practice	Absence of empirical practice: autonomous narrative

MODES OF MEDIATION

Besides proven competency, the scientific field asks of those who demand to be part of it recognition and respect for the rules that constitute it and define it uniquely as a specific field. Moreover, entry into a scientific career is dependent on strict mastery of these rules. The dual effort of learning and socialization occurs mainly at school (extending to university and the laboratory). School introduces a continuous learning process based on the progressive assimilation of connected information in order to build a coherent body of knowledge [see: 8—Table 22.2]. Each piece of information functions as an element of a whole. The preferred process of knowledge presentation in the school field is didactic transposition (Astolfi et al., 1978; Chevallard, 1985; Astolfi et al., 1998). That is, matching of the knowledge taught with extracurricular reference scientific knowledge produced in laboratories and research centres, but presented for purposes specific to the school. The program is canonic in form, followed step by step [9]. Also, the school speaks to groups that are homogeneous (age and competency) [10] and captive [11]. Its methods are coercive: the students are all required to attend the courses in which they are registered and, at least ideally, to confront problems with a comparable degree of preparation. Once they are registered, they can neither (theoretically) break nor abandon the learning "contract."

Things are presented differently in the exhibition field, as neither the formal competency sanctioned by the school, nor that acquired by practicing the profession or by direct contact with the culture, suffice to distinguish the social actors who intermingle in a shared space without really collaborating on a common project. That is why the exhibition is the meeting place par excellence of culture and taste. Moreover, the vision for non-formal dissemination of science, which consists of seeking to cultivate but without instructing, leaves no doubt on this matter [8]. "What we mean" by non-formal dissemination, to pick up the classic definition of Le Lionnais,

> is precisely this: all activity to explain and disseminate knowledge, culture and scientific and technical thought under two conditions, subject to two reservations: the first is that this explanation and dissemination of scientific and technical thought must be done outside official teaching or equivalent teaching (. . .) a second reservation is that these extracurricular explanations are intended neither to train specialists nor even to upgrade them in their own specialty, since we are claiming on the contrary to round out the culture of specialists outside their specialty. (Le Lionnais, 1958: 7)

In didactic communication, unlike scientific communication, the acquisition of knowledge is graduated, measured, and sanctioned. Even the most poorly designed didactic message is part of a process of which it marks only one step. The concept of continuity and progress governs the development

Table 22.2 Formal Education and Non-formal Education

CATEGORIES	FORMAL EDUCATION SCHOOL	NON-FORMAL EDUCATION EXHIBITION	INDICATORS (MUSEUM – EXHIBITION)[1]
Didactic communication and exhibition communication			
[8] Goal	Train specialists (or at least qualify the learners in a field through the appropriation of a coherent body of knowledge)	Cultivate without training specialists	
[9] Choice of themes	Elaborate system of official instruction (programs, syllabus, curricula)	Free from all constraints	Front-end evaluation
[10] Publics	Homogeneous groups	Diversified users (heterogeneous interests and training)	Audience survey Museum visitor survey
[11] Conditions	Captive: the students learn what the teacher's duty is to teach. Exposure to the message is mandatory for those (students, adults) committed to a training process.	Non-captive: taste for science, self-learners, curiosity, etc. The public voluntarily exposes itself to the message.	Museum/exhibition visit survey Leisure activity survey Tracking study Visitor behavior survey
[12] Evaluation	Validation system: control of pedagogical effectiveness (function of the exam which sanctions the learner)	No validation system: absence of control of message effectiveness (failing tests, quiz or evaluation taken voluntarily entails no sanction)	Contextual learning analysis/survey[2] Formative evaluation
[13] Time	School	Leisure	Leisure activity survey
[14] Challenge	Success	Pleasure and culture	Summative evaluation
Relationship to knowledge			
[15] Goal	Initiation to a thinking process and progressive entry into it	Familiarization with field of knowledge (or with certain of its aspects or implications)	

(continued)

Table 22.2 (continued)

CATEGORIES	FORMAL EDUCATION SCHOOL	NON-FORMAL EDUCATION EXHIBITION	INDICATORS (MUSEUM – EXHIBITION)[1]
[15] Goal (Continued)	Knowledge to know how: that is, knowledge aimed at "action"	Knowledge for its own sake: that is, knowledge aimed to be spoken (a knowledge of culture)	
[16] Content of the relationship	Deals with operations	Deals with generally concrete fragmentary elements	
[17] Strategy of appropriation	Development of critical spirit	Memorization of knowledge (encyclopedism).	
[18] Function	The knowledge is objectifying: i.e., the topic becomes "mediator"	The knowledge is objectified: i.e., knowledge is transformed into cultural objects Elaboration of a system of representations	

Non-formal dissemination practices

[19] Relationship to the discipline	Focused on the internal organization of disciplines	Concerned with real or anticipated interests and expectations of the public it addresses.	
[20] Supports	Generally prescribed	Free choice of supports	Exhibit evaluation Discourse analysis Semiotic analysis Semantic analysis Communicational model analysis
[21] Dissemination space	Closed, tangible space	Open, non-designated social space	
[22] Pedagogical time	Uncounted time	Portion limited to leisure time	

1. At present there is no single instrument in museum studies that would allow all the variables occurring in situations of informal learning to be included in one composite scale. For this reason, the studies in this field still combine a number of instruments, each one pertaining to a series of indicators, the interrelations of which yield an overall understanding of the situations studied. Our limited space here prevents us from describing even a small range of the instruments currently being used. In the spirit of our efforts to discern the characteristics of an informal learning situation (the exhibition), we here present the major strategies that researchers use to pinpoint the various indicators.
2. See: Falk & Dierking, 2000.

Suggesting Cultural Indicators Derived from Exhibitory Science 393

of this type of messages. Each one is defined in terms of prior knowledge that it seeks to go beyond and integrate. The didactic message implies a "déjà-là" (already acquired) knowledge within, *a fortiori*, the pretense of simulating the process of discovery undertaken by the student himself, placed in the position of a "researcher" because he must rediscover that which has already been discovered. The acquisition of a standardized knowledge orients and guides the process and will be assessed in these terms [12]. Inversely, from didactic communication that tends to reduce the "unknown" of the learner by approaching the already constituted "known" of science, scientific communication between researchers presupposes prior knowledge without which it is impossible: a "known" that opens on "the unknown" of the knowledge to be constituted.

It is different for the exhibition: the audience is neither homogeneous nor captive [11]. The messages are designed primarily to awaken, captivate, and constantly maintain the interest of visitors and, if possible, inform or cultivate them. For a designer, an exhibition's success is measured by public attendance numbers, just as it is evaluated by audience size for a producer of television programs.[5] In this context, the question of measuring the retention of concepts presented is simply not relevant. What occurs here is a fundamental torsion in establishing the relationship to knowledge that colours any non-formal dissemination project. Formal learning takes place during the school time [13], and the challenge is success in passing; pleasure and culture (Guichard, 2000), for their part, are for leisure time [14].[6]

THE RELATIONSHIP TO KNOWLEDGE

In exchanges between researchers or in the pedagogical relationship, the communication relationship enables the receptor to enter progressively into the transmitter's discursive framework through a series of interactions, the form of which is governed by the institutional and interpersonal context that modulates the exchange. The process of producing "meaning" thus remains open, because the goal is initiation into a certain speech and a certain practice. Even in the "cast-offs of lecture courses," as the classroom can occasionally be, the possibility for the receptor to enter into the transmitter's discourse is maintained, even if only as an alibi. Because the school seeks the learning of the thinking process at the very core of scientific "act" [15]. This is why, even in its most distorted forms, knowledge transmission is accompanied by the principles that preside over its creation. And if attaining this goal requires the mastery of concrete knowledge, this is because the thinking and acting process is anchored in and through it. They are its carriers. The content of the pedagogical relationship corresponds to the operations because "the specialist's knowledge is above all a knowledge, the linguistic expression of which is comprehensible only in reference to a practice (experimental or otherwise) that is essentially reproducible due to its underlying method (order of succession of operations to perform)" (Jurdant, 1973: 57, passim) [16].

In an exhibition layout, the relationship with knowledge rests mainly on tangible, fragmentary, and concrete elements [16]. The visitor can expect only a cultural type of collection, that is, a presentation of results de-contextualized (Schiele, 1984) [17] from results obtained by scientific research. The recounted experience is not the effective experience: the "facts" presented remain only effects of language.[7] Hence the knowledge elements, objectified and transformed into cultural objects through the media communication relationship of the exhibition, contribute to the creation of representations (Moscovici, 1961, 1976). The school, in contrast, promotes the integration of knowledge and know-how because it advocates the development of a critical mind [17], the acquisition of which is the very condition of progress in knowledge. "Scientific creativity presumes the possibility of refuting theories rather than constructing them" (Jurdant, 1973: 70). So knowledge, in this perspective, can be defined as a field of "coordination and subordination of statements where concepts appear, apply and are transformed" (Foucault, 1969: 238). Thus transformed into objects of operations,[8] knowledge takes hold and can then be manipulated by a subject. Objective, it is objectifying: the subject becomes the mediator [18].

NON-FORMAL DISSEMINATION

To carry out their disclosure plan, exhibition designers adopt the essential arguments and manner of PUS discourse. Among their arguments, they purport to demystify science by humanizing it: bring man closer to a science created by and for him, but which has become distant and abstruse, and break with the jargon to address the general public in simple terms. They intend to translate science into a language accessible to all and do so for the benefit of all. On the practical level, it must be noted that the relationship to the discipline, which is central in the didactic relationship [19] is not pertinent to non-formal dissemination. The real or anticipated expectations of the target public are what are important.

It is the designers who choose the supports for the messages. They are not prescribed, as is generally the case with the pedagogical relationship (visual-script supports: manuals, exercise books, computer courseware, etc.) [20]. The important point is that these messages are public-oriented and aimed at society as a whole. This gives them a specific character: film, newspapers, magazines, television, and museums divide up heterogeneous abstract spaces because they each speak indiscriminately to everyone from everywhere [21]. In other words, in addressing everyone, they are really targeting no one. In contrast, the school, which addresses a few, does so from institutionally designated places. Moreover, the school is characterized by uncounted pedagogical time as it indicates the "time required for the student to establish himself as a questioning subject in his capacity as subject through scientific discourse." In contrast, non-formal education is limited to leisure time (days, weekends, vacations). This time, closed and subjective, is determined by work patterns (Jurdant, 1973: 70–71, passim).

The exhibition is understood to initiate a complex receptive process. It is essentially apprehended by the dynamics of "seeing," propelled by the meaningful combination of a contrastive layout of planes and volumes. This combination serves as a greeting context and a structure to take charge of the visitor. At the content level, the science exhibition draws mainly on scholarly discourse and its imagery (Schiele and Boucher, 1994; Schiele, 2001). Created first for research purposes, they are accompanied by representations from the "world of everyone," a true system of benchmarks that connect the unknown of the scientist's knowledge to the "already-known" of the layman, and that of the "classroom world," which guides the layman to the "already-formulated" of science (Jacquinot, 1977).

KNOWLEDGE SUPPLY AND DEMAND

To contemplate the dynamics of the science museology field, its links with other fields must be recalled. Like PUS, of which it is a variation, the science museum deals with a body of already constituted knowledge that it seeks to disseminate. The carrier of a knowledge disclosure plan, with close ties to the school apparatus that provides it a stable and captive visitor-public, the science museum nonetheless dissociates itself from the school by rejecting its boundaries and strategies. On the other hand, hard-pressed to secure the loyalty of a fluid public, it must resort to the contrivances of media language. For this reason, the problem of its cultural contribution cannot be posed independently of the specific constraints imposed by the media form. The media's objective is less to transpose science for easier assimilation than to target population segments, and, to reach this goal, to compose messages that meet their expectations and interests as well as their listening or visiting habits. The museums subscribe to this perspective.

Let us parenthetically address the role of cultural industries (of which the mass media are part). We might be inclined to think that non-formal dissemination, because it is open and uncircumscribed, lacks coherence. Nothing is more misleading. Non-formal education, similar to the school, is based on a network of institutions (associations, groups, clubs, museums); dissemination circuits (book fairs, radio, television, multimedia, cultural animation); and financing methods (government assistance, communications industry). In fact, non-formal dissemination takes place from distinct and characteristic places. Two are traditionally distinguished: one that immediately enters its products in the marketplace (book industry, media, etc.); and the other, which assigns them to non-market circulation (scientific animation, volunteer networks, awareness activities, etc.). Clearly these two universes are not mutually impervious. Moreover, cultural industries in this area of activity grow through an encroachment of the first circuit on the second. And, insofar as the market circuit absorbs the non-market circuit, the ties it has developed with the school or science fields are stretched. More specifically, what fell under the order of symbolic exchange is reduced to market exchange. For example, the school as a place of non-formal activities mutates into a market for the products and services of non-formal disclosure. This transforms the knowledge offer of the actors of non-formal disclosure. It is in this perspective that the current trend in science museology is being questioned. It is less about wanting to present scientific and technical knowledge than about exploiting a segment of the culture consumption market.

In this context, whereas the school's knowledge offer is part of society's reproduction space (Bourdieu and Passeron, 1964, 1970), that of the exhibition falls under either the recreational or cultural space in non-market circuits, or the consumption space in the market relationship [23]. The inequality in principle between the school's knowledge offer and individual expectations must also be considered. The knowledge offer functions as a constraint to which the demand for knowledge is submitted [24]. The reciprocity of the exchange, to which the school strives, and which is the finished form of the pedagogical relationship—in which the student catching up with the teacher is his equal—requires at least tacit recognition of and submission to the fundamental inequality of the teacher-learner relationship. Agreeing to learn implies acceptance of wanting to move off-centre, agreeing to make the effort to change reference point.

> In learning to read and write, the child is led to confront a universe of signs that pre-existed him and that are not directly intelligible to him. Pedagogy is demagoguery when it denies this off-centering and the effort it implies, when it suggests that culture is only a simple response to the needs and wishes of young people. It is illusory if it leaves the belief that its methods could make the significations immediately accessible and transparent. In this sense, reading and writing necessarily include at the outset a "drudgery" dimension and the break with the logic of the supremacy of "me." Pedagogy seeks to help achieve this transition, but cannot make it disappear. (Le Goff, 1999: 64)

The universe of non-formal dissemination is *a contrario* governed by the demand to which it continually readjusts. It is the main regulator of the offer.

A final point: the science exhibition achieves its aim of creating a relationship of science dissemination only by re-activating the dynamics of social processes engaged in any transfer of cultural traits. It presupposes a set of processes that at the same time relay and put in contact individuals belonging to distinctive social groups, binding with the knowledge of differentiated appropriation relationships (expectations, familiarization, instruction, etc.). The designer of a science exhibition may well imagine that visitors will follow the imagined ideal path and thereby apprehend its meaning. In fact, this path is obvious only to him. The marks, indicators, and traces must be read and then integrated in a pattern by the receptor before coherence is imposed on it and, consequently, the meaning of his reading. The exhibition invites various readings and meanings because its appropriation strategies express the plural relationships of proximity or distance from the knowledge presented. The cultural appropriation practices manifest habitus and *ethos* (Bourdieu and Passeron, 1964, 1970) [25]. At least ideally, this is precisely what the school project is trying to deconstruct [26]. This is the meaning of the expression "achieve equality of opportunity" (*réaliser l'égalité des chances*).[9] Otherwise, how is one not to be constantly sent back to oneself? [26].

CONCLUSION

This contribution sought to characterize the social situations and communication situations in which scientific information circulates between social actors. We think we have shown that the current methods of measuring science literacy of the public, although they have their usefulness, are insufficient when it comes to describing and apprehending real situations of dissemination of scientific information. We also sought to show that a rich area of research has developed in recent years on these questions and that the theoretical and methodological tools developed are pertinent. Finally, the indicators introduced and described in this chapter represent only a first formulation. Nevertheless, we hope they will contribute to nourishing the debate on the measurement of science literacy.

NOTES

1. The analysis offered on this subject expands on thoughts on non-formal dissemination of knowledge. It reformulates and develops elements incorporated from two prior papers: Schiele, B., (1995), "La médiation non formelle", Perspectives, XXV (1): 95–107, and Schiele, B., (2001), "Cinq remarques sur le rôle pédagogique de l'exposition scientifique et un commentaire sur la réforme de l'éducation", in L. Julien and L. Santerre, (Edit.), L'apport de la culture à l'éducation—Actes du colloque Recherche: culture et communication, Montréal: Editions Nouvelles, 135–157.
2. For a critical analysis of the popularizing of science in the media, see Dornan (1990).
3. It is obvious that the social functions of the school overflow the strict learning relationship. The word "training" (*formation*) is used to emphasize the extended socialization effort of the school.
4. At present there is no single instrument in museum studies that would allow all the variables occurring in situations of informal learning to be

included in one composite scale. For this reason, the studies in this field still combine a number of instruments, each one pertaining to a series of indicators, the interrelations of which yield an overall understanding of the situations studied. Our limited space here prevents us from describing even a small range of the instruments currently being used. In the spirit of our efforts to discern the characteristics of an informal learning situation (the exhibition), we here present the major strategies that researchers use to pinpoint the various indicators.
5. But understand us clearly: this is not to deny that any situation can, at least potentially, give rise to learning or prove enriching for someone.
6. See: Falk & Dierking, 2000.
7. On this point, see Roqueplo (1974), p. 131 and following.
8. Which does not exclude, quite the contrary, a gradation of information items to arrive at operations: "knowledge of particular information, knowledge of the ways to use particular information, knowledge of abstract representations", etc. (Bloom, 1975).
9. Which does not exclude, of course, the search for social status and a convergence between the demands of graduation and the interests of society (Jurdant, 1973: 69, passim).

REFERENCES

Astolfi, J.-P., Giordan, A., Gohau, G., Host, V., Martinand, J.-L., Rumelhard, G., Zadounaïsky, G., (1978), Quelle éducation scientifique pour quelle société?, Paris: PUF.
Astolfi, J.-P., Peterfalvi, B., Vérin, A., (1998), Comment les enfants apprennent les sciences, Paris: Retz.
Bachelard, G., (1963), Le nouvel esprit scientifique, Paris: PUF.
Bally, C., (1951), Traité de stylistique française, Georg & Klincksieck.
Bloom, B. S., (1975), Taxinomie des objectifs pédagogiques, Montréal: Les presses de l'Université du Québec.
Bourdieu, P., (1997), Les usages sociaux de la science, Paris: INRA Editions.
Bourdieu, P., Passeron, J.-P., (1964), Les héritiers, Paris: Les Editions de minuit.
Bourdieu, P., Passeron, J.-P., (1970), La reproduction, Paris: Les Editions de minuit.
Charaudeau, P., (1991), "Contrats de communication et ritualisation des débats télévisés", in P. Charaudeau, (Edit.), La Télévison—Les débats culturels « Apostrophes », Paris: Didier Erudition, p. 11–35.
Chevallard, Y., (1985), La transposition didactique, Grenoble: Editions la Pensée sauvage.
Davallon, J., (1986), "Penser l'exposition comme rituel de représentation", in J. Davallon, (Edit.) Claquemurer pour ainsi dire tout l'univers, Paris: Centre Georges Pompidou, p. 269–279.
Dornan, C., (1990), "Some Problems in Conceptualizing the Issue of 'Science and the Media'", Critical Studies in Mass Communication, 7: 48–71.
Falk, J., Dierking, L., (2000), Learning from Museums, New York: Alta Mira.
Foucault, M., (1969), L'archéologie du savoir, Paris: Gallimard.
Granger, G.-G., (1967), Pensée formelle et sciences de l'homme, Paris: Aubier—Montaigne.
Guichard, J., (2000), Médiatique des sciences, Paris: PUF.
Guilbert, L., (1973), "La spécificité du terme scientifique et technique", Langue française, 17: 5–17.

Jacobi, D., (1999), La communication scientifique, Grenoble: Presses Universitaires de Grenoble.
Jacobi, D., Schiele, B., (1990), "La vulgarisation scientifique et l'éducation non formelle", Revue française de pédagogie, 91: 81–111.
Jacquinot, G., (1977), Image et pédagogie, Paris, PUF.
Jurdant, B., (1973), Les problèmes théoriques de la vulgarisation, Strasbourg: Université Louis Pasteur.
Kocourek, R., (1991), La langue française de la technique et de la science, Wiesbaden: Brandstetter Verlag, La Documentation française.
Le Goff, J.-P., (1999), La barbarie douce, Paris: La Découverte.
Le Lionnais, F., (1958), "La vulgarisation scientifique", Bulletin de l'Association de écrivains scientifiques de France, Paris: Palais de la Découverte.
Moscocivi, S., (1976), La psychanalyse, son image et son public (édition refondue), Paris: PUF.
Moscovici, S., (1961), La psychanalyse, son image et son public, Paris: PUF.
Roqueplo, P., (1974), Le partage du savoir, Paris, Editions du Seuil.
Schiele, B., (1984), "Note pour une analyse de la notion de coupure épistémologique", Communication—Information, VI (2–3): 43–98.
Schiele, B., (2001), Le Musée de sciences—Montée du modèle communicationnel et recomposition du champ muséal, Paris: L'Harmattan.
Schiele, B., (2008), "On and about the Deficit Model in an Age of Free Flow", in D. Cheng, M. Claessens, T. Gascoigne, J. Metcalfe, B. Schiele, Shunke Shi, (Edit.), Communicating Science in Social Contexts, Springer, pp. 93–117.
Schiele, B., Boucher, L., (1994), "Some processes particular to the scientific exhibition", in R. Miles, Zavala, L., (Edit.), Towards the Museum of the Future— New European Perspectives, London, New York : Routledge, pp. 177–191.
Ullmo, J., (1969), La pensée scientifique moderne, Paris : Flammarion.

23 Building a Science News Media Barometer—SAPO

Carlos Vogt, Yurij Castelfranchi, Sabine Righetti, Rafael Evangelista, Ana Paula Morales and Flavia Gouveia[1]

The importance of science, technology and innovation (ST&I) and their influence on the processes of transformation of contemporary societies are undeniable. The production of ST&I has a significant impact on several social dimensions, such as economics, politics community, culture and values. And, based on this premise, the need emerges for and is consolidated in the construction of indicators geared to scientific and technological production that measure, in some way, the impact of this production. Examples would be indicators of technological innovation, of scientific production, of human and financial resources for research and, more recently, indicators of the public perception of science—which examine in what way and to what extent ST&I topics are part of people's daily lives.

The development of indicators of the public perception of science, specifically, is fundamental for a society that would claim to be democratic, because they can assist in the participatory process of decision-making. Today, it is accepted that awareness of public opinion about ST&I is fundamental, and many pertinent decisions relating to the work of scientists are now taken with the participation of different actors, such as politicians, bureaucrats, business people, military officials, religious figures, social movements, consumers and patient associations.

This rather complex and yet barely studied phenomenon seems to have begun after the world wars, in Europe, and it deepened and became more widespread in the 1990s. In a scenario in which scientists, sometimes reluctantly, are increasingly learning to leave their laboratories and their universities to communicate with various social groups, a discussion shaped on the importance of scientific communication (and of indicators that deal specifically with scientific communication). And, today, it is generally acknowledged, within and outside the academic environment: there is no science without its due communication to society.

SCIENCE IN THE MEDIA

The torrent of information emanating from recent scientific discoveries has progressively spread out from the academic environment to the public in general through specialized magazines and the mass media. Polemical topics related to advances in the most varied scientific fields come to be discussed within and outside laboratory walls, and people, increasingly, are called on to reflect and give opinions on the benefits, risks and ethical, moral and social implications deriving from research. Consequently, society needs scientific information.

For the first time, there is talk of scientific and technological communication, and national and regional governments have supported the creation and the activities in the field of scientific culture.

In Brazil, in the last twenty years, the communication of science has advanced significantly. One of the reasons for this development has been the consolidation of national scientific research itself which, if it still has not reached a desirable stage, has certainly progressed in comparison to the recent past (Capozzoli *in* Oliveira, 2002). Nowadays, it should be noted that the mass media are no longer afraid to deal with science and technology and actually exploit them to clarify the current reality in general.

When one begins to pore over studies of scientific communication, it is necessary, first of all, to delimit the field. Several works go in the direction of understanding ST&I in the press as materials that devote themselves to explanations of terms and scientific concepts or that deal with new discoveries. In this work, science is understood as a broad concept, a **scientific culture**, that is, a large ecosystem of symbols, ideas, histories, facts and notions that circulate and stir up society and have, therefore, an extremely strong media reflection.

This means that, in the media, science is present not only in the science sections, but, for example, in the readers' letters that serve, not to disseminate or explain anything scientific, but to express opinions, values and beliefs with respect to an important subject in which science enters into the country's political and social life. Science is present too in the newspaper pages that express opinions, in economics columns or in political articles, concerning strategic decisions, for example, in the technological innovation sector.

As science gains ever more space in the mass media and, consequently, conquers space also in people's daily lives (increasing interest in ST&I), the need to develop mechanisms geared to measuring and analyzing this content, even with the aim of producing indicators, such as, in fact, SAPO does, grows stronger.

THE SAPO PROJECT

The SAPO project (Scientific Automatic Press Observer) emerged in the Laboratory for Advanced Studies in Journalism (Labjor), of the State University of Campinas (Unicamp), in 2003, with the aim of analyzing the manner and the extent to which science-related subjects frequent the media and, consequently, affect society.

The SAPO system is simple: it consists of a piece of software[2] that collects, selects, organizes and measures the content published in four important online and print (digitally available)[3] vehicles: the print newspaper **Folha de S.Paulo,** its online homologue—**Folha***Online*, the **Portal G1** and the online **Estadão.com.br.** The system enables the user to work also with a fifth vehicle, the print newspaper **O Estado de S.Paulo,** collected up to the start of 2009, when it changed its electronically available version to "flash". The decision to integrate online newspapers in the sample, in addition to representing a technical facility regarding the automatic collection of these newspapers, also made it possible to integrate in the scope this new journalistic format, which began life in the second half of 1990 and which is increasingly consumed, especially by young people. The newspaper, previously only available in print, today can also be read in digital versions on the Internet and, furthermore, a new version of the vehicle has emerged exclusively online (the content of which is produced exclusively for the web).

This phenomenon completely changed the relationship of newspaper consumption. If previously it was necessary to go to the newsstand, or the newspaper needed to be physically delivered to the consumer's house, today the information arrives directly through the computer. Where previously a newspaper attained a readership of up to five times the total number of copies in circulation,[4] today the number of readers reached by an online newspaper is much higher and practically incalculable, given the extinction of physical and geographical limitations made possible by the Internet.

To study online newspapers and incorporate them into SAPO is also interesting given the worldwide phenomenon of a general decline in the penetration of print newspapers. Increasingly, readers/consumers of information are migrating to the digital platform. In Brazil, the analysis of the circulation of copies shows a reduction of around 11 per cent in the overall daily figure, which dropped from 3.5 million copies each day, in 1995, to 3.09 million copies each day, 2005 (Righetti, 2008).[5] It is interesting that, in Brazil, the same "public" that had access to the print newspaper, today, receives information via the Internet.[6] In other words: the "new readers" on the Internet are the old readers of the printed paper.

Today, with a sample that focuses primarily on online vehicles that tend to grow in penetration, but that also considers printed vehicles, SAPO's sample may be considered robust: around 700,000 articles have been collected and stored in the database in 85 different sections—and this number is growing daily.

SAPO's Methodology for Classifying Materials

All the content published in the vehicles collected by SAPO is stored in the database and duly classified. The classification methodology is based on science-related keywords. The sum of these keywords determines a score that defines whether the text *is* of scientific content, *may be* of scientific content (and, in this case, the text is passed on to a base of human verification[7]) or *is not* of scientific content.[8]

Based on tests carried out at the start of the work using human coders, with the aim of analyzing the classification results of the software coders, we verified a high level of reliability of the system.[9]

The development of keywords for the classification of the content collected was delimited by a restricted number of words that could be updated and refined dynamically. This consists of a minimum set of terms *characteristic of the language of dissemination materials* or of those that deal with science, scientific policy, environment etc., and that are, at the same time, *less frequent* in materials that deal with other topics. Thus, obviously, "DNA" or "transgenic" may be important terms, but "electrophysiology" and "antiproton" are not necessarily (since they are not that common).

The SAPO coders, as mentioned above, select materials of science, technology and innovation from the **scientific culture** perspective, that is, based on the principle that science is present in the journalistic materials of science sections and also in other sections, articles and readers' letters. Thus, SAPO's classification selects, for example, materials on advances in **research-linked cutting edge technologies**—such as nanotechnology, molecular biology and aerospace technology, major **discussions on policies and the impact** of ST&I (for example, pollution, electromagnetism, transgenics and digital TV), materials on **health policies** (in the broad area of medicine) and on **environmental policies** (in the environmental coverage), among others.

The SAPO Indicators

In the first phase of the work four **quantitative indicators** were developed, the first three of which have been affirmed among media studies researchers and the fourth developed by the Labjor/Unicamp team.

In the formulation of the indicators, we call **N_Tot the total number of materials** that were published on a specific day in a specific vehicle, **P_Tot the total number of words** contained in a specific newspaper on a specific day, N_selected the number of materials selected for a specific day, in a specific vehicle, and P_selected the total number of words contained in these items. Based on these values, we defined the following numeric indicators:

> I. *Media "Mass" indicator,* M. This represents the *absolute number of ST&I items* **published** in each vehicle analyzed, for a specific day or period.
>
> $$M = N\_selected$$

Temporal analysis of this indicator makes it possible to identify peak moments of ST&I coverage, enabling it to indicate interesting case studies. For example, it is easy to identify, from the peaks of Indicator M, the recurrence of topics, anniversaries or small media epidemics.

> *II. Indicator of "frequency", f.* This represents the *relative quantity*, that is, the percentage of ST&I materials out of the total number of materials published in the vehicle.

$$f = M / N\_Tot$$

It may be considered as an indicator of the "attention" given by a specific vehicle to ST&I matters. It enables, in a better way than the absolute value (Indicator M), a comparison of the attention given, in the same vehicle, to other topics, such as sports, economics, leisure etc. This indicator, having a relative value, is fundamental for a comparison of the behavior of different newspapers.

> *III. Indicator of "density" in the media, d.* This represents the relative *space* that the materials of ST&I occupy in each vehicle analyzed, that is, the percentage of characters for the area.

$$d = P\_selected / P\_Tot$$

This indicator is especially significant in dealing with online vehicles, as it makes it possible to ascertain the space dedicated to a specific subject by a vehicle, without the need of the old metric and tactile methodologies—the so-called "comparative journalism" methodologies, formerly used in the analysis of media coverage (see example in Melo, 1972). The *d* indicator fulfils that function exactly.

> *IV. Indicator of depth in the media, A.* This represents the relative weight that the vehicle gives to ST&I matters in comparison to "average" material in the newspaper.

$$A = d/f$$

This is a typical index of the political and cultural policy of the newspaper. It indicates that the lower the value of Indicator A, the less "in-depth" and more dispersed are the items on ST&I.

OPERATION AND FUNCTIONING OF SAPO

Today, several studies can now be obtained using SAPO. For example: i) evaluation and measurement of general tendencies in the coverage of different themes;

ii) analysis of the coverage of new media cases (recent example: swine flu); iii) study of how a news story evolves over time; longitudinal coverage of "classic" topics (cancer, space, computing etc.); iv) study of the public's perception and response (for example, the Readers Letters sections); and v) the correlation between the type of coverage of a particular subject and other variables.

The database material may be exploited by means of a user-friendly interface, available on the Internet,[10] principally in two ways:

I) **Quantitative Indicators**—the four **Quantitative Indicators** of SAPO, previously mentioned, allow the user to generate graphics and obtain data based on the choice of vehicle(s), on the *status* of the content ("science," "non-science" and filtered items—for human verification) and for a predetermined period (day, month, year). The quantitative indicators, as the name implies, are of most use in *quantitative* research.

II) **Content search**—SAPO users may also search the SAPO database by entering specific terms (for example, the expression "stem cells"), by determining the vehicle to be consulted, and entering the time periods. Thus, it is possible for the user to verify which scientific topics are the object of greatest interest to the press, how these topics are handled and in what section of the newspaper they appear most frequently. When undertaking a search, by clicking on a specific item, the user has access to the item's metadata: title, subtitle, authors, section where found, vehicle, date of publication, date of collection, total number of words, total number of score and density filter matches—in addition to the complete content of journalistic items (in this case, only for registered users). Searches are of use above all for *qualitative* research on specific topics.

III) **Data crossing**—the user may, furthermore, cross the two tools—Quantitative Indicators and Content searches—carrying out a *quali-quantitative study*.

Exploration: How to Use SAPO

To expound and illustrate what the SAPO system can do, an exploratory exercise is proposed. Let us, therefore, conduct an investigation of peaks of science materials published in the year 2008 in the *Estado de S.Paulo* newspaper.

Initially, it is observed, through the SAPO *"Mass" Indicator (M)*, generated for the entire base of the aforementioned vehicle in 2008, that this year had five large peaks of ST&I items, on 14 May, 7 and 14 August, 2 October and 6 November, with the first peak occurring on a Wednesday and the others on Thursdays.

The peaks of science items identified on the days of the week, that is, excluding Sundays, have an interesting relevance, as in all the vehicles analyzed one observes an increase in science items on Sunday, due to both the increase in the total number of items published, and a greater participation of science-related subjects.

406   Carlos Vogt et al.

Table 23.1   Average of Total Items and Science Items Published over the Year 2008 and in May 2008—Mass Indicator in the Media

	EACH DAY	EACH SUNDAY	OTHER DAYS OF THE WEEK (MONDAY TO SATURDAY)
Average number of items published in 2008	189.16	206.56	186.36
Average number of science items published in 2008	10.49	15.82	9.64
Average number of items published in month of May	192.90	230.50	187.33
Average number of science items published in May	13.35	18.75	12.55

Source: SAPO, our extrapolation.

In the *Estado de S.Paulo* newspaper, in 2008, on average, 10.49 science items were published each day. When we analyze the Sundays in isolation, it is observed that an average of 15.82 items of scientific content were published. On the other days of the week, excluding Sundays, the average number of items with scientific content drops to 9.64 (Table 23.1).

It may also be observed from Table 23.1, that the total average of science items published each day in the month of May (13.35) is around 27 per cent higher than the total average of items in the same area published over the course of 2008 (10.49), which further underlines our interest in studying this month.

Peak Analysis by means of the Indicators

Once again by way of exploration, let us choose to study the first peak in the year 2008, found in the *Estado de S.Paulo* newspaper, in May.

By means of the **Media "Mass" Indicator (M)** generated for the month of May 2008, we see that there were, in this month, five peaks of science items: 14 May (Wednesday, 28 items); 11 May (Sunday, 24 items); 18 May (Sunday, 23 items); 29 May (Thursday, 20); and 28 May (Wednesday, 19) (Figure 23.1A).

The first peak, on 14 May, with 28 items, presented 16 per cent more science items compared to the second peak of the month (on Sunday, 11 May, with a total of 24 items published).

The **M Indicator** is very interesting for this type of analysis, in absolute terms. But it is also necessary to consider the percentage occupied by ST&I items out of the total number of items published during a specified period, that is, the *relative quantity* of ST&I items out of the total number published in the period. In this case, we resort to the "Frequency" Indicator (f), SAPO's second indicator (Figure 23.1B).

A   Media "mass" indicator (M) for O *Estado de S.Paulo*.

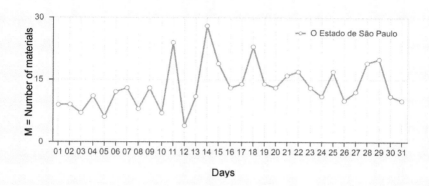

B   Indicator of "frequency" (f) in the media for O *Estado de S.Paulo*.

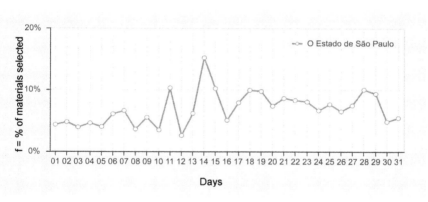

C   Media "density" indicator" (d) for O *Estado de S.Paulo*.

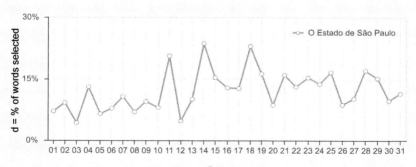

D Indicator of "depth" in the media (A)—*O Estado de S.Paulo*.

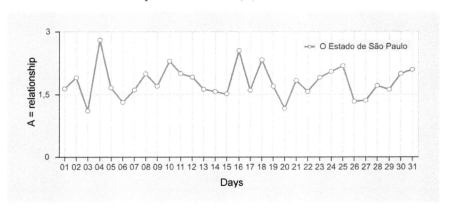

E Indicator (M) for *O Estado de S.Paulo* and *Folha de S.Paulo*.

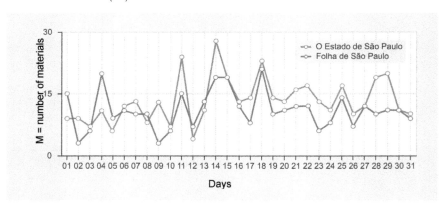

Figure 23.1A–E Quantitative indicators of SAPO.

A) By means of the media "mass" indicator (M), one can see five peaks of science items on 14 May (Wednesday, 28 items), 11 May (Sunday, 24 items), 18 May (Sunday, 23 items), 29 May (Thursday, 20) and 28 May (Wednesday, 19); B) The "frequency" Indicator (f) reinforces the peak of 14 May (on this day, 15% of the total number items published are science-related) and indicates that Sundays tend to carry a higher (absolute) number of science-related items; C) The media "density indicator" (d) shows that 23% of characters was dedicated to ST&I on 14 May; D) The Indicator of "depth" in the media (A) shows that the vehicle is publishing matters of science and technology that are, on average, of greater length than that of the general content, as A=1.56 (A> 1); and E). The comparison of the vehicles *O Estado de S.Paulo* and *Folha de S.Paulo* (both in printed versions published digitally) using the media "mass" indicator (M) shows that *Folha de S.Paulo* published less ST&I matter, in absolute numbers, than *O Estado de S.Paulo*, with the exception of three days: 01, 04 and 12 May. It is worth noting that on 14 May, with the peak of ST&I items previously analyzed, *O Estado de S.Paulo* had 33% more ST&I items than *Folha de S.Paulo*. *Source*: SAPO, May/2008.

The **f Indicator** reinforces the peak of 14 May. On this day, 15.22 per cent of the total number of items published is science-related—the second peak, on 11 May, had 10 per cent of items of scientific content. And, as we saw previously, Sundays tend to carry a higher (absolute) number of science-related items (Figure 23.1B).

Let us make our analysis more complex. Despite knowing the total number of ST&I items in the newspaper, in percentage terms, over the period analyzed, it is interesting to analyze also the *space* that these materials occupy out of the whole paper, that is, if they are smaller, similar or larger than non-science-related items.

In this case, to analyze the space, we use the **Media "Density Indicator"** (d), which shows the percentage of *characters* dedicated to ST&I (Figure 23.1C).

To extend the analysis of the space that the newspaper dedicates to matters of ST&I, it is interesting to make use of the fourth indicator, developed by the SAPO team, the **Indicator of depth in the media, A**, which relies upon two indicators (A=d/f) to analyze the relative weight that the vehicle gives to ST&I matters in comparison to the "usual" material in the newspaper.

With the peak in question, on 14 May, we see that A=1.56. Considering that A>1 (Figure 23.1D), we noticed that the vehicle is publishing matters of science and technology that are, on average, of greater length than that of the general content—and, consequently, may be more in-depth, more complete and more analytical.[11]

However, we see that despite the peak in absolute number (Indicator M), in frequency (Indicator f) and in density (Indicator d), the depth of the items (Indicator A) on the peak of 14 May 2008 was not significant. This demonstrates that the items published on that day were not of greater depth (longer).

Peak Analysis by means of Search System

Once again in an exploratory manner, we will analyze in this article the data from the 28 science items published on the peak of 14 May 2008. To do this, we are going to list them. Using the SAPO "Search" tool it is possible to select the exact date (14 May 2008), the vehicle (*O Estado de S.Paulo* newspaper) and the *status* of the materials ("about science") and check the content published.

In this list, we may confirm that the science content was found in readers' letters, reports and editorial columns in six sections: Agriculture, Supplement 2 (Culture), Cities, Politics, Economics and Life & (science)—out of a total of 26 sections in the paper.[12]

We can see, from the titles of the items, that a large number of them are related to coverage of environmental issues (Table 23.2). From a brief

410 Carlos Vogt et al.

Table 23.2 List of Materials Published on the Peak—*O Estado de S.Paulo*—14 May 2008

Content	TITLE OF MATERIALS	TOTAL
Directly related to the resignation	'Noisy' departure of minister irritates president Who left the government was the symbol Minister collected enemies Lula sounds out Carlos Minc, but PT prefers Viana From the Amazon forest to the center of power It's a disaster, say environmentalists Environment Minister felt disrespected Complete text of resignation letter Deforestation declines by 60 per cent in 3 years	9 materials
Not related to the resignation	Action on embryo cell halted two and half months ago New site offers "virtual telescope" * Sound trap attracts cicadas Waiting for Brazil * Essential to adopt technology to conserve Overspill cities for SP? * "Green" consumer seen as risk to businesses Bus fare replaces items of 1st necessity The history of peace diplomacy Direct from the source NASA probe will land on Mars on 25th * Where cattle and the forest aren't in conflict Study indicates Viagra for treatment of Dystrophy Letters * Sheriff of native Indian lands arrives Believing in extraterrestrials does not offend faith, says priest "Country violates biosafety treaty" Anvisa creates system for control of embryos Research obtains capillary vessels from umbilical cell	19 materials

* Items linked to the environmental question (environmental research, agricultural technologies related to the environment and native Indian questions), but not related to the Minister's resignation.

analysis of the content of the first item listed[13] ("'Noisy' departure of minister irritates president") it is clear that the text involves the resignation request of the then Minister of the Environment, Marina Silva, which occurred the previous day at the peak, 13 May 2008. This type of discussion, along with health matters, has special appeal for SAPO's selection, given the composition of the filter (as was mentioned in section *3.1 SAPO methodology for classifying materials*).

Based on an analysis of the other items listed by SAPO, we can separate the items related and not related to Marina Silva's resignation (Table 23.2).

It is concluded, based on the material analyzed by SAPO's "Search" system that the resignation of the ex-Minister of the Environment Marina Silva caused a significant surge in the production of scientific content in *O Estado de S.Paulo* newspaper.

However, we observed that, excluding the 9 items directly related to the political event in question, the remainder of the ST&I items published (19) already exceeds the average for the month of May, which is 13.35 items on weekdays and 18.75 on Sundays (Table 23.1). This signifies that, hypothetically, even if the newspaper had not published any material on the Minister's resignation, 14 May 2008, would have a higher average of published ST&I matter (very probably due to the items in the Agricultural Supplement, published on Wednesdays). However, without the items about the resignation of Marina Silva, 14 May would not have been a visible peak in the year or higher than the other peaks observed on the Sundays of that month.

Comparative Analysis

The composition of the SAPO sample with different vehicles, some of which even "compete" against each other, such as the *Folha de S.Paulo* and *O Estado de S.Paulo*, is particularly interesting, because it enables the comparison of data. To exemplify this, we will analyze again the month of May 2008, this time comparing the Media "Mass" Indicator of the vehicles *O Estado de S.Paulo* and the *Folha de S.Paulo*—both in the printed versions published digitally (Figure 23.1E).

We notice that the *Folha de S.Paulo* published less ST&I matter, in absolute numbers, than *O Estado de S.Paulo*, with the exception of three days: 01, 04 and 12 May (Figure 23.1E). It is worth noting that on 14 May, with the peak of ST&I items previously analyzed, *O Estado de S.Paulo* had 33 per cent more ST&I items than the *Folha de S.Paulo*. This percentage is three times higher than expected, given that the first newspaper, which has a daily average of 10.49 science items, publishes approximately 11 per cent more ST&I content than the second newspaper, with a daily average of 9.49 items in the area.[14]

Furthermore in the analysis using the "Search" tool, it is possible to establish that the *Folha de S.Paulo* published only 4 items directly related to the departure of the Minister ("Landowners lobby celebrates Marina leaving office", "Without support from Lula, Marina announces departure from government", "NGOs see government image harmed" and "Article: Marina under fire ever since Lula's first term"), as against the 9 items published in the other newspaper.

A series of comparative analyses could now be undertaken using SAPO, for example, comparisons of content of printed and online vehicles (such as the *Folha de S.Paulo* and *FolhaOnline*) or a comparison of an online newspaper (Estadão.com.br) with an online news portal (G1).

PERCEPTION OF SCIENCE

Work on analysis of ST&I publication in the media made possible by SAPO falls within the ambit of studies of public perception of science, the beginnings of which were incentivized by the post-war social movements. The objective of this type of research is to evaluate society's knowledge, interest and appropriation in matters of ST&I to guide decision-making or the formulation of public policies in the area. And media presentation of scientific matters is justly one of the factors that determines a region's scientific culture, as the media is one of the sources—if not the main one—of the access of individuals to ST&I.

The Ibero-American and adjoining countries have conducted national research into perception more recently. Brazil undertook national research of this type in 1987 (CNPq/Gallup), 2007 and 2011 (MCT/CDN). Since the beginning of the decade of 2000, the *Project for the Development of an Ibero-American Standard of Indicators of Social Perception, Scientific Culture and Citizen Participation in S&T* has brought together different countries in the Ibero-American region with the objective of constructing a methodology in the unique and comparable area.[15]

The SAPO methodology tends to complement and, what is more, to enter into dialogue with the research activities already undertaken in the area of the public perception of science, especially by means of surveys (questionnaires applied to populations). For example, in the last major survey of perception of ST&I carried out with 1,825 people throughout the State of S.Paulo, by Labjor/UNICAMP,[16] some interesting data may be processed in synchrony with the qualitative and quantitative analyses made possible by SAPO.[17]

With regard to levels of information about science and technology, 54.1 per cent of those interviewed declared that they feel barely or not at all informed about matters relating to ST&I. These respondents, when questioned why they felt this way, declared, above all, that they do not understand journalistic scientific items (36.6 per cent of those interviewed), that they are not interested in that type of information (17.8 per cent) or, yet again, that they do not know how and where to acquire information on ST&I (14.5 per cent).

In another analysis, we can see that, out of the total interviewed in the state of São Paulo, barely 7 per cent stated that they frequently read scientific news in newspapers, 26.3 per cent said that they sometimes read and **66.5 per cent never read this type of information in newspapers.** When asked about their consumption of scientific magazines, the figures are even worse: **79.8 per cent of those interviewed declared that they never read magazines that disseminate science,** 16.4 per cent read from time to time and barely 3.6 per cent frequently read (Table 23.3).

This raises some questions: these respondents, who never consume scientific news in newspapers, do they in fact consume any newspapers? If they consume other news, what are the reasons for their lack of interest?

Table 23.3 Index of Consumption of Scientific News in Different Media (%)

	FREQUENTLY	SOMETIMES	NEVER	DON'T KNOW/ DID NOT REPLY	TOTAL
NEWSPAPERS	7.0	26.3	66.5	0.2	100.0
MAGAZINES	3.6	16.4	79.8	0.2	100.0
TV	16.1	55.9	27.7	0.3	100.0
RADIO	2.4	14.2	83.1	0.3	100.0

Source: Research into public perception of S&T carried out in the State of São Paulo (Labjor/Unicamp).

And when addressing TV especially, a very strong presence in Brazilian scientific culture (it is estimated that 95 per cent of the country's homes has a set), **27.7 per cent of those interviewed said that they never watch programs or documentaries about science and technology** or about nature, and 55.9 per cent declared that they "sometimes" watch (Table 23.3). Is there little science on the TV? If there is little science, is the reason for the lack of interest on the part of viewers or is the lack of interest due to the paucity of scientific programs?

Research into the public perception of science using surveys and using new methodologies such as SAPO are complementary and mutually enriching. The figures indicated in the *survey* of perception in the State of São Paulo, for example, crossed with the information that may be obtained from SAPO, may highlight phenomena, suggest new studies and indicate public policies. In fact, they may reflect what type of dissemination we are engaged in—and for whom.

NEXT STAGES AND CONSIDERATIONS

As approached in this work, the much-needed discussion about the impact of science, technology and innovation in the various fields of human activity—and the manner in which ST&I comes to be a living part of our culture—should be based on suitable instruments in order to permit a deep analysis. In this sense, the SAPO proposal arises out of an innovative perspective and contributes to the development of cultural indicators of ST&I in the line of studies of the public perception of science.

As we have shown, it is possible to generate a very interesting body of statistical information capable of identifying "chronicities" and "epidemics" relating to the frequency of the ST&I theme in the media.

In the next steps of the work, new vehicles should be added to the sample in order to broaden the scope of comparison. Also, minor systems adjustments are already being done in order to enable the improvement of search and indexing tools. In the near future, these adjustments should permit the categorization of the texts in large areas within the major topic of ST&I. The development of new indicators and new ways of presenting the data are already being studied and should also be deployed as a next stage of development of SAPO.

Ultimately, the index created by SAPO does not pretend to be an exact reflection of how much science is disseminated in the media. It offers a specific snapshot based on arguable, albeit fixed, criteria, which once they are adopted and measured historically, take on a meaning and succeed in highlighting interesting fluctuations and processes.

SAPO has already proven to be a fairly valuable instrument for reflection and may offer the research community, from a variety of areas, daily information on the media coverage of ST&I, making it possible to glimpse not only how much this subject finds the reader, but also how the reader finds it in newspapers.

NOTES

1. Assisted by Giovana Martineli and Eden Miki Suetake.
2. The *software* was developed by the SOLIS Company, with funding from the State of São Paulo Research Foundation (FAPESP), and was internally refined by the team.
3. According to the latest data from the Instituto Verificador de Circulação (IVC, 2008) national circulation is led, respectively, by the following print titles: 1st Folha de S.Paulo (SP), 2nd Super Notícia (BH), 3rd Extra (RJ), 4th O Globo (RJ) and 5th O Estado de S.Paulo (SP).
4. The methodologies for calculating the penetration of print newspapers usually multiply the total number of copies in circulation by five, based on an estimate that, on average, five people have access to the same printed newspaper.
5. The analysis of the circulation specifically of two newspapers collected by SAPO—*Folha de S.Paulo* and *O Estado de S.Paulo*—that are among the five papers with the largest circulation in the country also shows a reduction in readership. In the case of the *Folha de S.Paulo*, the average number of copies fell 49 per cent in ten years (1995–2005). The *O Estado de S.Paulo* newspaper suffered a 39 per cent decline in circulation over the same period (Righetti, 2008).
6. This means that the Internet does not promote, necessarily, a distribution of access to information in the different levels of society. In absolute numbers, the portion of society that has access to Internet is still small: approximately 30 per cent, according to recent data from Ibope//NetRatings and from the Comitê Gestor da Internet Brasileira (Cgi.br). And, undoubtedly, this percentage with access to the web is concentrated in the most comfortable classes.
7. Entered in this category are those texts the scores of which are located in a "grey zone", which makes it impossible to affirm with a sufficient level of reliability whether they should have selected "about ST&I". This concerns, therefore, items (around 5 per cent of the total) for which the supervision of

a human coder is necessary. At the same time, these are materials that, normally, human coders have difficulties in classifying.
8. Initially, to verify the correspondence between the material collected by the system and that which is published in the newspapers, we compared the correspondence between the items contained in the database with the material also available in the printed version of the newspapers over three "constructed weeks": seven days of the week, although chosen at random over a six-month period of availability of the newspaper. After initial adjustments in the collection vehicles (some newspapers altered their production, introduced or eliminated supplements etc.), we checked that all the supplements of the newspapers of the sample are collected with an excellent level of correspondence between the material in the database and that which appears on the news-stands, with the exception of sections such as caricatures, comic strips, images, advertisements, classified ads and brief notes.
9. It should be pointed out that some refinements to the software are being made with a view to improving further the level of reliability of SAPO. Currently, for example, there is an increased incidence of error in the items that are classified as "of scientific content" with low scores, that is, those that are in the "transition zone"—very close to the "may be of scientific content" classification. Based on a sampling study undertaken on the database, we verified a margin of 10 per cent in materials classified as of scientific content due to items in the "transition zone". For the purposes of the present article, in which the operations and the data obtained are illustrative in nature, the results do not alter the demonstrative character of how the system works.
10. http://www.sapo.labjor.unicamp.br
11. As we saw previously, A>1 signifies that the vehicle is publishing items on science and technology that are, on average, longer than those of its general matter.
12. The 26 sections of O *Estado de S.Paulo* are: Supplement 2, Cities, Economics, Sport, International, Politics, Open Space, Life&, Travel, Agriculture, Estadão Sul, Estadão Norte, Estadão Leste, Estadão Oeste, Guide Supplement 2, Children's, Feminine, Alias, House&, TV & Leisure, Cars & Acessories, Construction, Opportunities, Notes and Information, Link, Gastronomy.
13. The arrangement of the items may be determined by the user by: alphabetical order of the titles, score, density, date of publication, date of collection, number of words of the text or number of words in the filter. In this case the chosen order was ascending scores.
14. Data from SAPO's database.
15. At the start of this decade, Labjor/Unicamp joined the *Ibero-American Project*, created in 2001, which began with a collaboration between the Organização dos Estados Iberoamericanos (OEI) and the Rede de Indicadores de Ciência e Tecnologia (RICYT). Until the emergence of the OEI-RICYT network, in the region there had been no attempt to find a common methodology, or one with the objective of constructing quantitative indicators that would enable an international comparison. To find out more about these works, consult Vogt & Polino (2003) and Vogt et al. (2005).
16. The paper will be published in the next edition of FAPESP's *ST&I Indicators in the State of São Paulo*, before the end of 2009.
17. The questions in the questionnaire applied in the survey undertaken in the State of São Paulo, the results of which are discussed in this work, are: Q **11**. You stated that you had little or no information on topics of science and technology. Why? Q **12**. I'm going to read some sentences about different information habits. Please tell me in each case whether you have information

Frequently, Sometimes, or Never? The *survey* consists of 39 questions common to the Ibero-American questionnaire, in addition to five additional questions by the São Paulo team. The two questions discussed in this work (Q11 and Q12) are common to the questionnaire applied by the other countries in Ibero-America.

BIBLIOGRAPHY

Bauer, M.W.; Gaskell, G. (2002): *Pesquisa qualitativa com texto, imagem e som. Um manual prático.* Petrópolis, RJ: Vozes.
Belda, F.R. (2003): *Alimentos transgênicos e imprensa—Um estudo do discurso jornalístico de divulgação científica.* 342 p. Dissertação (Mestrado em Ciências da Comunicação), Universidade de São Paulo, São Paulo.
Castelfranchi, Y. (2002): Scientists to the streets: Science, politics and the public moving towards new osmoses. *Jcom*, Vol. 1, n. 2, Trieste: Junho.
CNPq/Gallup (1987): **O que o brasileiro pensa da ciência e da tecnologia?** Relatório. Rio de Janeiro. Mimeografado.
CNPq/Ibope (1992): **O que o brasileiro pensa da ecologia?** Relatório de pesquisa, Brasília, 1992.
Figueiredo, S.P.; Vogt, C.A.; Knobel, M. (2005): **Percepção pública da C&T: um importante instrumento de apoio a políticas públicas.** In: XI Seminario Ibero-Americano de Gestión Tecnológica—Innovación Tecnológica, cooperación y desarollo, 2005, Salvador—BA. Anais. São Paulo: PGT-USP. v. 1.
Greco, P. (2002): Communicating in the post-academic era of science. **Jcom**, Vol. 1, n. 1, Trieste, Março 2002. Disponível em: <http://jcom.sissa.it/editorial/edit0101.pdf>.
IVC (2008): Instituto Verificador de Circulação.
MCT (2011): Percepção Pública da Ciência e Tecnologia no Brasil. Relatório de pesquisa. Brasília, 2011.
MCT/CDN (2007): Percepção Pública da Ciência e Tecnologia no Brasil. Relatório de pesquisa. Brasília, 2007. Disponível em: http://www.mct.gov.br/index.php/content/view/50877.html Acesso em novembro de 2008.
Melo, J.M. de. (1972): *Estudos de Jornalismo Comparado.* São Paulo, SP. Ed. Pioneira.
Meyer, P. (2004): *The Vanishing Newspaper—Saving Journalism in The Information Age.* Missouri: University of Missouri Press.
Mídiadados (2008): *Relatório anual publicado pelo Grupo de Mídia.*
Oliveira, F. de. (2002): **Jornalismo Científico.** Ed. Contexto.
Polino, C. (2007): Regional efforts toward an Iberobarometer on public perception, scientific culture and citizen participation. Conference paper presented to International indicators of science and the public workshop, 5–6 November 2007, London, England. See abstract at http://royalsociety.org/page.asp?tip=1&id=7200.
Polino, C., Cerezo, J.L., Fazio, M.E., Castelfranchi, Y. (2006): Nuevas herramientas y direcciones hacia una mejor comprensión de la percepción social de la ciencia en los países del ámbito Ibero-americano. In: Albornoz, M.; Alfaraz, C.; Arber, G.; Barrere, R.; Kataishi, R. (Org.). *El Estado de la Ciencia. Principales Indicadores de Ciencia y Tecnologia Ibero-americanos / Interamericanos.* Buenos Aires: REDES, 2006, Vol. 1, p. 50–60.
Righetti, S. (2008): *Inovação, formação de competências e diversificação no setor de comunicação: a exploração da internet em dois grupos brasileiros de mídia impressa.* Dissertação de mestrado defendida na UNICAMP em fevereiro de 2008.

Vogt, C., Knobel, M., Evangelista, R., Pallone, S., Castelfranchi, Y. (2005): "Percepção Pública da Ciência e Tecnologia: uma Abordagem Metodológica para São Paulo". Em: Landi, F. (Org.). **Indicadores de Ciência, Tecnologia e Inovação do Estado de São Paulo.** São Paulo, FAPESP: 2005, cap. 12.

Vogt, C., Marques De Melo, J., et al. (2001): "C&T na mídia impressa brasileira: tendências evidenciadas na cobertura nacional dos jornais diários sobre ciência & tecnologia (biênio 2000–2001)". Em: Guimarães, A.. (Org). *Produção and Circulação do Conhecimento: Política, Ciência, Divulgação.* Campinas: Pontes Editores.

Vogt, C.A.; Polino, C. (Orgs.) (2003) *Percepção pública da ciência: Resultados da pesquisa na Argentina, Brasil, Espanha e Uruguai.* Campinas, SP: Editora da UNICAMP; São Paulo: FAPESP.

Ziman, J. (2000): *Real Science: What It Is, and What It Means.* Cambridge: Cambr. Univ. Press.

24 Validating Survey Measures of Scientific Citizenship

Niels Mejlgaard and Sally Stares

DUAL DIMENSIONS OF SCIENTIFIC CITIZENSHIP[1]

The spread and growth of science and technology, including its risk and benefits, are main features of a globalised world. The governance challenges that accompany such dispersal have seen the deployment of mechanisms of public and stakeholder education and participation; and competent, active citizen engagement in science and technology is increasingly considered key in legitimate science governance. In this chapter, we use survey data to explore patterns of citizen competence and participation in science across Europe, and tentatively propose a typology of what we call 'scientific citizenship' based on latent class analyses. In order to assess the robustness of the results, we further qualify the findings by examining contextual information for selected exemplary countries.

The idea of 'scientific citizenship', i.e. full and active citizen membership in the knowledge society, has recently become an aim in science and society policies and appears to involve a rethinking of the field of science communication. Maya Horst (2007: 151) recently conceptualised the emerging notion of scientific citizenship in the following way:

> The notion of scientific citizenship (Irwin 2001) points to an increasing awareness of the intermingling between science and society. It implies not only that scientific knowledge is important for citizenship in contemporary society but also that citizens can lay a legitimate claim about accountability on scientific research. As such, the notion can be perceived as a normative ideal concerning the appropriate form of democratic governance in a society that has become increasingly dependent on scientific knowledge.

Horst argues that scientific citizenship is two-dimensional, combining what we might call a competence perspective on the one hand and a participation perspective on the other hand. Scientific competence is considered a prerequisite for effective human agency in modern societies. In an increasingly complex world, where science and technologies extensively shape the everyday lives of the public and affect social practices, citizens are in need of particular competences, knowledge and skills, to navigate effectively and define their own role within the system. We might say, in line with traditional conceptions of citizenship, that citizens in modern societies have a *right* to be informed, by means of appropriate dissemination schemes, about the developments, potentials and risks alike, in science and technology, in order not to be marginalised from social systems. This competence dimension has been a central concern to the PUS movement and the practical science dissemination efforts that it has involved, ever since the Royal Society report on the public understanding of science in the mid-80s set in motion a wave of effort to bring science to citizens and fill the public knowledge gaps that the report had identified.

But, as Horst argues, knowledge or competence is not an exhaustive parameter of scientific accountability. Scientific competence may facilitate human action and cultivate an enlightened citizenry; however, there is also a need for mechanisms to ensure that citizen concerns are in fact fed into decision-making processes. If modern societies are to be considered legitimate, citizens should thus also actively make use of their competence to lay claim on scientific practices and take part in public debate about scientific and technological developments. The inherently normative notion of (republican) participatory citizenship stresses the importance of full citizenship in terms of both certain rights and privileges, which serve to protect and empower the individual on the one hand, but also an ideal of societal obligation or *duty,* in which civic participation is central. Participatory citizenship is not simply about enjoying the right to enter the sphere of decision-making, but rather about actually entering it. This participatory dimension has been central to what has been called the PES (public engagement with science) approach in science communication. The PES approach is based on a belief that science communication should be a two-way exchange where active citizens 'speak back' to science (Gibbons, 1999), make their concerns heard to scientists as well as science policy-makers, and contribute to setting the agenda for research. To a large extent, the PES approach has defined itself in opposition to PUS and deficit understandings of the public.

We find, then, that the notion of scientific citizenship conceptually combines two separate, or even competing, paradigms within the field. It stresses the importance of citizen competence (which has traditionally been the focus of PUS) as well as the need for active involvement (which has been the aim of the PES approach). In this paper, we have two intentions. First, we wish to present measures of scientific competence and participation in science based on survey material, in order to examine the *empirical* interrelatedness between them, and, ultimately, point towards dominant types or modes of scientific citizenship across Europe. Second, we wish to assess the robustness of the quantitative analyses by contextualising the results on the basis of qualitative data.

In so doing, we hope that the paper might contribute to bridging two divides within the field: between PUS (focusing on citizen competence, literacy or understanding) and PES (focusing on citizen participation), but also a divide between qualitative and quantitative approaches. We contend that methodological triangulation, particularly the combination of quantitative and qualitative approaches, is something worth exploring a little more within this field of research. For the present chapter, the analyses are limited to including survey-based measures of citizen competence, i.e. the objective and subjectively felt abilities that citizens hold in relation to science and technology, and participation, i.e. citizen performances or practices related to science and technology. We use these to develop a tentative typology of scientific citizenship. Subsequently, we offer a post hoc validation of the typology by examining existing qualitative material for selected countries. We conclude with some thoughts on related studies that could be carried out using a complementary or similar approach.

SURVEY-BASED DATA ON PARTICIPATION AND COMPETENCE

We make use of the recent European survey on 'Europeans, Science and Technology' (EB 63.1), fielded in 2005 in 32 countries, with a total sample size of 31,390 and focus on two sets of variables, capturing participation and competence. Table 24.1 below gives frequencies for these items for the complete data set, with each country's contribution to the total weighted according to its population size.

We employ latent class models (Lazarsfeld and Henry, 1968) to construct summary measures of competence and participation, and to test for their cross-national equivalence, in statistical terms. We infer that the associations that we find between people's responses to these survey items can be explained by some underlying, general variable characterising participation (for the first set of items) and competence (for the second set of items). These general variables cannot be observed directly: they are hypothesised *latent* variables, and we infer their existence from the analysis. In a latent class model, the latent variable is categorical, with a class for each category of the variable. The observed items are typically also categorical. In this chapter we treat both latent and observed variables as nominal, that is, unordered categorical variables. We then use a logistic link function to model the conditional response probabilities for every item—e.g. the probability that if one were a member of class j, one would respond in category s to item i. We also model the overall probabilities of belonging to each of the classes, and from these we can say, for example, what proportion of our target population we would expect to belong to each of the classes.

In order to arrive at the cross-national models, we begin by running latent class models separately within each country and informally

Table 24.1 Items Capturing Participation and Competence in Science

Survey question	% responses		

PARTICIPATION ITEMS
How often do you...?
Responses recoded: Regularly/occasionally/hardly ever into 'Yes', and Never into 'No'

	Yes	No	Don't know
Read articles on science in newspapers, magazines or on the Internet	78.3	21.3	0.4
Talk with your friends about science and technology	70.8	28.7	0.5
Attend public meetings or debates about science or technology	28.4	71.0	0.6
Sign petitions or join street demonstrations about nuclear power, biotechnology or the environment	24.3	74.8	0.9

COMPETENCE ITEMS

Let us talk about those issues in the news which interest you. For each issue I read out, please tell me if you are very interested, moderately interested or not at all interested in it.

	Very Interested	Moderately interested	Not at all interested	Don't know
New inventions and technologies	28.9	46.9	22.9	1.3
New scientific discoveries	28.7	46.6	23.1	1.6

I would like you to tell me for each of the following issues in the news if you feel very well informed, moderately well informed or poorly informed about it?

	Very well informed	Moderately well informed	Poorly informed	Don't know
New inventions and technologies	10.9	50.3	36.4	2.4
New scientific discoveries	9.7	48.8	39.1	2.4

Knowledge	% answering 9 or more items correctly	% answering less than 9 items correctly
Derived variable from responses to 13 statements about science and technology[1]	50.2	49.8

[1] The thirteen items are the following: (1) The Sun goes around the Earth; (2) The centre of the Earth is very hot; (3) The oxygen we breathe comes from plants; (4) Radioactive milk can be made safe by boiling it; (5) Electrons are smaller than atoms; (6) The continents on which we live have been moving for millions of years and will continue to move in the future; (7) It is the mother's genes that decide whether the baby is a boy or a girl; (8) The earliest humans lived at the same time as the dinosaurs; (9) Antibiotics kill viruses as well as bacteria; (10) Lasers work by focusing sound waves; (11) All radioactivity is man-made; (12) Human beings, as we know them today, developed from earlier species of animals; (13) It takes one month for the Earth to go around the Sun.

422 *Niels Mejlgaard and Sally Stares*

comparing the similarities and dissimilarities between them. On the basis of these initial explorations, we then run a single model across all data sets, with country as a covariate. This allows the proportion of people expected to belong to each class to vary between countries, but crucially fixes the conditional response probabilities to be the same between countries. Where we can do this without compromising the fit of the model too much,[2] we claim that the underlying variables of participation and engagement can be characterised in broadly the same way between countries, so that it is reasonable to make comparisons between countries in terms of these constructs. This analysis only provides evidence of *statistical* comparability. The second part of the chapter begins to address the question of *substantive*, context-sensitive comparability, by examining contextual information for selected exemplary countries.

Participation

Latent class analyses of these items show that responses mirror each other for the first two items: respondents who say they have read articles about science are also likely to say that they talk about science with friends. These items might be taken as indications of *horizontal* participation in science: that is, engagement in activities that enhance 'scientific culture' and intersubjective learning, through 'non-political' modes of participation. These have an integrative quality and primarily represent a citizen–citizen perspective on issues of science and technology. In a similar way, responses to the second two items in the set often mirror each other: those who say they sign petitions or join street demonstrations are also likely to say they attend public meetings and debates about science and technology. These items might be taken as indicators of *vertical* participation, which aims at influencing policy agendas and represents to a wider extent a citizen–system perspective on issues of science and technology.

This pattern holds quite consistently between countries, although in many countries, even those in the most highly participating group will be unlikely to engage in signing petitions or joining street demonstrations. We might say, then, that engaging in petitions and demonstrations is the highest hurdle in this set of items on participation. The variation between countries on this last item is too great to be reconciled in a cross-national model, even if the number of classes is increased. If *petition* is dropped from the set, however, a three-class model fits satisfactorily,[3] and is therefore our preferred cross-national measure of participation.

This model is described in Table 24.2a, which shows the probabilities of each response to each item, conditional on class membership. Each class is represented by a column and each item response by a row. Notably high probabilities are highlighted in grey. For example, conditional

Table 24.2a Conditional and Prior Probabilities for a Joint Cross-national Model of Participation in Science

Item/response	Response probabilities for categories of items, conditional on class		
	Horizontal & vertical	Horizontal only	Non-participative
Read articles on science			
Yes	0.98	0.97	0.20
No	0.02	0.03	0.80
Talk with friends about science and technology			
Yes	0.99	0.80	0.13
No	0.01	0.20	0.87
Attend public meetings/debates about science or technology			
Yes	0.67	0.01	0.01
No	0.33	0.99	0.99
Estimated proportion in each class (pop. weighted)	0.42	0.33	0.25

on membership in the first class in the table (looking at the first column of figures), a respondent has a 0.98 probability of saying he or she has read articles about science, a 0.99 probability of having talked about it and a 0.67 chance of having attended a meeting or public hearing on the topic. Given such a pattern of likely responses for people in this class we could characterise it as one of both horizontal and vertical participation. This suggested label is included at the top of the column of figures, alongside suggested labels for the other two classes. The last row of the table gives the estimated probabilities of belonging in each class, weighted according to the relative population sizes of the 32 countries in the data set. For example, this model estimates that 42 per cent of this European public is highly participative.[4,5]

Competence

We have tried to include items from the EB 63.1 that could potentially constitute a compound measure of competence that involves not only 'objective' knowledge of science, but also 'subjective' competence, indicated by respondents' interest in science and what we could call internal 'techno-scientific efficacy', that is, the extent to which people subjectively *feel* well-informed on issues of science and technology. In the questionnaire, respondents are asked separately about their subjective competence with regard to new inventions and technologies, and new scientific discoveries. The first striking finding from our analyses is that respondents tend to report very similar levels of interest and informedness for the former and the latter. These items work together so closely, in fact, that for our purposes in finding a joint cross-national model for competence, we consider from this point just those items relating to 'scientific discoveries'.

The joint cross-national model using these two items, and a third to denote levels of factual knowledge, is described in Table 24.2b. To achieve a well-fitting model[6] six classes are needed—perhaps more than is ideal, but still a very much less complex solution than the 18 possible response patterns that the items might yield. Statistically speaking the classes are unordered (the latent variable is nominal), but for the purposes of substantive interpretation we can roughly order them, in common-sense terms, from high to low competence. We can label them in terms of self-reported, subjective, competence (high, mid and low) and in terms of high or low textbook, objective knowledge (+ or -).

In the six classes, the patterns of responses for interest and informedness mirror each other fairly closely, but with ratings of informedness lagging behind those of interest: even those who are very interested in new scientific discoveries are most likely to report feeling moderately rather than very well informed about them ('High+' class), whereas

Table 24.2b Conditional and Prior Probabilities for a Joint Cross-national Model of Competence in Science

Item/response	Response probabilities for categories of items, conditional on class					
	High+	High-	Mid+	Mid-	Low++	Low-
Interest in new scientific discoveries						
Very interested	1.00	0.85	0.14	0.18	0.12	0.00
Moderately interested	0.00	0.14	0.80	0.82	0.60	0.07
Not at all interested	0.00	0.01	0.06	0.00	0.28	0.93
Informedness about new scientific discoveries						
Very well informed	0.31	0.38	0.05	0.01	0.00	0.00
Moderately well informed	0.66	0.54	0.95	0.63	0.00	0.08
Poorly informed	0.02	0.08	0.00	0.37	1.00	0.92
Knowledge						
9 or more correct	0.91	0.27	0.73	0.14	0.69	0.08
less than 9 correct	0.10	0.73	0.27	0.86	0.31	0.92
Estimated proportion in each class (pop. weighted)	0.14	0.08	0.27	0.15	0.17	0.19

those who feel poorly informed might be moderately or not at all interested ('Low++' and 'Low-' classes). These items on 'subjective' competence are only weakly correlated with 'objective' competence, as captured by the knowledge variable. It is, as for example in the case of the High- class, possible to feel rather competent even though the level of factual knowledge is probably low and vice versa. Nonetheless, the magnitudes of the response probabilities for these items give the suggestion of a positive relationship between objective and subjective competence: in the High+ group the probability of answering nine or more knowledge items correctly is 0.91, whereas in the Low++ group it is only 0.69; likewise the probability of having low objective knowledge is 0.73 in the High- class, but 0.92 in the Low- class. Europeans are fairly evenly spread between the classes, although with notably very few in the High- class.

INTERRELATEDNESS OF COMPETENCE AND PARTICIPATION

Using correspondence analysis we examine the relationship between citizen competence and citizen participation. The bi-plot in Figure 24.1 shows the relative strengths of associations between the classes: those classes that appear close together in the plot are relatively more strongly associated with each other than with classes which are further away. The plot suggests a positive association between participation and competence: those in the horizontal and vertical participation class are

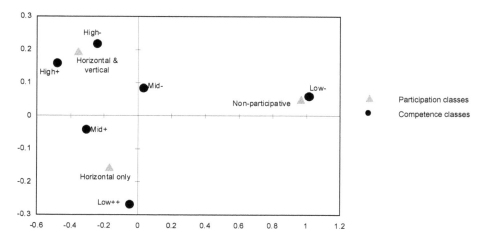

Figure 24.1 Bi-plot from correspondence analysis of measures of participation and competence.

relatively more likely also to belong to the high-competence classes than to the low-competence classes, and those in the non-participative class are relatively more likely to belong to a low-competence class than a high-competence class.

We tentatively suggest that three types of scientific citizenship can be identified: One group of citizens, in the top left corner of the correspondence plot, are likely to be both highly participative (vertical and horizontal) and high on competence. We could label this cluster 'involved' citizens. It is worth noting that both the High+ and the High- clusters are located here, which seems to suggest that the objective aspect of competence—textbook knowledge—is not an absolute prerequisite for action, rather, vertical participation appears to be associated with subjective competence (or self-confidence). Another group of citizens, at the right-hand side of the plot, might be labelled 'detached' citizens. This cluster of citizens do not get involved, neither horizontally nor vertically, and they appear to be neither subjectively nor objectively competent. Effectively, these citizens are marginalised in the knowledge society. It is a bit harder to try to name and distinguish between those in the middle of the plot. Relatively speaking, the 'horizontal only' class of participation is most closely associated with the Low ++ competence class, which is characterised by feeling poorly informed, even if they are somewhat interested and possess relatively higher levels of textbook knowledge. These citizens might in some way resemble what Miller (1983) has called the 'attentive' public for science, that is, people who are scientifically literate and rather interested, and who engage, horizontally, in active patterns of knowledge acquisition in terms of reading articles on science and talking to friends about science, but who are not necessarily politically mobilised.

DISTRIBUTIONS OF SCIENTIFIC CITIZENSHIP TYPES, AND EXEMPLARY CASES FOR POST HOC QUALITATIVE VALIDATION

Table 24.3 shows for each country the percentage of its population expected to belong to each of the classes of participation and of competence, with the table ordered by levels of non-participation on the left hand side and by levels of high+ competence on the right hand side. In this initial study into developing indicators we would not attach too much importance to the rank ordering of the countries. But it is interesting to note some general intuitive patterns emerging, and to identify countries where the majority of the population can be identified as falling broadly into one of our citizenship types.

For example, in countries such as Portugal, Turkey, Bulgaria and Romania, the estimated proportions of citizens who are non-active and have low subjective and objective competence are high. In Portugal we find the highest estimated proportion of people who belong to the 'non-participative' class across all European countries, combined with a very high percentage of people estimated to belong to the 'low-' competence class, i.e. people who are neither objectively nor subjectively competent in matters of science and technology. As such, Portugal is an exemplary case for what we termed 'detached' citizenship.

In comparison, citizens in countries such as the UK, Estonia and Norway to a higher extent tend to resemble what we tentatively called 'attentive' citizens. In these countries, the estimated proportion of people in the 'horizontal only' participation class is largest, in combination with high proportions of people in the middle and low++ classes of competence. Other countries, such as Iceland and the Netherlands, also have very high proportions of people in the horizontal only class, but in these countries competence levels are generally estimated to be somewhat higher.

Finally, in countries such as Sweden, Germany, Switzerland and Greece large proportions of the populations are estimated to belong to the 'horizontal & vertical' class, and high proportions of people also are in the high-competence classes. These countries thus appear to come close to the 'involved' type of scientific citizenship. Greece has a remarkably high proportion of people in the 'horizontal & vertical' class, and also an estimated 41 percent of Greeks have high competence, but among these, two out of three have low objective competence. In comparison, all of the 36 percent of the Swedish respondents who are estimated to belong to high-competence classes have subjective *and* objective competence in matters of science and technology.

How might we explain these cross-country variations in 'the typical scientific citizen'? We do not have space here for a comprehensive country-by-country account. Instead, we focus on three countries: Portugal, the UK and Sweden. These countries clearly each represent a type from the tentative typology, and crucially, ample qualitative material is available for analyses. A major six-country comparative study—Optimizing Public Understanding of Science (OPUS)—finalised in 2003 (Felt, 2003) provides extensive, detailed information about the national context for science and PUS policies, science dissemination and the media landscape, state-, university- and local-level initiatives regarding PUS, public consultation and foresight exercises and much more for the selected countries. Portugal, the UK and Sweden were also covered in an eight-country comparative study on Science, Technology and Governance in Europe (STAGE) that produced 29 case studies on public engagement in science and the governance of science and technology. The final report of the STAGE project group was published in 2005 (Hagendijk et al.), the same year in which the survey was conducted.

Table 24.3 Estimated Percentages in Each Class of Participation and Competence, By Country

	Hor. & vertical	Hor. only	Non		High+	High-	Mid+	Mid-	Low++	Low-
Sweden	46	49	4	Sweden	36	0	29	0	35	0
Finland	52	43	5	Netherlands	31	3	27	8	27	4
Iceland	30	63	7	Germany	26	4	28	9	25	7
Slovenia	40	53	8	Norway	24	0	37	5	30	4
Norway	38	54	8	Denmark	23	1	30	2	37	7
Netherlands	25	64	11	France	22	13	37	9	9	9
Switzerland	60	29	11	Switzerland	22	12	35	9	18	5
Luxembourg	38	50	11	Luxembourg	21	9	38	8	15	8
Estonia	30	57	13	Iceland	19	6	16	15	38	6
Germany	55	31	14	Croatia	19	7	26	20	14	15
Denmark	46	40	14	Hungary	19	2	27	12	22	18
Croatia	47	37	16	Belgium	19	7	36	6	23	9
Slovakia	51	31	18	Finland	17	0	30	2	47	4
Latvia	34	48	19	UK	15	8	34	8	22	14
Belgium	36	46	19	Greece	14	27	23	20	0	16
Austria	68	13	19	Slovakia	13	0	24	13	34	16

Country			
Czech Republic	44	35	21
Cyprus	41	38	21
France	31	47	22
Lithuania	35	43	22
UK	33	43	24
Greece	76	0	24
Hungary	50	24	26
Poland	26	47	27
Ireland	46	26	28
Italy	54	17	29
Romania	28	39	33
Spain	43	19	38
Bulgaria	45	17	38
Turkey	36	23	41
Malta	26	31	43
Portugal	21	33	46

Country						
Austria	12	9	24	17	22	16
Czech Republic	12	0	50	0	31	7
Ireland	11	9	20	21	18	21
Slovenia	10	7	47	0	31	5
Spain	8	9	17	24	21	20
Estonia	8	4	20	18	33	18
Italy	7	3	45	13	14	18
Poland	7	4	23	20	19	26
Malta	7	24	2	35	5	28
Latvia	6	4	7	46	9	28
Bulgaria	5	8	10	30	8	39
Romania	4	4	19	24	7	42
Portugal	3	4	16	37	1	39
Lithuania	1	1	11	26	18	43
Cyprus	0	51	3	34	0	12
Turkey	0	21	4	30	0	46

Contextualisation

The three countries in question differ markedly with regard to the historical relationship between science and society. In Portugal, political and economic conditions have kept science in isolation from society, and the diffusion of scientific knowledge has not been encouraged until quite recently (Goncalves and Castro, 2003a). For decades, the political authoritarian regime of the Estado Novo (1926–1974) looked at scientific research with suspicion (Castro, 2009), and until the 1974 revolution, Portuguese science was comparatively under-developed, with only minor investments and efforts going into research in the areas of the natural and exact sciences. Areas that would adhere to the social sciences were almost non-existent, with the exception, according to Nunes and Matias (2004), of those disciplines directly linked to the legitimisation of the regime and its social engineering, such as law, history and economics.

In contrast, the UK and Sweden have long traditions for science popularisation, and science has had a prominent position in society. Scientific institutions have been supported by the state, and popularisation efforts have generally benefited from a favourable educational and political climate. The roots of modern PUS activities in Sweden, according to Nolin et al. (2003), go back more than a hundred years when democratic movements in Sweden sought legitimisation by referring to contemporary scientific knowledge and scholarship, and as such, explicitly made connections between appropriate social and political processes on the one hand and scientific advice and knowledge on the other. Science has played an important role in the development of the social-democratically led development of the welfare state, and the state has, at an early stage, formally obliged universities to contribute to societal progress, for example by legally committing universities to 'third mission' dissemination activities in the 1977 university act. The British tradition for science popularisation dates back centuries, with prominent institutions such as the British Association for the Advancement of Science, founded in 1831, and the Royal Society from 1660, which from the very start were attentive to science dissemination and the establishment of intermediaries between science and society (Healey, 1999).

Although the role of science in society is slowly changing in Portugal, and even if initiatives such as the *Ciencia Viva* is slowly introducing a new educational mode of science–citizen interaction, the historical context plays a major role for understanding the present governance of science and the role of citizens. Portugal is, according to Castro (2009), still significantly characterised by an immature civil society with no tradition for either public science education or public involvement, and by secretive institutions and secretive institutional habits. In the STAGE project, Portugal was discussed as an example of what was termed 'discretionary governance', in the sense that science and technology policy-making takes place with virtually no explicit interaction with the public. Decisions are taken with very little input to the policy process by any group outside the government institutions directly responsible for science and technology policy (Hagendijk & Kallerud, 2003).

In the light of these descriptions of Portuguese science policy, the results of the latent class analyses seem plausible. We characterised Portugal as a country in which large proportions of the population would be 'detached' in the sense that people would have relatively weak scientific competence and little active involvement. The contextual information, highlighting the low levels of science education and dissemination as well as discretionary, secretive governance practices, is entirely congruent with the observations and analyses based on the survey material.

Among the major conclusions of the STAGE project was that different national styles of governance had to be considered non-stable, something in periodic transition, and that although there seemed to be patterns of mutual influence between countries, clearly the trajectories that different countries had followed historically differed significantly (Hagendijk et al., 2005). This observation may help us understand why the results of the latent class analyses for the UK and Sweden diverge, even if these countries, at least compared to Portugal, seem to share certain features, such as a historic tradition for state acknowledgement and support of science.

In broad terms, The UK trajectory over the last two to three decades, which has been influential in science communication and in the broader discussion over science governance in EU and other European countries, provides a story of a shift from PUS to PES strategies and visions, i.e. from focusing one-way dissemination of knowledge towards increased attention to active public involvement. The House of Lords' report on 'science and society' (2000) had a strong focus on public engagement with science and technology, and signalled a shift away from the science dissemination efforts of the British PUST movement in the 1980s and 1990s (Irwin, 2006; Bauer, 2003). Since the turn of the century, dialogical, deliberative methods have been at the core of British initiatives and policies within this area. In the UK, the turn towards more dialogical methods in science communication, or what the House of Lords Select Committee on Science and Technology called the emerging 'mood for dialogue' (2000), developed in the context of the BSE crisis and the wider unease with expertise or crisis of confidence between science and citizens that the BSE affair seemingly reflected. As Goncalves & Castro (2003b) explicitly point out, this 'mood for dialogue' has more ancient roots in Sweden, where the nuclear energy debate of the 1970s invoked a strong sense of need for public involvement in technology debate and decision-making processes. According to Glimell (2004) Sweden earned a reputation as a pioneer of a socially responsible science policy and wider societal concerns for collaborative modes of governance. Educational and participatory practices were developed against the backdrop of the specific controversy over nuclear energy (which is very similar to what happened in Denmark), but also in the context of wider technical advances and automation that reshaped production and working conditions of the engineering industry in the 1960s and 1970s. Building on the corporatist labor market structures of collective agreement practices and stakeholder primacy, a variety of new, inclusive methods for handling techno-social conflicts, such as issues of electronic surveillance and the digital divide in the successive years (Glimell, 2004), emerged in Sweden.

It seems that participatory culture or tradition simply has been nurtured longer in Sweden than in the UK when it comes to the relationship between science and society. This may be part of the explanation why our quantitative analysis identify people in Sweden as likely to be 'involved', horizontally and vertically and with high competence, in science, whereas UK citizens are more likely to be 'attentive', i.e. moderately active, primarily horizontally, and also less competent. Interestingly, at the turn of the century, science policies also seem to be changing hugely in Sweden. Not towards participatory governance and dialogical formats of technology assessment, as in the UK, but rather towards a re-emphasis of traditional ideas of academic autonomy and less focus on socially responsible science. In Glimell's words (2004: 2), 'Sweden appears no longer home to industrial democracy, but rather a re-purified techno-scientific authority in government: A re-purified authority sitting uncomfortably with the current European level concerns with the public accountability of science and the democratisation of expertise.' Sweden is turning too, but away from, not towards, intensified public accountability procedures, and in the longer run, it will be interesting to see how this change of paradigm in Swedish science policy will come to shape the practice of scientific citizenship in Sweden.

It is not entirely straightforward to connect the contextual, qualitative data on Sweden and the UK with the survey-based identification of different types of scientific citizenship. It does seem reasonable to say that the dominant 'involved' mode of scientific citizenship in the Swedish public appears to reflect a long tradition of a lively civil society, where the public is actively involved in political processes related to science and technology. Likewise, the British PUS movement, primarily expert-led and highly concerned with translation and passage of knowledge from science to passive, spectator citizens, might plausibly have enhanced a mode of scientific citizenship in which policy-oriented involvement is modest, but attention to the world of science generally is present among citizens. But trends are turning, in Sweden as well as in the UK, and it is very difficult to assess the extent to which the emerging British orientation towards active engagement and the recent Swedish return to sound science orthodoxy and traditional values of academic autonomy might be reflected in the 2005 survey-based measure of scientific citizenship. What kind of time lag would we expect with regard to the impact of civil society procedures, public policies and PUS initiatives on the observed practice of scientific citizenship?

In this chapter we have focused on measures of competence and participation, but it is worth stressing that other dimensions or elements would be equally interesting if we were to present a comprehensive picture of scientific citizenship. A thorough examination of the notion of techno-scientific rights, for example, would need to involve not only individual perceptions but also extensive analyses of regulatory frameworks at national and international levels, and cross-country comparisons or even benchmarking would need to score legal records as well as other institutional structures supporting such rights. Likewise, citizen participation, or performed scientific citizenship, is not independent of institutional arrangements and political culture, which provides the opportunity structure for public involvement. Systematic attempts to develop databases containing comparable information on formal and informal national procedures for public engagement, actors involved in governing science and technology, the use of science in policy-making and activities and trends in science communication would provide an important insight into such opportunity structures. The upcoming European project on 'Monitoring Policy and Research Activities on Science in Society in Europe' (MASIS) involves an attempt to build such a database for 38 European countries, and might offer valuable input for contextualising and supplementing survey-based results on scientific citizenship.

As a first step towards a more broad-ranging approach to exploring scientific citizenship in cross-national context, we have presented a combination of survey analyses and qualitative contextualisation of the results as an illustration of how historical trajectories might be used as a way of illuminating and invigorating raw statistics. We hope that the paper can contribute to bridging the divide between PUS and PES concerns by showing that these dimensions interact positively and could be thought of as integrated dimensions of scientific citizenship, and also to bringing quantitative and qualitative methodologies together more often in studies of the relation between science and society.

NOTES

1. This chapter is a moderated version of the paper 'Indicators of Scientific Citizenship' that we presented to the Royal Society workshop on international indicators of science and the public, 5–6 November 2007, and later published as 'Participation and Competence as Joint Components in a Cross-national Analysis of Scientific Citizenship' in *Public Understanding of Science*. We refer to the article (Mejlgaard and Stares, 2009) for an extended argument and particularly for technical details related to the latent class analyses.
2. In our assessments of model fit we rely mainly on two-way marginal residuals, drawing on Bartholomew et al. (2002), Bartholomew and Knott (1999) and Jöreskog and Moustaki (2001). For responses to each pair of items, we create a two-way marginal table, by collapsing over responses to the other variables. We then compare O, the observed frequency in a single cell of such a table, with E, the expected frequency for that same cell. The residual for each cell is calculated in standardised version, as $(O-E)^2/E$, where values greater than 4 are taken to indicate poor fit (Bartholomew et al., 2002). The

greater the number of large residuals, the worse the model is, and we take as our fit statistic for each model the percentage of standardised marginal residuals greater than 4, overall as well as conditional on country, for cross-national models. For fuller details of fit statistics for the models in this chapter, we refer the reader to Mejlgaard and Stares, 2009.
3. Overall, 4.4 per cent standardised marginal residuals are greater than 4. Conditional on country, the range is from 0.0 per cent (in Belgium, Spain, France, Luxembourg, the Netherlands, Lithuania, Malta, Poland and Slovenia) to 50.0 (in Greece), with a mean of 13.5.
4. Note that this is much higher than the 26 per cent of people who would be classed as highly participative according to the model, which includes petition as a criterion of high participation.
5. The thirteen items are the following: (1) The Sun goes around the Earth; (2) The centre of the Earth is very hot; (3) The oxygen we breathe comes from plants; (4) Radioactive milk can be made safe by boiling it; (5) Electrons are smaller than atoms; (6) The continents on which we live have been moving for millions of years and will continue to move in the future; (7) It is the mother's genes that decide whether the baby is a boy or a girl; (8) The earliest humans lived at the same time as the dinosaurs; (9) Antibiotics kill viruses as well as bacteria; (10) Lasers work by focusing sound waves; (11) All radioactivity is man-made; (12) Human beings, as we know them today, developed from earlier species of animals; (13) It takes one month for the Earth to go around the Sun.
6. Overall, 6.5 per cent standardised marginal residuals are greater than 4. Conditional on country, the range is from 0.0 per cent (in Belgium, Denmark, Croatia, Switzerland and Norway) to 52.4 (in Greece), with a mean of 13.4.

REFERENCES

Bauer, M. (2003). The vicissitudes of "Public Understanding of Science": From "literacy" to "science in society". In Fundacâo Callouste Gulbenkian, *Science Meets Society* Lisbon: Fundacâo Calouste Gulbenkian.

Castro, P. (2009). *STEPE Portuguese expert interviews—1st report*. Unpublished working paper from the fp7 project 'Sensitive Technologies and European Public Ethics'.

Felt, U. (ed.) (2003). *O.P.U.S. Optimising public understanding of science and technology—final report*.

Gibbons, M. (1999). Science's new social contract with society. In *Nature*, 402(6761): C81–C84.

Glimell, H. (2004). *Restored Scientific Authority and Orchestrated Deliberation: A Review of the Current Swedish S&T Policies*. STAGE discussion paper 27.

Goncalves, M. E. & Castro, P. (2003a). Policy–public interface in Portugal. In Felt, U. (ed.) *O.P.U.S. Optimising public understanding of science and technology—final report*.

Goncalves, M. E. & Castro, P. (2003b). PUS policies—introduction. In Felt, U. (ed.) *O.P.U.S. Optimising public understanding of science and technology—final report*.

Hagendijk, R. et al. (2005). *STAGE final report—science, technology and governance in Europe: challenges of public engagement*.

Hagendijk, R. & Kallerud, E. (2003). *Changing Conceptions and Practices of Governance in Science and Technology in Europe: A Framework for Analysis*. STAGE discussion paper 2.

Healey, P. (1999). Popularising science for the sake of the economy: The UK experience. In Miettinen, R. (ed.) *Biotechnology and public understanding of science—Proceedings of the UK-Nordic co-operative seminar, Helsinki October 25–27, 1998*. Helsinki: Academy of Finland.

Horst, M. (2007). Public expectations of gene therapy: Scientific futures and their performative effects on scientific citizenship. In *Science, Technology and Human Values*, 32(2): 150–171.

House of Lords Select Committee on Science and Technology (2000). *Science and Society*. London: The Stationery Office.

Irwin, A. (2001). Constructing the scientific citizen: Science and democracy in the biosciences. In *Public Understanding of Science*, 10(1): 1–18.

Irwin, A. (2006). The politics of talk: Coming to terms with the 'new' scientific governance. In *Social Studies of Science*, 36(2).

Lazarsfeld, P. F. & Henry, N. W. (1968). *Latent Structure Analysis*. Boston: Houghton Mifflin.

Mejlgaard, N. & Stares, S. (2009). Participation and competence as joint components in a cross-national analysis of scientific citizenship, *Public Understanding of Science*, prepublished (online first) June 26. 2009. doi: 10.1177/0963662509335456.

Miller, J. D. (1983). *The American People and Science Policy: The Role of Public Attitudes in the Policy Process*. NewYork: Pergamon Press.

Nolin, J. et al. (2003). PUS-policy: The Swedish context. In Felt, U. (ed.) *O.P.U.S. Optimising public understanding of science and technology—final report*.

Nunes, J. A. & Matias, M. (2004). *Science, Technology and Governance in Portugal*. STAGE discussion paper 22.

Royal Society of London (1985). *The Public Understanding of Science*. London: Royal Society.

Stein, J. A. (2003). Public understanding of science and the policy context in the United Kingdom. In Felt, U. (ed.) *O.P.U.S. Optimising public understanding of science and technology—final report*.

25 Benchmarking Climate Indicators for Science Communication and Public Engagement across Europe

Steve Miller

If one is trying to buy a new car or a new super-computer, one can run a series of standard tests to distinguish which is the best model to buy, which gives best value for money, and which should not be touched with the proverbial bargepole. This sort of inter-comparison goes under the general heading of "benchmarking," a term quite literally derived from carpentry, whereby carpenters measure one piece of wood against another on their workbench by means of a series of marks—bench-marks. One test for a car might be the time it takes to accelerate from a standing start to 100 kilometres per hour. For a computer one might have a series of programmes designed to get behind the manufacturers' claims of gigaflops per processor and find out the actual time it would take to run the job(s) that needed running—conceptually straightforward, if not necessarily straightforward in practice.

But how do you benchmark a country? How do you tell which of Europe's member states is the most efficient, the most productive, the most effective at turning scientific discovery into economic advantage and hard cash, the best at retaining its researchers in research jobs, particularly in academia? And how do you work out which country is best at keeping its citizens "on board" when it comes to the possibilities and pitfalls of new scientific discoveries and technologies?

That was the task facing the European Union at the turn of the century as it tried to face up competition from traditional rivals—the USA and Japan—and newly emerging powerhouses, such as China and India. At its summit in Lisbon in 2000, the European Union adopted a strategy goal of making Europe the most dynamic "knowledge-based" society in the world by the end of the decade. But first it needed to know just how far it had to go to make the rhetoric reality.

To answer that question, the European Commission came up with a series of Working Groups to "benchmark" performance in a number of areas. When the original criteria to be examined were set out in 2000, the chosen areas were "Human Resources in Research, Technology and Development (RTD)," "Public and Private Investment," "Competitiveness and Employment," and "Productivity." But the Portuguese Science Minister,

Jose Mariano Gago, took advantage of his country's presidency of the European Union to add "Promotion of RTD Culture and Public Understanding of Science (PUS)."

So it was that a group of six "experts" in the general area of science communication, RTD culture, and PUS was assembled in 2001, together with a number of advisors in specific areas, such as the role of women, to benchmark this rather nebulous subject. The results of the five working groups were published in summary form by the EC at the end of 2002, and the one looking at RTD culture and PUS [1] is available at: ftp://ftp.cordis.europa.eu/pub/era/docs/bench_pus_0702.pdf.

BENCHMARKING JELLY

For those groups trying to benchmark the more straightforwardly economic areas of EU performance, figures from the Commission itself and from the Organisation for Economic Cooperation and Development provided some reliable numbers. Benchmarking the "Promotion of RTD Culture and PUS" turned out to be akin to nailing jelly to a wall—very slippery. Inputs for the RTD culture and PUS benchmarking exercise came in the form of a new Eurobarometer survey of public knowledge of, and attitudes towards, matters scientific, as well as information that should have been supplied by a High Level Group of government representatives. Additionally the Working Group made use of its own sources of information, web-available materials, and some presentations from invited speakers.

During the course of the Working Group's activities, the European Commission issued a new set of Eurobarometer data that included questions about knowledge of, and attitudes towards, science, technology, and medicine [2], more or less standardised across the Members State. At the same time, and rather independently, the European Commission also drew up its own *Science and Society Action Plan* [3]. In some ways, this provided background for the benchmarking activity; in others, it pre-empted any recommendations the Working Group might come up with.

In a simple "input-output" model for economic performance, figures may be available for how much has been invested for what return. Again, in the area of RTD culture and PUS, this is far less straightforward. The Eurobarometer data could be seen as providing a snapshot of the "outputs" of activities to raise the general educational level, as far as science is concerned, as well as efforts to direct the attention of European citizens towards matters scientific, without making any assumptions as to whether those efforts produce positive or negative results. In the surveys, respondents give answers to the questions posed at the time they were posed. But citizens are not laboratory rats and they do not give answers to questions about science, technology, and medicine in a vacuum.

To a greater or lesser extent, those answers are influenced by government activities and media coverage, at long and short terms. Other actors, such as those identified by the Benchmarking Working Group, play their part too. These all form part of the "input" side of the equation, but their mutual interactions form a complex web and disentangling this so as to perform some kind of analysis necessarily leads to (over)simplifications. Moreover, there is no unique way of separating out the various input streams; although one scheme may appeal more than another, in the end "you pay your money and take your choice." In the end, the Working Group set up a "matrix of actors" to gauge what they might be doing in the relevant area—governments, the scientific community, the education system, science museums and centres, the mass media, and—last but not least—industry. This subdivision was designed to get the temperature for science communication activities aimed at European citizens, and to gain some insight into the climate for initiatives in the area.

The available data on inputs were very variable in extent and quality. Moreover, the Expert Group identified at least three broad approaches: the UK-driven PUS approach; the approach typified by Germany and some of the Nordic countries, which included the social sciences within "science" (Public Understanding of Science and the Humanities—PUSH); and the concept of "culture scientifique," less knowledge based and more about the inclusion of science in general culture, thought to be prevalent in France and southern Europe. These different approaches made it more difficult to define what activity was occurring that might come under the heading of "promoting" RTD culture and PUS.

As a result, it was not possible to carry out a strict, numerical benchmarking exercise. Instead, the Working Group had to make use of the much looser type of comparison that involved looking at some detailed case studies [4]. This (from a strictly comparative point of view) laxer methodology made a virtue out of necessity by concentrating attention on practices and initiatives in one or a limited group of countries that might prove useful more widely across Europe.

USING THE EUROBAROMETER DATA AS OUTPUTS

The results of the 2001 Eurobarometer were given to the Working Group as soon as they were available. As the task given to the Working Group was to benchmark at the country level, no attempt was made to subdivide the results further, and it resisted the temptation to try to use the data—particularly the figures on citizen knowledge of science—as performance indicators. Instead, the Eurobarometer was used to develop "climate indicators," designed to be sensitive to the public mood for science communication or scientific literacy initiatives. The indicators used were calculated relative to averages across the European Union's then 15 member states.

They included "knowledge," based on the number of correct answers to the 12 quiz questions that have become standard in such surveys, "interest" based on respondents own claims as to how interested or otherwise they were, "activity" in terms of visits to science museums and centres, and "esteem" for scientists as a profession.

In this and previous Eurobarometers, it had been noticeable that interest levels for non-science subjects, such as sports, correlate quite well with self-reported "informedness." This is not the case with science, technology, and medicine subjects. So a new "indicator" was proposed: the ratio between the level of knowledge (expressed in units of the EU average) and interest (again expressed in unit of the EU average)—K/I. This indicator was intended to reflect a potential gap between knowledge levels and interest, or a "surfeit" of knowledge compared with interest. Analysis of the 1992 Eurobarometer [5] by Durant et al. [6] had shown that some highly industrialised countries such as Germany had a tendency for their citizens to develop a post-industrial ennui as far as science was concerned. K/I > 1.0 might indicate that citizens of that country would express some measure of the ennui identified by Durant and coworkers; conversely, K/I levels *below* 1.0 might indicate that more

Table 25.1 Measure of Knowledge, Interest, Activity, and Esteem Derived from the 2001 Eurobarometer Survey (in units of the EU average)

Country	Knowledge	Interest	K/I*	Activity	Esteem
Austria	1.00	0.83	1.20	1.04	0.86
Belgium	0.92	0.93	0.99	0.86	1.07
Denmark	1.11	1.34	0.83	1.50	0.97
Finland	1.12	1.17	0.95	0.91	1.02
France	1.02	1.20	0.84	0.71	1.10
Germany	0.98	0.66	1.48	1.09	0.93
Greece	0.85	1.34	0.64	0.45	1.07
Ireland	0.84	0.70	1.20	0.36	0.75
Italy	1.05	1.13	0.93	0.77	0.99
Luxembourg	1.01	1.15	0.88	1.23	1.11
Netherlands	1.16	1.30	0.89	1.21	1.07
Portugal	0.81	0.84	0.96	0.79	0.93
Spain	0.91	0.94	0.97	1.00	1.01
Sweden	1.25	1.42	0.88	1.72	1.13
U.K.	1.01	1.04	0.97	1.40	1.01

* K/I is *not* normalised to the EU average for this quantity. The EU average for K/I is 0.97.

science communication or engagement activities really were required, in line with self-reported interest and informedness.

Policy-makers who took note of the K/I indicator prior to embarking on initiatives to raise interest in or knowledge of science, could deduce that the climate for doing this would be quite favourable in Denmark (K/I = 0.83) and France (K/I = 0.84), where interest outstripped knowledge, but much less so in Germany (K/I = 1.48), Austria (K/I = 1.20), and Ireland (K/I = 1.20), where the reverse was true. With knowledge levels around the EU average in Austria and Germany, this might not worry policy-makers too much, but Ireland's rather low knowledge level, compared with its European partners, might make this environment especially challenging.

Further analysis of the 1992 Eurobarometer enabled changes over time to be identified, as outlined in Table 25.2. Once more, a negative change in K/I indicated a climate for science communication becoming more "favourable" in terms of unfulfilled interest, whereas positive changes in this indicator might make policy-makers less inclined to try something new—or, possibly, even more desperate to halt the "slide" towards ennui. Although the absolute figures are probably not particularly significant, trends may be. Germany and Ireland once more showed up as countries in which the climate was not only difficult, but also becoming increasingly so. The 1992 figures had shown Germany above EU average for knowledge, but as the archetypal "disenchanted" country. Comparison with the 2001 data might be interpreted as indicating a country sliding towards disinterested mediocrity. Similarly, Ireland, already low on the 1992 knowledge scores, could be seen to be on a downward spiral of lack of interest leading to even lower knowledge levels.

Conversely, with levels of K/I decreasing (interest outstripping knowledge), the Netherlands could be interpreted as showing a pleasing increase in knowledge accompanied by rapidly increasing interest, perhaps the best climate for an initiative in public engagement with science. Denmark, France, and Greece, with small decreases in knowledge but large negative changes in K/I, might all show that initiatives were both needed and were likely to be positively received by their citizens. The K/I indicator could be considered as a *relative* "deficit/surplus of knowledge," not in terms of some *absolute* knowledge scale made up of items drawn from within the scientific community, but in terms of the self-professed desires of citizens. This, in turn, could be set in a wider context by additional parameters.

These could include, for example, the "activity" scale—do citizens do anything about their professed interest in science—or the esteem scale. If the former were used, then one might be inclined to believe Danish citizens really were as interested in science as they claimed, as they were well above average in terms of visiting museums and science centres. On the other hand, Greece's above average reported interest might have more to do with the perceived "worthiness" of science, or with a national sense of history, than real personal interest. Luxembourg's interest would be supported both

Table 25.2 Changes in Knowledge and Interest Levels for 11 EU Member States in Terms of European Averages*

Country	Knowledge 2001–1992	Interest 2001–1992	K/I 2001–1992
Belgium	-5%	-2%	-4%
Denmark	-1%	+32%	-28%
France	-7%	+14%	-19%
Germany	-12%	-29%	+23%
Greece	-3%	+25%	-16%
Ireland	-9%	-21%	+18%
Italy	+4%	+9%	-4%
Netherlands	+9%	+25%	-13%
Portugal	+2%	-5%	+7%
Spain	-6%	-6%	-
U.K.	-5%	-	-5%

* This table relates only to those 11 EU member states for which data from 1992 were available.

by high levels of activity and esteem. Ireland's low knowledge and interest correlates well with low activity and low esteem for scientists. In France, interest and esteem were quite well correlated but not so activity, despite the fact that Paris boasts one of Europe's big four science centres—la Cite des Science et de l'Industrie.

In all of this analysis the conditional has been used—might, could, etc. The aim here is not to assert that "this is how it is (was)" but to suggest how data from surveys could be used to work out a climate for science communication and public engagement in science, based on combining the responses different (types of) questions. Unfortunately for the European Commission, time did not allow for a more thorough and sophisticated analysis, or for testing just how robust indicators such as K/I might be as a measure of climate and trends. It seems to be a feature of surveys that more time, effort, and money goes into collecting data than analysing. And once the headlines have been cherry- picked, little attention goes into getting beneath the surface to tease out a more complicated and more "realistic" picture of what is happening.

GOVERNMENTS: WHO WAS DOING WHAT?

The table given below is a snapshot of where government-sponsored activity could be found, under what heading, across Europe. Note that at the time of the benchmarking activity—2001–2002—the EU had 15 member states. Later Eurobarometers, e.g. 224 *Europeans, Science and Technology*, provide updated information, including the 10 new member states. The table is an indication that activity was identified, but not the level at which it was being undertaken, which was very hard to judge.

Leadership was an indication that governments were at least making public statements about the importance of [improved] relationships between their citizens and "science"; *Policies and Activities* referred to identifiable policy documents and initiatives; and a tick under resourcing. *Resourcing* meant that the Working Group was able to find some evidence either that governments were spending money in the area, or that they were making funding available for others to spend.

The Commission's own *Science and Society Action Plan* drew attention to the perceived mismatch between the hopes and aspirations of European citizens, and what the science and research community was delivering. Some of the areas it highlighted included an emphasis on bringing more women into science, and more citizen participation. So the Working Group included promotion of RTD culture and PUS specifically to women in its enquiries, and activities that went beyond a strictly "deficit model"

Table 25.3 Government Activity in the Promotion of RTD Culture and PUS

Country	Leadership	Policies & Activities	Resourcing	Women	Internet	Participation & Dialogue
Austria	✓	✓	✓	✓	✓	✓
Belgium	✓	✓	✓		✓	
Denmark	✓	✓	✓	✓	✓	✓
Finland	✓	✓			✓	✓
France	✓	✓			✓	✓
Germany	✓	✓	✓			✓
Greece	✓	✓	✓			✓
Ireland	✓	✓	✓			✓
Italy	✓	✓	(✓)			
Luxembourg						
Netherlands	✓	✓	✓			✓
Portugal	✓	✓	✓	(✓)	✓	
Spain	✓	✓				(✓)
Sweden	✓	✓			✓	✓
U.K.	✓	✓	✓	✓	✓	✓

approach to include some notion of participation and two-way communication. Finally web-based initiatives were felt to be increasingly important, notwithstanding warnings not to be taken in by "cyberbole" [7]. Hence the headings: Women, Internet, and Participation and Dialogue. (The report itself and its appendices provided the evidence for the ticks.)

Most countries (Luxembourg made no response to repeated requests for information) provided evidence of government leadership and policies. One of the most striking findings was how quite difficult it was to find out any concrete information about what resources were being allocated to "promotion": governments simply did not know, or did not care to tell about, how much they were investing in science communication.

A key issue was that if the education system, or at least that part of it that had to do with science education, was included, then governments could argue that they were committing very large resources. Indeed, educational funding entirely swamped any other money specifically allocated to initiatives (e.g. the German PUSH project). The Working Group took the view, however, that normal in-school science education should **not** be included as part of a government's contribution, and that only initiatives outside of this should count—including after-school science clubs, for example. Instead, the education system was looked at separately.

So far, so good. But that still left a lot of scope for argument as to what did count. For example, one might include public health campaigns on AIDS and HIV, and in some countries it would be appropriate to include that. For the EU member states, however, health/social-behaviour awareness-raising was not included. Nor was, for example, expert advice to specific groups, such as farmers or fishing communities. Instead, more general initiatives, such as the use of consensus conferences to inform policy decisions and government support for science festivals was identified as "promotional" activity over and above what might be thought of as normal government responsibilities.

This discussion shows how difficult it is to even define what was supposed to be being benchmarked. Another question was to what extent it was "fair," in terms of producing an accurate picture of what was happening, not to make comparisons that failed to go below the country level. Traditionally, France has been a model of a centralised state, so the national government might well be the best choice for the actor in this category. But in neighbouring Germany the state governments—the länder—could be more important than the federal government for some activities, including science communication initiatives. It was difficult for the Working Group to capture these differences in national composition, given information aggregated by central governments.

One easily identifiable and quantifiable initiative came from Portugal. There the government had decided that 5% of the budget available for publicly funded research should be allocated to the *Ciençia Viva* initiative, which involved the setting up of several new science centres across the country, and a drive to get the country's research community involved in outreach projects, including opening up laboratories to young people and organising summer schools for all ages. So *Ciençia Viva* became a "model" against which other European countries, particularly those with little or no tradition in science communication activities, could measure their governmental response to this aspect of building the European Research Area and the *Science and Society Action Plan*. To our knowledge, however, this was not taken up within the *European* Community: it was a goal of the government of South Korea when it hosted the 2006 biennial conference of the Public Communication of Science and Technology network.

INVOLVEMENT OF THE SCIENTIFIC COMMUNITY

Since the mid-1980s when a series of reports and initiatives were produced, there has been an expectation that members of the scientific community—individually and through their institutions and professional bodies—should be engaged in media, outreach, and engagement activities aimed towards the general public and towards school students (both within and outside of the main curriculum). The European Commission's *Science and Society Action Plan* set out 38 actions that were required to make scientific research better aligned with "the needs and aspirations of Europe's citizens."

The Working Group found considerable levels of activity promoting RTD culture amongst the research communities of all of the EU's member states. For example, all countries, with the exception of Greece and Luxembourg, had their own science festivals or science weeks, or participated in the European Week of Science. One interesting trend was that those festivals that had been long established were run by scientific organisations, whereas those that had only been on the scene for a few years relied more on governmental agencies. The Working Group pointed out that the scientific community had its own interests that did not necessarily align with those of governments and that, although cooperation might be fruitful, care should be taken not to compromise the independence of scientists themselves.

It was also clear that little evaluation of the outreach and communication activities of scientists, such as science festivals, had been carried out. In the one clear example where it had, for the 2001 *Science@Austria* event, the results showed that citizens enjoyed meeting scientists face-to-face, and discussing issues that related to everyday life in a relaxed setting. For their part, the scientists involved in the event seemed unclear just who they were addressing, often equating "success" with whether or not they received any media coverage.

The Working Group also found that individual researchers at the time felt they had little community and institutional support. Typical complaints were that being involved in outreach and engagement activity carried little to no weight with promotion or remuneration committees, and that too little training for public communication—directly or through the media—was available.

THE EDUCATION SYSTEM

Science education inside the organised curriculum had been the subject of several international studies at the time of the EU's benchmarking activity, and many of the data generated by those exercises could properly be made use of. These studies showed that within Europe there was a trend away from purely formal, theoretical teaching, towards a more hands-on approach that emphasised the importance of lab work. There were clear differences in when "science" was introduced into children's studies. Most countries started with children at age 6 or 7, but Germany, Greece, and Ireland delayed this until pupils were 10 years old, and Sweden until age 12. At the same time, "after-school" science activities and visits to museums and science centres were becoming more popular, a trend that the Working Group commended on the basis of results from Holland and Germany.

One area where the Working Group could find real room for development was in the teaching of the more social and "public understanding" aspects of science. This would have to be included in teacher training if it were to make its way into schools: so too would training for teachers in how to work with science museums and centres.

SCIENCE MUSEUMS AND CENTRES

Since the end of the Enlightenment, museums have played an important role in *culture scientifique*, the francophone notion of science as part of wider culture. More recently, there has been a growth in interactive science centres, and some governments are even looking to the museum and science centre sector to provide forums for public engagement along the lines of holding "dialogue and debate" events. Figures provided for the Working Group by the European Network of Science Centres and Museums, known as ECSITE, showed that some 35 million Europeans visited these attractions every year. Of more than 160 individual science museums and centres in Europe, the "Big Four"—the Natural History Museum and the Science Museum in London, la Cite des Sciences et de l'Industrie in Paris, and the Deutsches Museum in Berlin—accounted for some 20% of visitor numbers and nearly 40% of the total budget across the sector. For this reason, support was urged for the medium and smaller centres, to enable them to play their role in national policies for increasing PUS, and cooperative projects were encouraged as a way of sharing the burden of developing exhibits and exhibitions.

ECSITE's own survey showed also that Eurobarometer data have to be set into a nuanced context. If "activity" were being measured by visiting science museums and centres, then those institutions had to be available. The UK had well above average levels of "activity," but not only does it have two of the "Big Four," it also had a major programme of building science centres as part of its Millennium Commission programme. At least some of these came on stream in time for the 2001 survey. At the time of writing, the UK has 77 ECSITE members. In contrast, Ireland, with much lower than average "activity" currently has just two, and—in 2002— would not have had more. So "activity" as measured by the Eurobarometer is in part a measure of "availability." That said, iSCAN—the Irish Science Centres Awareness Network—currently lists 65 "places" across the whole island (including Northern Ireland) associated with science and technology that accept visitors. Once more, this raises questions of "comparability" of data—what does, or does not, count as a "science museum or centre."

THE MEDIA

Given the importance that Europeans place(d) on the mass media—at that time, television in particular—for informing them about science, technology, and medicine, the Working Group felt that it had no option but to include them in the benchmarking exercise as important actors. Given the heterogeneity of the mass media—medium, ownership, audience, etc.—it also proved almost impossible to carry out any genuine country-by-country comparison. One issue that arose out of observations of the media at work, rather than any comparative study, was the lack of "local voices" in many science stories. This was becoming more of an issue as reliance increased on key journals, such as *Science* and *Nature*, for the weekly science news pages or broadcasts.

INDUSTRY

The final actor chosen by the Working Group was industry, an actor all too often forgotten in discussions about public understanding of science that tend to emphasise public policy-making and information campaigns. There, the boundary between PR and advertising, as against informing and being part of a genuine societal dialogue, was hard to draw. Risk communication was seen as one of the areas where industry had an important part to play, but trust was a key factor in limiting the role that private companies could play in public debates. Benchmarking multinationals in an exercise based on country-by-country comparisons proved next to impossible, however, and all that could be done at this level was to point to examples and case studies.

REFLECTIONS

Given that one could not even identify the inputs to the effort to promote RTD culture and PUS, the outcome that the EU had hoped for—that one could identify the "most" productive/efficient country and learn from, and generalise, their experience—was pretty much a non-starter. The Working Group decided not to use the Eurobarometer survey data as a measure of "success," even if ministers had wanted to: there were, and still are, many well-known and well-rehearsed arguments against using such surveys as performance indicators [8]. As this chapter has tried to show, however, that in no way makes the results of surveys such as Eurobarometer worthless. Used as "climate indicators" and as measures of "trends," this long-time series of European surveys are worth their weight in gold both to researchers and to those with the responsibility for drawing up science and society policies and programmes.

Another way of looking at the exercise was to abandon entirely the concept of benchmarking. Instead, what the Working Group was able to do was to gather a lot of information—albeit partial—to draw up a snapshot of the climate for RTD culture and PUS across the European Union. That might not deliver the "magic bullets" beloved of sound-bite policy-making. But it would give policy-makers an indication of the (comparative) environment in which they were working. And if that could be combined sensitively and non-judgmentally with survey data, such as that provided by the Eurobarometers, then some really useful climate indicators could be produced. So the Working Group was able to point to "good" practices that could be applied more generally—such as the *Ciençia Viva* initiative, or the Danish consensus conference approach to citizen involvement.

A final word of (probably obvious) warning: six people, plus a few support staff, working less than one day per week for nine months cannot do this effectively for 25-plus countries, particularly when government representatives are unwilling to provide the necessary inputs. Developing genuine climate indicators for science communication and citizen involvement takes time, effort, and—above all—commitment at the top.

REFERENCES

Barré, R. 2001. "Sense and nonsense of S&T productivity indicators", *Science and Public Policy* 28, 263–264.

Bauer, M. and Schoon, I. 1994. "Mapping variety in public understanding of science." *Public Understanding of Science* 3, 141–156.

Commission of the European Communities, 2002. *Report from the Expert Group benchmarking the promotion of RTD culture and Public Understanding of Science.* (Brussels, Belgium).

Commission of the European Communities, 2001. *Eurobarometer 55.2: Europeans, science and technology.* (Brussels, Belgium).

Commission of the European Communities, 2001. *Science and Society Action Plan*. (Brussels, Belgium).

Commission of the European Communities, 1992. *Basic Statistics of the Community*. (Eurostats, Brussels, Belgium).

Durant, J., Bauer, M., Gaskell, G., Midden, C., Liakopoulos, M. and Scholten, L. 2000. "Two cultures of public understanding of science and technology in Europe", in *Between Understanding and Trust: The public science and technology*. (Eds. M. Dierkes and C. von Grote, Harwood Academic Publishers, Amsterdam), 131–156.

Woolgar, S. (Ed), 2002. *Virtual Society?—technology, cyberbole, reality*. (Oxford University Press).

26 Monitoring Science in the Public Sphere
The Case of Italy

Massimiano Bucchi and Federico Neresini

The debate on the science and society relationship in Italy has been quite intense during the past two decades. However, this debate has for a long time pre-eminently nourished itself with stereotypes: stereotypes of science and scientists among the public, stereotypes of the public among scientists and policy makers, and widespread stereotypes of the media.

Until the early 2000s, substantial and comprehensive data and studies on public perception and media coverage of science were significantly scarce and had modest impact on the wider societal debate. It was in this context that the non-profit research centre Observa Science in Society started its activities in 2001. The main mission was to offer to key audiences—scientists, policy makers, journalists—, but also to citizens at large, research tools that could feed a more constructive and informed science/society debate. The choice of establishing a non-profit entity was linked with two key factors that were considered necessary in order to fulfil such a goal. The first was independence. Studies were going to touch several issues—like nanotechnology or biotechnology—in which many existing research organizations (including public and academic bodies) had legitimate interests and investments running. Independence was to be preserved in order to achieve credibility among a broad range of actors.

The second element is more specific to the Italian context. Because research was intended to make an impact on current debates, it had to be closer to practice and ongoing debates than is usually possible through academic research, without of course losing quality and scientific rigour. With this purpose, an international scientific committee was established that at present comprise scholars from different fields (life sciences, social sciences, humanities). The first research programme launched was the Science in Society Monitor, a regular monitoring of public perception of science and technology in Italian public opinion. In 2005, the publication of *Annuario Scienza e Società* (*Science in Society Facts and Figures*) was started with a view to providing key data on this topic in an accessible and compact form, based on information from recognised national and international sources. This publication is now in its sixth edition, and is published by the leading Italian social science publisher, Il Mulino. Observa website (www.observa.it, www.science-insociety.eu), offering full access to reports and data, has become a reference point for several audiences, with approximately 2,000 subscribing to the monthly newsletter in Italian and English and 3,000 visiting the website on a monthly basis. Active collaborations have resulted with some of the key European and international networks, including Macospol (Mapping Controversies in Science for Politics), Rose (Relevance of Science Education), Esconet (European Science Communication Training Network).

THE SCIENCE IN SOCIETY MONITOR

The Science in Society Monitor was launched in 2002 with the aim of providing a regular and updated source of information on public perception and attitudes towards science and technology. Surveys are conducted through CATI (Computer Assisted Telephone Interviews) approximately twice a year on a representative sample of around 1,000 citizens aged over 15, stratified by gender, age, and area of residence. Many items have been designed in order to make comparison possible with other key national and international studies in this area (e.g. Eurobarometer). Questionnaire sections in areas such as scientific literacy, media exposure to science and technology content, priorities in research investments, trust in and credibility of science, scientists, and research institutions are run on a regular basis. Thematic sections are run in connection with specific issues that acquire particular salience in the policy agenda or public debate.

Here is a selective list of issues covered and years of data collection:

- Aging (2005)
- Assisted reproduction (2002, 2005, 2006, 2009)
- 'End of life' decisions (2005, 2007, 2009)
- Climate change (2007, 2009)
- DNA tests and archives (2006, 2010)
- Embryo stem cell research (2005, 2006, 2009)
- Evolution theory (2006, 2009)
- GMOs (2003, 2004, 2005, 2008, 2010)
- Homeopathy (2006, 2009)
- Intellectual property and access to knowledge products (2007, 2009)
- IT and digital media technologies (2010)
- Nobel science laureates (2005)
- Nuclear energy (2003, 2005, 2007, 2009)

- Priorities of research (2005, 2007, 2009)
- Problems of Italian research/brain drain (2006)
- Renewable energy sources (2005, 2007)
- Risk perceptions (2003, 2004, 2005, 2010)
- Science festivals (2007)
- Urban pollution and traffic (2005, 2007)
- Waste disposal technology (2008)
- Young women and science (2006, 2008)
- Youth and science university curricula (15–18 year-olds sample, 2006)
- Youth and technology (2006)

TRENDS IN ITALIAN PUBLIC ATTITUDES: THE CASE OF NUCLEAR POWER

Nuclear power stands out as one of the science and technology issues that have recorded a significant change in public attitudes across the years in Italy.

In 1987, one year after the Chernobyl disaster, a referendum vote marked the substantial demise of nuclear power in Italy. Both research into the development of nuclear power for civil use and the building of new plants were halted, and the few that were in existence were decommissioned. Subsequently, opposition to nuclear power remained high for more than a decade.

More recently, however, Italian attitudes have shifted more and more in favour of investing in nuclear power. From 2003 to 2009, those who think investments in this area should be pursued have risen from 22% to almost 52%; in the same period, those against have decreased from 56% to 39%. Thus, for the first time, after more than twenty years since the referendum marking the Italian referendum on nuclear power, positive attitudes outnumber negative attitudes. One Italian in five remains uncertain, but the percentage of those with no opinions has slightly decreased during the past two years—probably shifting towards favourable attitudes, as the percentage of those against has remained stable during the same period. It should be noted that nuclear power moderately increases in relevance also in terms of citizen priorities for research investments: research on nuclear energy now ranks third in this list of priorities, together with biotechnology and after research on climate change and renewable energy sources.

What are the reasons for this change? On one hand, the current economical-political scenarios seem to play an evident role: the need to reduce Italy's dependence on countries controlling oil sources is the first motivation of those favouring nuclear investments. The risk of exhausting current energy sources also appears relevant to interviewees. During the past couple of years, however, the international dimension also seems to gather the attention of the public: the fact that other countries in Europe are investing in the production of nuclear energy has become decisive for 28% among those in favour (it was 19% in 2007). It is also possible that the recent strategies of transnational collaboration involving Italian business as well as political actors in this field have had an impact on this perception. On other policy issues, too, it is a recurrent element that Italian public opinion is more sensitive to themes and policy initiatives being shaped in a European and international context, compared to those emerging in a national context. Among those against nuclear power investments, the weight of those considering more appropriate investing in renewable sources is now overwhelming (from 45% to 56% in two years), and concerns on the plants' localisation now seem less relevant (from 17% to 7%).

The proportion of critical attitudes is highest among the youngest (reaching 53% among those 15–19 years old) and among those living in northwest Italy (46%). However, the most interesting result in this respect seems to be the relationship between these opinions and scientific literacy. As long as scientific literacy grows, the propensity to invest in nuclear energy also grows, but only up to medium-high levels. Among those scoring the highest levels in terms of literacy (and partially also among those with the highest educational level), critical attitudes towards nuclear power prevail again.

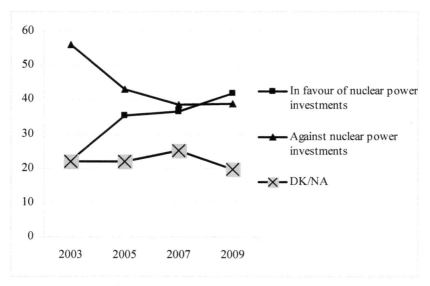

Figure 26.1 Attitudes to investments in nuclear energy—Science in Society Monitor 2003–2009. (%, 2003 n=1000; 2005 n=1021; 2007 n=998; 2009=1020)

SCIENCE IN THE ITALIAN DAILY PRESS

Most studies of the relationship between science and the media consider relatively short time periods or focus on relevant yet specific issues or domains.[1] The Science in the Media Monitor was developed by Observa Science in Society with the aim of monitoring, on a comprehensive and regular basis, the presence of articles in the Italian media that deal with science- and technology-related issues. The method can be described as an automatic system that gathers, analyses, and files articles published by the main Italian newspapers, and which, for the time being, is limited to articles published on their websites.[2]

The decision to focus, at least initially, on the web version of the daily newspapers was taken on the basis of two particular considerations. First, the web is an increasingly popular source of news for large sections of the public. Second, it is quite plausible that there is very little difference between digital and paper editions when we consider media coverage of science and technology. In addition, articles published on the web are relatively easy to trace and are produced in a format that makes them more readily available for (partly automated) analysis.

One significant feature of the Science in the Media Monitor concerns the way in which it has tried to answer a crucial question. Analysing media coverage of scientific matters means deciding whether to consider only articles devoted explicitly to science and technology or also those that are in some way related.

In this case, it was decided that it is undoubtedly more appropriate to include all articles that are judged to be relevant, regardless of the section of the newspaper into which they are inserted, because many articles in which science and technology occupy a role of prime importance are published in the News, Business, Politics, or Sport sections.

This, however, leads to the problem of how to define criteria that are sufficiently clear for the purpose of distinguishing relevant articles from irrelevant articles.

The solution adopted by the Science in the Media Monitor is to use a software which constantly gathers articles as they are put on the websites by the four daily newspapers under examination and analyses their contents to select those that are relevant to science and technology.

Selection is based on a wide thesaurus of approximately 200 keywords, each weighted according to its relevance for discriminating scientific and technological contents in opposition to others. The system assigns a score of 'science and technology relevance' to each published article. Articles above a certain threshold are selected as relevant; articles below a certain score are discarded; articles falling below the threshold but within a certain range are stored to be manually inspected by researchers.

The Science in the Media Monitor (SMM) then uses this information to produce a series of indices tracing the presence of science and technology in the media (i.e. coverage of science and technology, ratio of science and technology articles on total published articles and most relevant themes; each index is available for defined periods of time and for each newspaper). From the general database, which constantly grows through the SMM systematic daily monitoring of online newspapers, it is also possible to extract subsets of articles according to ad hoc criteria. Thus, more specific analysis can be carried out on emerging issues, e.g. nuclear energy, nanotechnoogy, stem cells, climate change.

SCIENCE AND TECHNOLOGY IN ONLINE DAILY NEWSPAPERS

During 2008, a total of 46,896 articles were published on the websites of *Il Corriere della Sera, La Repubblica, Il Sole 24 Ore,* and *La Stampa*: 5,221 relevant articles were selected, equivalent to 11% of the total number of articles published. It is difficult to establish whether this percentage is high or low, as it is still not possible to make comparisons with longer time periods or other national contexts. However, it seems relevant to notice that one article in ten was found to deal significantly with science.

Observing the presence of relevant articles by month, we note that both overall results and those relative to each newspaper are quite similar. The tendency towards a general increase during the last months of the year, preceded by a slight peak between the months of May and July, is particularly evident. These results are in part determined by the fact that the four newspapers redesigned their sites towards the end of 2008 and started adding ever more articles, some of which were on the subject of science and technology. It is, however, necessary to point out that certain unique events such as the Englaro case, the beginning of the LHC experiment at CERN, and the awarding of the Nobel Prizes significantly influenced article distribution.[3]

This is even more obvious if we examine the ratio of relevant articles to the total number of articles published, i.e. the salience of themes connected to science and technology among the entire range of subjects dealt with by the media.

Table 26.1 Distribution of Relevant Articles and the Index of Salience by Newspaper (2008)

	Relevant articles	*Total articles*	*Index of salience (% of relevant articles from total)*
Il Sole 24 Ore	1713	12636	13.5
Il Corriere della Sera	1536	17412	8.8
La Repubblica	1115	10231	10.9
La Stampa	857	6617	13.0
Total	5221	46896	11.1

When one considers the average values of the index of salience, one observes a rather discontinuous pattern over the course of the year, fluctuating between 10% and 15%, with a clear reduction in the number of relevant articles in the last few weeks of the year.

There are obviously differences between the newspapers. Although *Il Sole 24 Ore* and *La Repubblica* share very similar results in the various indices, values for the former are consistently above those for the latter. *Il Corriere della Sera* shows a more linear pattern that is generally intermiediate between the two aforementioned newspapers, with the exception of the middle part of the year.

Science and Technology Everywhere?

The publication of articles in online daily newspapers is signalled by means of RSS, which covers various sections including for example home pages, environment, politics, news, and so on. Collected data show that, among these macro categories, more than half of the articles (67.1%) are to be found at some point in the home page.[4]

This is extremely interesting, both because it unequivocally underlines the importance that the media attributes to science and technology, and because it justifies our decision not to limit analysis exclusively to specific science and technology sections within the newspapers.

As it is still necessary to establish whether we are dealing with a peculiar characteristic of online information, which can only be clarified by comparison with paper versions, it is worth analysing the distribution of the articles excluding the home page section of the website. This adjustment reveals that most of the relevant articles (47.6%) come from the Science and Technology section, 12.4% from the News section, and 10.1% from the Culture section.[5]

Table 26.2 Distribution of Relevant Articles among Sections of Online Daily Newspapers Excluding the Home Page (values expressed in %)

SECTION	
Science and Technology	47.6
Other Sections	16.9
News	12.4
Culture	10.1
Entertainment	5.3
Business	4.8
Sport	2.0
Politics	0.9
TOTAL	100.0

Even when the homepage is excluded from the analysis, there is a considerable distribution of content devoted to science and technology among the other sections.

Finally, comparisons of the distribution of relevant articles among the various sections demonstrates the prevalence of articles in sections specifically devoted to science and technology, although there is a consistently notable presence of articles with a high level of content devoted to science and technology in the home pages as well as in other sections.

NUCLEAR POWER IN THE ITALIAN DAILY PRESS

The problem of the social effects of the media has been the main stream of a research tradition going back hundreds of years, which has been extensively studied and widely debated by scholars studying mass communication. Although this research field may have not provided definitive answers, some elements have nonetheless been clarified (McQuail, 1987; Bruhn Jensen, 2002). It is, for example, generally agreed among researchers that the media tend mainly to reinforce pre-existent attitudes and opinions.

However, empirical research into the effects of the media with regard to science issues is still relatively limited, the studies by Mazur at the beginning of the 80s being one of the most notable exceptions.

The objective of these studies was to analyse whether, and to what extent, media coverage of certain controversial issues related to science and technology may be responsible for effects in terms of public attitudes. Its initial hypothesis can be summarised as follows: "when media coverage of a controversy increases, public opposition to the technology in question (...) increases; when media coverage wanes, public opposition falls off" (Mazur 1981, p.109).

In the context of this hypothesis, Mazur does not forget to consider an aspect of crucial importance, i.e. the difference between how much attention the media devotes to a particular matter and how they actually deal with it. Consequently, it is not sufficient merely to dedicate a great amount of space to an issue in order to influence public opinion, it is also necessary to comment on it, either favourably or unfavourably, in order to produce the same effect among the public.

As is well known, if the media deal extensively with an issue over a long period of time, they are likely to be able to focus our attention on such an issue and thereby modify the order of importance on the agenda of issues that are discussed in our daily and institutional lives (McCombs and Shaw, 1972).

However, is the fact that there are many arguments for or against a subject of research sufficient to affect the possibility of those arguments significantly orienting public opinion? If most articles and television reports tend to underline the risks of a technology, will public opinion display attitudes which are in line with such a tendency?

Empirical evidence on this aspect offers contradictory indications.

In 1979—the year of the accident at the Three Mile nuclear power station and the bankruptcy of Skylab—media coverage was certainly not favourable to science and technology, yet, despite some fluctuations, public attitudes towards science remained generally positive. In the same way, around the middle of the 1980s, the media widely reported the tragedy of Bhopal, the explosion of the Challenger shuttle, and the Chernobyl accident, with several negative comments about science and technology. However, even in this case, the public overcame an initial shock and eventually general opinion was in favour of further advances in technology, with the exception of nuclear power (Nelkin 1987, p.72).

According to Mazur, however, "people are affected more by the quantity of coverage, especially the repetition of simple images (. . .), than by its substantive content" (Mazur 1998, p.459). Furthermore, given that the public seems to favour a "conservative" structural preconception, meaning that in situations of uncertainty, familiarity is preferable to the unknown (Mazur 1981, p.114), the original hypothesis of Mazur is reinforced: the more the media cover a controversial technical or scientific issue, the more likely the general public is to assume a critical attitude towards it.

The debate on nuclear energy is a case study of particular interest, not just because it has been the subject of many studies, but also because extensive data collected by both the Science and Society Monitor and the Science in the Media Monitor allow further analysis.[6] Despite the fact that we are dealing exclusively with data about Italy, the case appears as a rare opportunity to understand more about the relationship between public opinion and the media when it comes to scientific and technical issues. Furthermore, some aspects are of particular interest in the specific case of Italy.

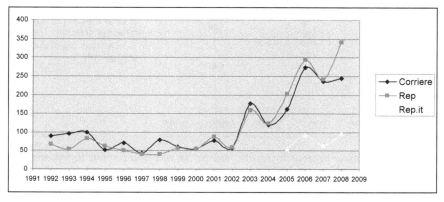

Figure 26.2 Patterns of coverage of the issue of nuclear power in *Il Corriere della Sera* and *La Repubblica* (no. of articles per year, 1992–2008).

As has been detailed above, Italian citizens' attitudes to nuclear energy have significantly changed during the last decade. How much of the responsibility for this change is to be attributed to the media? Coverage of nuclear energy related issues by the two most widely circulated dailies (*Il Corriere della Sera* and *La Repubblica*) has, despite a few fluctuations, tended to increase in the long term[7]. Furthermore, patterns for the two newspapers have been rather similar. In addition, patterns displayed by the digital and paper editions of *La Repubblica* were also rather similar. For this reason we consider it appropriate to concentrate on this newspaper, developing further analysis, including comparative analysis, using data from the Science in the Media Monitor.

Data for the most recent period (2006–2008) indicate that the web version of *La Repubblica* generally deals with the issue of nuclear power in a constant manner, though in 2007 a blip was recorded. If, however, we restrict analysis to articles that deal specifically with the issue of nuclear power for home energy needs, therefore excluding articles referring to both military use (as in the cases of international events regarding Iran and Pakistan) and for the study of the physical properties of particles, we find that there is an upward trend, rising from 19 articles in 2006 to 33 in 2008 (Figure 26.3).

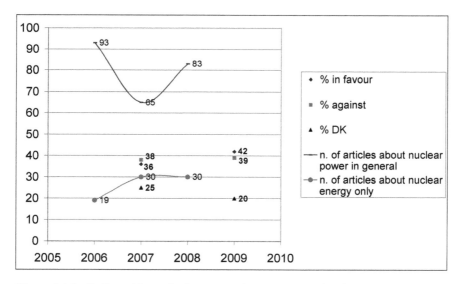

Figure 26.3 Italian citizens in favour, against or uncertain about investments in nuclear power (2007* and 2009*) and patterns of media coverage. (Articles published in *La Repubblica* online in 2006**, 2007**, and 2008***.)

* *source*: Observa, Science and Society Monitor
** *source*: Repubblica.it archive
*** *source*: Observa, Science in the Media Monitor

According to Mazur's hypothesis, greater media coverage of a scientific and technical matter should have fostered public concern. However, public opinion in Italy seems to have followed a somewhat opposite trend.

On the other hand, available data are by no means sufficient to claim that the media do not influence patterns of public opinion. It is, in fact, possible to hypothesise that the way in which the media deal with nuclear power has changed and that this may have in turn influenced public opinion.

If, however, we analyse the content of the articles in the period under examination, we find rather consistent and balanced views of the issue; actually, negative tones were slightly more prevalent than positive attitudes in 2008 coverage. It is also worth pointing out that coverage of nuclear accidents does not disappear from the Italian media: 2006 was the 20th anniversary of the Chernobyl disaster and the media were keen to recall it with articles and dedicated pages; the following year, an earthquake caused another major accident at the Kashiwazaki-Kariwa plant in Japan, an event which received a great deal of coverage; in 2008 there was a technical fault at the Krsko plant in Slovenia, just across the border from Italy, causing a radioactive leak, which also received a certain amount of media coverage. In addition, many of the recent articles which deal with nuclear energy do so in relation to the Iranian decision to develop a nuclear weapons capability, a framework which clearly favours the association between nuclear power and danger. Overall, the italian media is characterised by its relatively balanced coverage of nuclear power. If anything, negative opinions are slightly prevalent given the repeated nature of accidents at nuclear power plants and constant references to the nuclear threat in military affairs. However, despite this, public opinion seems to see nuclear power in increasingly positive terms.

It is therefore necessary to bring other factors into play that, along with the media, may have contributed to shape this shift in public opinion.[9]

Some hypotheses have already been sketched above, particularly in terms of increased public sensitivity to the broader energy issues. More specifically, it should be remarked that the price of oil rose continuously from 2006 onwards, reaching a peak of $147 a barrel in July 2008. This undoubtedly influenced the way in which the media dealt with nuclear power, which is often presented as a possible alternative energy source to oil: but this increase is not perceived by the public solely because they read about it in the press; most people have direct experience of higher prices when buying fuel for their vehicles and paying their heating bills.

Besides, it is important to remember that it was in the very period under examination that a notable fracture opened up within the environmental movement in Italy, of which one part began to see nuclear power as a valid energy alternative for dealing with global warming caused by the excessive use of fossil fuels. This inversion is due to several factors, and the media seem to have accepted and represented it rather than promoting it. For example, in 2006 *La Repubblica* hardly ever mentioned nuclear power in articles dealing with climate change, but in subsequent years nuclear power began to appear as a possible "eco-compatible" energy source on a quasi-regular basis.

Finally, another way to reinterpret Mazur's hypothesis in the face of these data would be to limit it to the initial emergence of a science/technology issue: i.e., the relationship between amount of coverage and mounting negative attitudes would hold for the initial stages of appearance of an issue on the public agenda. Because nuclear energy has now been—albeit with significant cycles—on the agenda for several decades, variations in media coverage might not necessarily affect public concerns but rather reflect and reinforce, for example, the cycle itself of public interest and attention.

More generally, the significant shift in attitudes towards investments in nuclear energy in Italy during the last five years does not appear to be linked to specific events or media coverage, nor to a specific reappraisal of risks related to advancement in nuclear technology. Rather, attitudes towards nuclear power need to be understood within wider trends in terms of public sensitivities for energy issues and their global scenarios, which now seem to provide a different, more favourable, framework for public perception.

CHALLENGES FOR THE FUTURE

Both the Science in Society Monitor and the Science in the Media Monitor face significant challenges in terms of future activities and further development.

In particular, the Science in Society Monitor should aim at regular data collection on core general themes (scientific literacy, access to information, attitudes to science) as well as on more specific issues, strengthening international collaboration and comparisons through key networks in which Observa and the Science in Society Monitor researchers are already active members.

Being comparatively a more recent endeavour, the Science in the Media Monitor still needs to consolidate, validate, and fine-tune its methodological basis and operational tools, particularly in the face of continuous technical changes and updates in its key sources (online newspapers). Extension to other web sources (blogs, search databases, twitter) is also being sought. Contacts and discussions are also under way in view of international collaborations and comparative research in monitoring media coverage of science,[8] as well as in view of exploring the potential for adapting the system to other national and media contexts. Content analysis with regard to specific issues is being developed, using methods and tools from statistical text analysis.

Both activities share common organisational challenges, e.g. in terms of ensuring regular and continuous funding and appropriate staffing, feeding equally both into scholarly discussions and policy/public debates.

Finally, efforts should be made to better integrate the two instruments; together, they offer a significant opportunity for comprehensive study and understanding of the dynamics of public discourse and perception about science and technology issues, as well as for putting to empirical test several hypotheses and theoretical discussions within the fields of public understanding of science and science and technology in society.

NOTES

1. Notable exceptions include Bauer et al. (1995) and for Italy Bucchi and Mazzolini (2003).
2. The Science in the Media Monitor was inspired by the SAPO project (Science Automated Press Observation), developed by Labjor at the University of Campinas (www.labjor.unicamp.br). The software for the Science in the Media Monitor was developed by Observa Science in Society (www.observa.eu) in collaboration with Weird Studio (www.weirdstudio.com).
3. Eluana Englaro was a young woman who entered a comatose state in 1992, following a car accident. Since 1996 her father had been advocating suspension of feeding and hydration therapies. After several court decisions and an intense public debate that became particularly heated in late 2008, feeding and hydration therapies were suspended and Eluana was declared officially dead on 9 February 2009.
4. Despite the fact that each daily makes its own decisions about where to place its articles, it is still possible to group the various sections into homogeneous macro categories.
5. The "other" section includes articles that come from sections that deal with diverse subjects, e.g. governance and security issues.
6. About the public debate on nuclear power see, for example, Gamson and Modigliani (1989) and Wiegman et al. (1995).
7. Data were collected by consulting the archives of the two dailies using "nuclear power" as the search term.
8. See for example the classical study on the UK daily press by Bauer et al. (1995) or indicators like the PEW news index developed for the US by the PEW research centre for people and the press http://people-press.org/
9. In a U-turn national referendum of mid June 2011, of 57% Italians, 94% voted against the restart of a nuclear energy programme in Italy. This was a vote in reaction to the nuclear accident in Fukushima (Japan) of March 2011 and a vote against the current government of PM Berlusconi.

REFERENCES

Bauer M. et al., 1995, *Science and Technology in the British Press, 1946–1990*, London, The Science Museum.
Bruhn Jensen K. (ed), 2002, *A Handbook of Media and Communication Research*, Routledge, London.
Bucchi M., Mazzolini R., "Big Science, Little News: Science Coverage in the Italian Daily Press, 1946–1997", *Public Understanding of Science*, 12, 2003, pp.7–24.
Gamson W. A., Modigliani A., 1989, "Media Discourse and Public Opinion on Nuclear Power: A Constructionist Approach", *The American Journal of Sociology*, 95, 1, pp. 1–37.

Mazur A., 1981, "Media Coverage and Public Opinion on Scientific Controversies", *Journal of Communication*, 31, 2, pp.106–115.

Mazur A., 1998, "Global Environmental Change in the News—1987–90 vs 1992–6", *International Sociology*, 13, 4, pp.457–472.

McCombs M.E., Shaw D.L., 1972, "The Agenda-Setting Function of the Press", *Public Opinion Quarterly*, 36, pp.176–187.

McQuail D., 1987, *Mass Communication Theory*, Sage, London.

Nelkin D., 1987, *Selling Science*, Freeman, New York.

Wiegman O., Gutteling J. M., Cadet B., 1995, "Perception of Nuclear Energy and Coal in France and the Netherlands", *Risk Analysis*, 15, 4, pp. 513–521.

Contributors

ALLUM, Nick, Essex University, UK

BAUER, Martin W., London School of Economics, UK

BOY, Daniel, Science Po, Paris, France

BUCCHI, Massimiano, University of Trento, Italy

CASTELFRANCHI, Yurij, University of Bela Horizonte, Brazil

CRETTAZ-von, Roten Fabienne, University of Lausanne, Switzerland

EVANGELISTA, Rafael , University of Campinas, Brazil

GODIN, Benoit, UQAM, Montreal, Canada

GOUVEIA, Flavia, University of Campinas, Brazil

HE, Wei, CRISP, Peijing, China

KEETER, Scott , PEW foundation, Washington, USA

KIM, Hak-Soo, Sogan University, Seoul, South Korea

LIU, Xuan, CRISP Beijing, China

LOSH, Susan, Florida State University, USA

MASCI, David Masci, PEW Foundation, Washington, USA

MATSUURA, Takaya, University of Hiroshima, Japan

MEJLGAARD, Niels, Aarhus University, Denmark

MILLER, Jon D., North Western University, Chicago

MILLER, Steve, University College London, UK

MORALES, Ana Paula, University of Campinas, Brazil

NERESINI, Federico, University of Padua, Italy

PARDO, Rafael, Fundacion BBVA, Madrid, Spain

PETKOVA, Kristina, Bulgarian Academy of Sciences, Sofia

POLINO, Carmelo, REDES-RICYT, Buenos Aires, Argentina

RAI, Anil, Indian Agricultural Statistics Research Institute, New Delhi, India

RAZA, Gauhar, NISTADS, Delhi, India

REN, Fujun , CRISP, Peijing, China

RIGHETTI, Sabine , University of Campinas, Brazil

SCHIELE, Bernard, UQAM, Montreal, Canada

SCHREINER, Camilla, University of Oslo, Norway

SHIMIZU, Kinya, University of Hiroshima, Japan

SHUKLA, Rajesh, NCAER Delhi, India

SINGH, Surgit, NISTADS, Delhi, India

SJØBERG, Svein, University of Oslo, Norway

SMITH, Gregory Smith, PEW foundation, Washington, USA

STARES, Sally, London School of Economics, UK

STONEMAN, Paul, Essex University, UK

TANG, Shukun, Hefei Technical University, China

TODOROV, Valery, Bulgarian Academy of Sciences, Sofia

VOGT, Carlos , University of Campinas, Brazil

WANG, Ke, CAST Peijing, China

ZHANG, Chao, CRISP, Peijing, China

Index

Note: Page numbers ending in "f" refer to figures. Page numbers ending in "t" refer to tables.

A

adolescents, 200–213; ACE question, 202–203, 203f, 209–210, 210f; attitudes toward science, 200–211; cluster analysis, 203–204, 203f; education similarities, 202–204, 203f; HDI scale, 5, 203f, 204–212, 206f, 207f, 208f, 209f, 210f, 210t, 212n5; ROSE, 5, 200–205, 209–210; school curriculum, 200; science career interest, 206–211, 208f, 209f; science/technology interest, 204–211, 206f, 207f, 208f; scientist interest, 206–207, 208f
Adorno, Theodor, 304–305
Adult Civic Science Literacy, 55–71. *See also* Civic Science Literacy
Allum, Nick, 1, 6, 301
America, 55–75; Civic Science Literacy (CSL), 55–71; cohort factors, 57–58, 61–72; multivariate analyses, 66–68, 67t; pseudoscience beliefs, 55–60, 60t, 64f, 65–68, 65t, 67t; research methods, 58–61; research results, 61–71; science education in, 56–58; science knowledge in, 55–57; science literacy in, 55–71. *See also* United States
animal cloning, 7, 325–326, 329–332, 329t, 330t
Annuario Scienza e Società, 449
ANOVA models, 62, 81, 89t, 90t, 121, 144t, 150f, 264, 267
Anthony, R. N., 26
astrology: astrological predictions, 301–302, 304; authoritarianism and, 304–306, 313, 320; beliefs about, 301–321, 312t, 314t, 315f; data and measures, 306–307; horoscopes, 301–312, 308f, 310f, 320–321; hypotheses, 302–305; religious beliefs and, 347; science and, 301–308, 308f; understanding of, 303–305; variation in beliefs, 311–315, 312t, 314t, 315f
"authoritarian personality," 304–305, 313
authoritarianism, 304–306, 313, 320

B

Bauer, Martin W., 1, 4, 40, 92, 139, 179
benchmarking, 436–447
biological literacy, 355–359, 360t, 363, 364t–365t, 368
biomedical research, 217–218, 362, 367
biotechnology: human-animal boundaries, 323–324, 332; perceptions of, 332, 353, 355, 359, 366; scientific knowledge and, 242, 243t–244t, 248–259, 249f, 251t, 253f, 254t, 255f
Bishop, George F., 355
Boy, Daniel, 3, 5, 39
Broad, William J., 348
Brooks, Harvey, 39
Bucchi, Massimiano, 9, 449
Bulgaria, 76–90; ANOVA models, 81, 89t, 90t; attitude variables, 79; cohort variables, 79–81, 85–86, 88–89, 89t, 90t; "country" variables, 81, 89t, 90t; daily science relevance in, 83f, 86; distrust of scientists, 84f, 86; faith versus

466 Index

science, 84f, 86; generational attitudes in, 78–90; historical background of, 77–78; image of science, 76–90; interest in science, 82f, 85; science benefits, 83f, 85; science literacy, 79, 82–85, 82f, 83f, 84f, 85f; scientific research support, 85f, 86

C

campaign guidance system, 9–10
Castelfranchi, Yurij, 4, 5, 158, 400
Chernobyl disaster, 77, 205, 451, 457, 459
China, 126–135, 139–156; access to science information, 131–132; attitudes toward science, 133; cluster analysis solution, 146–149, 146t, 148t; Confucianism and, 140; Europe and, 139–156; national surveys in, 126–129, 127t, 134; questionnaire changes, 127–128; questionnaire wording, 154, 155t–156t; science attitudes, 144–145, 144t, 151–152; science interest, 142, 151–152; science knowledge, 141–142, 141t, 151–152; science literacy, 139–156; science museums/centers, 143; scientific literacy, 126–134, 127t, 129t, 131t, 132t, 133t; scientific rationalism, 144–145, 150–152, 150f; survey samples, 152–154; understanding of science, 139–156
Ciençia Viva, 443–444, 447
Civic Science Literacy (CSL), 55–71, 217–237; age factors, 57–62, 63t; in America, 55–71; cohort factors, 68–71; education factors, 69–70; generation factors, 57–62, 63t, 66–68; impact of, 217–237; quest for, 5; sources of, 217–237; understanding of, 60t
civic scientific literacy, 217–237. See also Civic Science Literacy
climate change, 339t, 340t, 343–348, 349t, 350
cognitive polyphasia levels, 316–317, 317f
comparative research, 2–3, 140, 460
competence: by country, 426–427, 428t–429t; measures of, 425f; in science, 420–426, 421t, 424t

creationism, 55, 69, 337–338, 340, 345, 348, 354
Crettaz von Roten, Fabienne, 7, 323
cross-cultural perspectives, 323–333
cross-national comparisons, 2–3, 12, 137
cross-national models: competence in science, 420–426, 424t, 425f; participation in science, 420–426, 423t, 425f; of scientific knowledge, 241–259
cultural appropriation, 353–355
cultural differences, 323–333
cultural distance: defining, 285–295; measuring, 286–288, 290t; scientific concept of, 290–291; shift in, 293–294, 294t; of various groups, 290–291; of various provinces, 291–293, 292t
cultural indicators: cross-cultural perspectives, 323–333; from exhibitory science, 385–397; global validity and, 1–17; of science, 10–11
cultural industries, 395–396
cultural systems, 9–10
culture of science: activity indicators of, 12; defining, 18–19; diffusion and uses of, 23–24, 34; impacts of, 24–25; institutions and, 21; measurements of, 19–22, 22f; in Old Europe, 92–108; perceptions of, 12; production and, 21–22; statistics on, 19–24, 20t
Curie, Marie, 42, 51
Curie, Pierre, 42, 51

D

Darwin, Charles, 336
data streams, 2, 7–10, 158, 373
Dawkins, Richard, 348
Durant, John, 220, 241, 284, 439

E

education: education gaps, 98–103, 100f, 101f, 106; formal education, 391t–392t; non-formal education, 391t–392t; science knowledge and, 66–70, 67t; science literacy and, 55–71; science museums/centers, 143, 445–446; similarities in, 202–204, 203f; socialisation in, 286–287, 287f
Elizabeth II, Queen, 1

Englaro, Eluana, 454, 461n3
Estado de S.Paulo, 402, 405–406, 407f–408f, 410t, 411
Europe, 139–156, 301–321, 436–448; astrology beliefs, 301–321; benchmarking, 436–447; China and, 139–156; cluster analysis solution, 146–149, 146t, 148t; education system, 445; government-sponsored activity, 441–444, 442t; public engagement across, 436–447; questionnaire wording, 154, 155t–156t; RTD culture, 436–438, 442–443, 442t; science activity, 439t, 440–441; science attitudes, 144–145, 144t, 151–152, 439t; science interest, 142, 151–152, 439t, 441t; science knowledge, 141–142, 141t, 150f, 439–440, 439t; science knowledge in, 441t; science literacy, 139–156; science museums/centers, 143; scientific community involvement, 444–445; scientific rationalism, 144–145, 150–152, 150f; stem cell research, 353–368; survey samples, 152–154; understanding of science, 139–156. *See also* Old Europe
Europeans, Science and Technology, 442
Evangelista, Rafael, 400
evolution, 337–341, 339t, 340t, 346–350, 349t, 354
exhibitory science indicators, 385–397

F

faith versus science, 84f, 86. *See also* religious beliefs
Folha de S.Paulo, 402, 408f, 411
FolhaOnline, 402, 411
France, 39–52; attitudes toward science, 39–52; generational attitudes, 39–51, 42f, 43f, 45f, 47f, 49f, 52t; image of researchers, 42–43; irrational attitudes, 48–49, 49f; judgment of science, 52t; scientific culture, 49f, 50–51; scientific development, 43–48, 45f, 46f, 47f; state presence, 41–42
Frascati manual, 25, 31–33, 179
future research considerations, 9–12

G

Gago, Jose Mariano, 437
Galileo, 336, 348
gender predictors, 61–62
global PUS research, 11–12
global validity, 1–17. *See also* cultural indicators
global warming, 339t, 340t, 343–348, 349t, 350
God Delusion, The, 348
God Is Not Great, 348
Godin, Benoît, 2, 18
Gould, Stephen Jay, 347–348
Gouveia, Flavia, 400
Gross Domestic Expenditures on Research and Development (GERD), 21

H

"hard science," 309–311, 310f, 315–317
He, Wei, 126
Hitchens, Christopher, 348
homosexuality, 340t, 341–343, 342t, 347–348, 349t, 350
Horst, Maya, 418–419
Human Development Index (HDI), 5, 184, 189, 193, 203f, 204–212, 206f, 207f, 208f, 209f, 210f, 210t, 212n5
human-animal boundaries, 323–333. *See also* nature

I

Iberoamerica, 158–174; attitudes toward science, 158–174; benefits from science/technology, 167–168, 167t, 169t, 175n10; comparative indicators by country, 160t–161t; countries of, 158–159, 160t–161t; ICIC index, 165t, 166t, 171–173, 172t; public policies, 168–171; PUS "movement," 158–159, 174n1; risks from science/technology, 167–168, 167t, 169t, 175n10; science interest, 164t, 175n8; scientific research, 164t; social participation, 168–171; survey of, 162–174, 175n6
"idea of progress," 40
"ideology of progress," 51
Il Corriere della Sera, 454–455, 458
Il Sole 24 Ore, 454–455
"immunisation" hypothesis, 302–305

index of democratisation (id), 286, 295n6
Indian National Science Survey, 184, 192, 194–195, 263–264, 268–274
"intelligent design," 69, 338, 354
irrationality, 48–49, 49f
Italy, 449–462; attitudes toward science, 451–452; monitoring science, 449–461; nuclear power issues, 451–452, 456–460; science in newspapers, 454–456, 454t, 455t; science in press, 453–460, 454t, 455t; society and science debate, 449
Item Characteristic Curves (ICCs), 118, 119f, 246, 248–252, 249f, 251t, 253f, 255f, 257f
Item Response Theory (IRT), 4, 119–124, 120f, 122t–123t, 220–221, 245

J

Japan, 110–124; Item Characteristic Curves (ICCs), 118–119, 119f; Item Response Theory (IRT), 4, 110–124, 120f, 122t–123t; J-SCITEK quiz analysis, 110–124; public understanding of science and technology (PUST), 110–112, 118, 123–124; quiz questions, 113–115, 116t–118t, 121

K

Keeter, Scott, 7, 336
Kim, Hak-Soo, 7, 375
knowledge: dissemination of, 385, 389t, 394–395, 397, 397n1; "formal dissemination" of, 385, 389t; measure of, 439t; "non-formal dissemination" of, 385, 394–396, 397n1; relationship to, 393–394; supply and demand for, 395–396; transmission of, 393–394

L

La Repubblica, 454–455, 458, 460
La Stampa, 454–455
laboratories, purpose of, 386–390, 389t, 401
latent trait models, 241–259. *See also* scientific knowledge
Law of People's Republic of China on Popularization of Science and Technology, 126

Lily, William, 301
Liu, Xuan, 4, 12, 139
logistic regression analysis, 263–264, 267–268, 271t–272t, 276–280
log-linear models, 264–267, 274–276
longitudinal analysis: in Bulgaria, 76–79; comparisons and, 2–3, 12; in Iberoamerica, 158; in Old Europe, 92, 103; in United Kingdom, 76–79
Los Angeles Times, 304
Losh, Susan Carol, 3, 55

M

MANOVA model, 98, 108, 150f
Masci, David, 336
MASIS, 433
Matsuura, Takuya, 4, 110
measurement issues: research and, 5–6, 19, 215; scientific knowledge, 241–242; scientific literacy, 219–223
Measurement of Scientific and Technical Activities, The, 28
media, science in, 8, 400–401, 406t, 413t, 446, 453–460, 454t, 455t
mediation, modes of, 390–393
Mejlgaard, Niels, 8, 12, 418
Miller, Jon D., 5, 58, 127–129, 217, 220, 241, 246
Miller, Steve, 9, 436
moral beliefs, 357, 358t, 359t
"moral" progress, 44
Morales, Ana Paula, 400
museum centers, 143, 385–386, 445–446
museum studies, 392t, 397n4

N

National Institute of Science and Technology Policy (NISTEP), 110–111, 245
National Science Foundation Surveys, 58–61
National Science Survey, 184, 192, 194–195, 263–264, 268–274
national surveys: in Bulgaria, 76; in China, 126–129, 127t, 134; in Iberoamerica, 159, 162–163; improving, 1–2; of science literacy, 13t–15t, 219–220; in United Kingdom, 76
nature, 328–332, 329t
Neresini, Federico, 9, 449

New York Times, 348
NOMA, 347–348
nuclear power issues, 451–452, 452f, 456–460, 457f, 458f

O

Obama, Barack, 347
OECD, 21–25, 22t, 27, 33
Ogburn, William F., 24
Old Europe, 92–109; changing culture of science, 92–108; cohort analysis, 94–108; education gaps, 98–103, 100f, 101f, 106; gender gaps, 98–103, 100f, 101f, 106; generational trends, 94–98, 95t, 100f, 102–103; knowledge rankings, 94–96, 95t; MANOVA model, 98, 108; postindustrial hypothesis, 102–103, 106; science attitudes, 95–98, 97f, 100f, 101f, 102–108, 104t–105t; science interest, 95–98, 102–108, 104t–105t; science knowledge, 95–98, 97f, 100f, 101f, 102–108, 104t–105t; scientific literacy, 94–97, 103; temporal trends, 92–94, 93f, 106. *See also* Europe
Outline of The National Scheme for Scientific Literacy, 126, 128, 130

P

"parasciences," 48–49, 49f
Pardo, Rafael, 7, 353
participation: competence and, 420–426; by country, 426–427, 428t–429t; measures of, 425f; in science, 420–426, 421t, 423t
Pasteur, Louis, 42, 51
Pearson, Ian, 1
P-engagements, 378–382, 379t, 380t
PEP/IS, 375–383; explanation of, 7, 376; key points of, 376–377, 382; measuring, 375–383; social problems, 378, 379t, 380t, 381t
Petkova, Kristina, 3, 76
Polino, Carmelo, 4, 5, 158
Power Rangers, 71
"Promotion of RTD Culture and Public Understanding of Science (PUS)," 437
pseudoscience beliefs: age and, 65t, 66–68; in America, 55–60, 60t, 64f, 65–68, 65t, 67t; astrology and, 302, 306, 309, 313, 320; education and, 66–70, 67t; generations and, 66–68; regression effects on, 66–68; science knowledge and, 55–71, 64f
public engagement: across Europe, 436–447; explanation of, 7, 376, 419; focus on, 431; investigating, 127, 174n1; participation in, 420–426; procedures for, 433; in science communication, 419; studies on, 427. *See also* PEP/IS
public engagement activities, 9–10
public understanding of science and technology (PUST), 110–112, 118, 123–124
Public Understanding of Science and the Humanities (PUSH), 438, 443
public understanding of science (PUS), 1–12; in America, 55–71; astrology and, 301–302; in Bulgaria, 86–89; in China, 139–156; cultural distance, 285–295; cultural model, 288–289; cultures of, 282–295; data analysis model, 268–274, 270t; dimension of, 19; in Europe, 139–156; global research base, 11–12; historical overview of, 282–283; in Iberoamerica, 158; improving, 375–376; Indian National Science Survey, 268–274; logistic regression analysis, 263–264, 267–268, 271t–272t, 276–280; log-linear models, 264–267, 274–276; measure of association, 273t; "movement" perspectives, 158–159, 174n1; notion of public and, 284–285; in Old Europe, 102–109; opposition to, 419; postindustrial hypothesis of, 102–103, 106; questioning, 376; researching, 1; science communication, 294–295; scientific method and, 283–284; statistical models of, 262–280, 289t, 290t; typologies of, 12; in United Kingdom, 86–89

R

Rai, Anil, 6, 262
Raza, Gauhar, 6, 11, 282
Reagan, Nancy, 69
"related scientific activities," 28–32

470   Index

religious beliefs, 336–352; clash of, 345–346; climate change, 339t, 340t, 343–348, 349t, 350; evolution, 337–341, 339t, 340t, 346–350, 349t; global warming, 339t, 340t, 343–348, 349t, 350; homosexuality, 340t, 341–343, 342t, 347–348, 349t, 350; science and, 56, 336–351; scientists and, 336–337; stem cell research, 354–357; in United States, 345–346
Ren, Fujun, 126
representations, alternative, 27–28
representations, cultural, 19, 24–28
research: comparative research, 2–3, 140, 460; future considerations for, 9–12; science as, 24–27; state presence and, 41–42; support for, 85f, 86
Research, Technology and Development (RTD) culture, 436–438, 442–443, 442t, 447
researchers, image of, 42–43, 42f
Righetti, Sabine, 400
Righter, Caroll, 304
ROSE (Relevance of Science Education), 5, 200–205, 209–210
Rousseau, Jean-Jacques, 44
RTD culture, 436–438, 442–443, 442t, 447

S
SAPO project: classifying materials, 402–403, 410; comparative analysis, 411; explanation of, 8, 400–414; indicators of, 403–406, 406t, 407f–408f; operation of, 404–409; peak analysis, 409–412, 410t; reliability of, 415n9
Schiele, Bernard, 8, 385
Schreiner, Camilla, 5, 200
science: alternative representations of, 19, 27–34; astrology and, 301–308, 308f, 315–321; attitudes in Bulgaria, 76–90; attitudes in France, 39–52; attitudes in Iberoamerica, 158–174; attitudes in United Kingdom, 76–90; attitudes of adolescents, 200–211; attitudes toward, 325–328, 327t, 336–351; characteristics of, 25–27; climate change, 339t; cognitive polyphasia levels, 316–317, 317f; communication of, 294–295, 375–383, 419, 436–447; communicative effectiveness about, 375–383; competence in, 420–426, 421t, 424t, 425f, 428t–429t; cultural indicators of, 10–11; cultural representations of, 19, 24–27; engagement with, 317–319, 318f, 319t; evolution and, 337–341, 339t, 347, 349t, 350; exhibitory science, 385–397; generational attitudes, 39–51, 42f, 43f, 45f, 47f, 49f, 52t; governance of, 325, 328, 328t, 331, 427, 430; on human relations, 43f; impacts of, 24–25; irrational attitudes, 48–49, 49f; judgment of, 51, 52t; measuring, 5–6, 19–22, 22f, 215; in media, 8, 400–401, 406t, 413t, 446, 453–460, 454t, 455t; modelling knowledge of, 262–280; monitoring in Italy, 449–461; participation in, 420–426, 421t, 423t, 425f, 428t–429t; perceptions of, 323, 348, 400, 412–413, 449; politics of numbers, 18–34; in public sphere, 449–461; R&D and, 25–33; as research, 24–25; reservations about, 346, 361–363, 368; scientific activities, 28–32; scientific and technological potential (STP), 27–32; scientific culture, 49f, 50–51; scientific development, 43–48, 45f, 46f, 47f; state presence, 41–42; technological services, 26–33; understanding in America, 55–71; understanding in China, 139–156; understanding in Europe, 139–156. *See also* culture of science
Science, technology and innovation (ST&I), 140, 400–414
Science and Society Action Plan, 437, 442, 444
science communication: climate indicators for, 436–447; effectiveness of, 375–383; public engagement in, 419; understanding of, 294–295
science culture index (SCI): benchmarking, 182–185, 186f; construction

of, 179–197; cultural indicators, 179–181; framework of, 183f, 191–194; indicators percentages, 188–189, 188t; methodology of, 182–185, 191–194; PUS database, 181–182, 186f, 187f; PUS indicators, 194–196, 196t–197t; validation of, 185–194; variability estimates, 190f
science exhibitions, 385–397
Science in Society Facts and Figures, 449
Science in Society Monitor, 449–450, 457, 460–461
Science in the Media Monitor (SMM), 453–454, 457, 460–461
science knowledge: education and, 66–70, 67t; in Europe, 141–142, 141t, 150f, 438–440, 441t; in Japan, 110–124; measure of, 439t; pseudoscience beliefs and, 55–71, 64f; regression effects on, 66–68, 67t. *See also* science literacy
science literacy: in America, 55–71; assessing, 241–259, 243t–244t; in Bulgaria, 82–85, 82f, 83f, 84f, 85f; cohort factors, 68–71; education factors, 69–70; measuring, 386–397; science education and, 55–71; in United Kingdom, 82–85, 82f, 83f, 84f, 85f. *See also* Civic Science Literacy
"science literacy scale," 241, 246, 256–259, 257f
science museum studies, 392t, 397n4
science museums/centers, 143, 385–386, 445–446
Science News Media Barometer, 400–414
scientific activities, 28–32
scientific and technological activities (STA), 30–34, 32f
scientific and technological potential (STP), 27–32
scientific and technological services (STS), 30–34, 32f
Scientific Automatic Press Observer (SAPO), 8, 400–414, 415n9
scientific careers: entry into, 390–393; interest in, 206–211, 208f, 209f
scientific citizenship: contextualism, 430–435; distributions of, 426–433; dual dimensions of, 418–419; society and science, 8–9, 418–419, 430–435; survey measures of, 418–433
scientific community involvement, 444–445
scientific competence, 418–419, 431
scientific culture index, 49f, 50–51
scientific development, 43–48, 45f
scientific discoveries interest, 82f, 85
scientific field, purpose of, 386–390, 389t, 401
scientific knowledge: acquiring, 388; assessing, 241–259, 243t–244t; biotechnology, 242, 243t–244t, 248–259, 249f, 251t, 253f, 254t, 255f; future survey design, 256–259; general science, 241–243, 255–256, 259; need for, 375–376; problem of, 388; scaling methods, 244–248; science literacy, 243t–244t; statistical models of, 262–280; survey measures of, 241–242; trait models of, 241–259. *See also* science knowledge; science literacy
scientific literacy: in China, 126–134; concept of, 218–219; by country, 222–223, 222f, 227–233, 229t; cultural appropriation, 353–355; democratic societies and, 217–218, 223, 232–235; dimension of, 19; impact of, 217–237; measurement of, 219–223; path models, 224–226, 225f, 226f; predicting, 224–226, 225f, 226f; science knowledge factor analysis, 236t–237t; score calibrations, 220–221; sources of, 217–237; stem cell research and, 353–357; in United States, 221–224, 221f. *See also* Civic Science Literacy
scientific method, 283–284
scientific research, support for, 85f, 86
scientists: distrust of, 84f, 86; opinion toward, 42f
Scopes, John Thomas, 336
"Scopes Monkey Trial," 336
sensitive topics, 6–7, 299
Shelley, Mary, 42
Shimizu, Kinya, 4, 110
Shukla, Rajesh, 1, 4, 6, 179, 262
Silva, Marina, 410, 411
Simpson, O. J., 347
Singh, Surjit, 6, 11, 282

Sjøberg, Svein, 5, 200
Smith, Gregory, 336
social problems: decision-making issues, 170, 172t; PEP/IS and, 7, 378–381, 379t, 380t, 381t
"soft science," 309–311, 310f, 315–317
SPS-engagements, 378–382, 379t, 380t
Star Trek, 71
Stares, Sally, 6, 8, 12, 241, 418
stem cell research, 353–368; across Europe, 353–368; attitudes toward, 354–356, 362–363, 364t–365t; controversy over, 217–218; moral beliefs, 357, 358t, 359t, 363; perceptions of, 353–368; religious beliefs, 354–357; scientific literacy, 353–357; trust of, 361–362, 368; worldviews of, 360–368
Stoneman, Paul, 6, 301
survey measures, validating, 418–433
Switzerland, 323–335; animal cloning, 325–326, 329–332, 330t; animals, 328–329, 329t; human-animal boundaries, 323–335; nature, 325–326, 328–332, 329t, 332; science, 325–328, 327t; technology, 325–326

T

Tang, Shukun, 139
technical development concepts, 43–48
technological activities concepts, 30–32
"technological enthusiasm," 361
technological optimism, 325–326, 327t, 330t
technology: assessing knowledge of, 241–259; attitudes toward, 325–326; in Iberoamerica, 158–174; in Japan, 110–124. *See also* biotechnology; science
Telos, 304
Todorov, Valery, 3, 76

TOPS Round Table, 383
trait models, 241–259. *See also* scientific knowledge
Twilight Zone, 71
typology constructions, 12

U

UNESCO, 25–34; conception of science and, 19, 25–34; scientific activities, 28–32; scientific and technological potential (STP), 27–32
United Kingdom, 76–90; ANOVA models, 81, 89t, 90t; attitude variables, 79; cohort variables, 79–81, 85–89, 89t, 90t; "country" variables, 81, 89t, 90t; daily science relevance in, 83f, 86; distrust of scientists, 84f, 86; faith versus science, 84f, 86; generational attitudes in, 78–90; historical background of, 76–77; image of science, 76–90; interest in science, 82f, 85; science benefits, 83f, 85; science literacy, 82–85, 82f, 83f, 84f, 85f; science literacy variables, 79; scientific research support, 85f, 86
United States, 336–352; attitudes toward science, 336–351; religious beliefs in, 336–351; scientific literacy, 221–224, 221f. *See also* America

V

Vogt, Carlos, 8, 400

W

Wang, Ke, 4, 126

Z

Zhang, Chao, 126